COMMUNICATION TECHNOLOGY

COMMUNICATION TECHNOLOGY

ROBERT BARDEN
MICHAEL HACKER

Glencoe
McGraw-Hill

New York, New York   Columbus, Ohio   Woodland Hills, California   Peoria, Illinois

Formerly published by Delmar Publishers Inc.®

NOTICE TO THE READER

Publisher does not warrant or guarantee any of the products described herein or perform any independent analysis in connection with any of the product information contained herein. Publisher does not assume, and expressly disclaims, any obligation to obtain and include information other than that provided to it by the manufacturer.

The reader is expressly warned to consider and adopt all safety precautions that might be indicated by the activities described herein and to avoid all potential hazards. By following the instructions contained herein, the reader willingly assumes all risks in connection with such instructions.

The publisher makes no representations or warranties of any kind, including but not limited to, the warranties of fitness for particular purpose or merchantability, nor are any such representations implied with respect to the material set forth herein, and the publisher takes no responsibility with respect to such material. The publisher shall not be liable for any special, consequential or exemplary damages resulting, in whole or in part, from the readers' use of, or reliance upon, this material.

COVER PHOTO CREDITS
Satellite dish photo courtesy of GTE Spacenet Corporation.
Workstation courtesy Clough, Harbour & Associates, Albany, New York;
      Photo by Tom Carney.
Compact disk photo courtesy of General Electric Company.
Register guidance system scanner photo courtesy of QUAD/TECH.

Glencoe/McGraw-Hill
A Division of The McGraw-Hill Companies

Send all inquiries to:
Glencoe/McGraw-Hill
3008 W. Willow Knolls Drive
Peoria, IL 61614

ISBN 0-8273-3225-4

Printed in the United States of America

2 3 4 5 6 7 8 9 10   003   01 00 99 98 97

CONTENTS

(Courtesy of TRW, Inc.)

v

(Courtesy of Prime Computers)

(Courtesy of Brodock Press Inc./James Scherzi Photography)

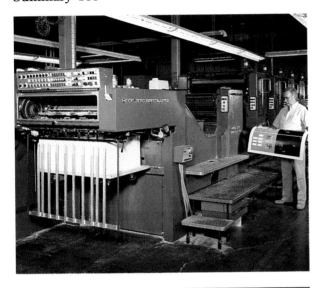

(Courtesy of Heidelberg Eastern, Inc.)

SECTION FOUR

PHOTOGRAPHY AND MOTION PICTURES 195

(Courtesy of Pentax Corporation)

(Courtesy of AMP Incorporated)

SECTION FIVE
ELECTRONIC COMMUNICATION 285

(Courtesy of Sony Corporation)

(Courtesy of Sony Corporation)

SECTION SIX _____

THE FUTURE OF INTEGRATED
COMMUNICATION SYSTEMS 405

FOREWORD

Communication is central to the human experience. Personal communication is an interchange of ideas and information involving words, gestures, body language, and speech patterns. It is an essential element in our society when two or more people interact.

Communication becomes much more complex, however, when we are not doing so in person. The need for instant access to information, the long distances that might separate people, the need to record a multitude of ideas, and the existence of an ever-exploding information base all play a dramatic role in our need to develop communication technologies. Technology permits us to ''extend our human potential'' to almost unlimited possibilities in dealing with information between people and machines. Today, we have an enormous capability for communication through technology. An example of this is the laser system that pulses 22 billion times a second over a fiber-optic channel; this permits us to send ten complete sets of the *Encyclopedia Britannica*, word for word, through a wisp of a glass thread in one second!

This book addresses the basic principles of communication technology, first reviewing the nature of communication technology, and then providing instruction in the fields of technical drawing, printing, photography, electronics, and computers. It culminates by focusing on communication systems and taking a peek into the future.

With an understanding of communication technology, people will be able to function more effectively in this technological age. As consumers and users of this technology, we find that it is a part of everyday life. Communication technology is used in all dimensions of the social setting, including electronic mail, FAX systems, and satellite transmission, as well as in a wide assortment of systems used in entertainment. Instantaneous communication to all parts of the world is a common activity.

In the workplace, effective communication technology is recognized as critical to the expansion of productivity. When we find people communicating with machines and, in turn, machines with people, the processes become even more complex. Machine-to-machine communication provides new information processing capacities for us. For example, scanners can digitize whole libraries of data in seconds; this data can be stored in a blink of an eye to be retrieved at will, and used in business and industry.

So, we call attention to the importance of the contents of this book, as a ''primer'' for the basics . . . and for things to come. The more effectively we can communicate, the more satisfying and productive our lives will be.

M. James Bensen, Dean
School of Industry and Technology
University of Wisconsin-Stout
Menomonie, Wisconsin

PREFACE

Communication is a basic human need. The need to communicate more quickly and more efficiently has become a central focus in our technological society. Communication has become an important and integral part of other technological areas, including manufacturing, construction, and transportation. In addition, the combination of computers and communication has produced a whole new information industry which both satisfies peoples' needs for better communication and drives the need for further improvements.

The development of this new field of technology, information systems, has had a tremendous impact on our society. Jobs that are information based, in which workers create, store, modify, and use information rather than things, now outnumber production and transportation jobs in this country. Information jobs are found in all areas of the economy. Many refer to the present era as the "information age."

TECHNOLOGY EDUCATION

Technology Education, distinct from "technical education" or "vocational education," is a program of activity-oriented study through which an appreciation of technology is given to all students as part of their general education. The goals of other technically or occupationally oriented programs are intended to provide students with job-specific skills. Technology Education is significantly different in that it is designed to provide technological literacy as part of every student's fundamental, liberal education.

In many established technology sequences, students are given an exposure to technology at the primary level, and are then given a more comprehensive overview of the various aspects of technology at the middle school level. The middle school course is often a cursory introduction to general technological concepts, followed by an exploration of various aspects of technology, often including communication, construction, manufacturing, transportation, and energy (often, construction and manufacturing are combined into a study of production technology). In the secondary school, each of these and other specific areas can be investigated through more in-depth courses that concentrate on each of these fields. This book is intended to be used in the secondary level course that concentrates on communication technology.

ORGANIZATION OF THIS BOOK

Communication Technology is divided into six sections. The first section serves as an introduction to communication technology, providing the student with a general base of understanding from which to delve into the more specifically targeted later sections. The first chapter presents some fundamental communications concepts.

Recognizing that computers are now widely used in technical drawing (CAD), printing, photography, and electronic communication, the second chapter presents an introduction to computer technology that will be referred to in later chapters.

The second section (Chapters 3 through 6) deals with technical drawing techniques, with emphasis on both traditional methods and newer computer (CAD) methods. Section 3 (Chapters 7 through 11) covers printing methods from Gutenberg's press to electronic paste-up and press. Chapters 12 through 15 make up Section 4, which describes still photography and motion picture photography. Section 5 (Chapters 16 through 21) deals with electronic communications, including an overview of electrical/electronic basics, telephone, radio, TV, microwave, satellite, fiber optics, data communications, and recording technologies.

Section 6 provides a glimpse into where current and pending developments may lead communication technology in the near future.

Each section has a set of culminating activities that will serve to reinforce the major concepts presented in the chapters within the section.

SPECIAL FEATURES

Communication Technology includes many special features to assist the reader to understand modern technology.

- The use of four color printing in the more than five hundred photographs and illustrations throughout the book helps to clarify the explanation of concepts.
- Each chapter begins with a set of major concepts. These prepare the reader to be aware of key ideas while reading.
- Key words introduced within the chapter are presented at the beginning of the chapter and are highlighted in the text.
- A series of activities provides a basis for action-oriented learning.
- Mathematics and science concepts are explained as they are introduced throughout the text.

- Complete summaries are included at the end of each chapter. These review the major concepts within the chapter.
- The text includes boxed inserts containing features of special interest.
- A glossary of terms with definitions is provided.
- A timeline of the development of communication technology is provided.

ACKNOWLEDGEMENTS

The development of a book covering the range of technologies presented here is a task to which many people have contributed. Special thanks is given to the manuscript reviewers, who provided helpful feedback to the authors and contributing authors:

Dennis Gallo
O'Fallon Township High School
O'Fallon, Illinois 62269

Bill Windham
Department of Technology
Southwest Texas State University
San Marcos, Texas 78666

Joseph Russo
Washburn High School
Washburn, WI 54891

Mike Harmon
West Point High School
West Point, Virginia 23181

Dennis Simpson
Fairfield Freshman Schools
Fairfield, Ohio 45014

Chuck Goodwin
Department of Technology,
Union-Endicott High School
Endicott, New York 13760-5271

Jim Payne
Victoria High School
Victoria, Texas 77903

Fred Posthuma
Westfield High School
Westfield, Wisconsin 53964

Ethan Lipton
California State University, Los Angeles
Los Angeles, California 90032

We wish to express deep appreciation to the Delmar staff, and particularly to Cynthia Haller and Christine Worden, who served as developmental editors for this work, and Christopher Chien, who served as project editor. Their dedication helped keep the project on track. Their expertise and professionalism grew into friendship, and a shared commitment to high standards.

We would also like to thank the many educators with whom we have spoken and corresponded, who have encouraged us in this effort, and who have contributed their advice and perspectives to it. They have earned our thanks and our respect.

ABOUT THE AUTHORS

Michael Hacker is a twenty-year veteran of secondary school teaching in Long Island, New York. As early as 1969, his technology-based junior high school program had received national attention. He has authored a dozen articles in national journals, consulted nationally, and has presented at numerous state and national conferences. He is past president of the New York State Technology Education Association and has actively served on various International Technology Education Association (ITEA) committees. In 1985, he was named an Outstanding Young Technology Educator by the ITEA. In his present capacity as Associate State Supervisor for Technology Education, New York State Education Department, his responsibilities include the development and implementation of Technology Education curricula.

Robert A. Barden is an electronics engineer specializing in the design and development of high capacity data, voice, and video communication systems. His work includes the integration of fiber-optic, microwave, and satellite technologies. Mr. Barden has served as a member of the National Advisory Council of the ITEA and has published articles and presented seminars on technology education and communication technology for teachers and teacher trainers. He is a Senior Member of the Institute of Electrical and Electronics Engineers (IEEE), a former member of its committee on Precollege Mathematics, Science, and Technology Education, and other committees, and a recipient of the IEEE Centennial Medal. He has published numerous papers and presented seminars in the fields of Local Area Networks and microwave communications.

CONTRIBUTING AUTHOR ACKNOWLEDGEMENTS

Special thanks are due the following contributing authors. Each is an expert in his area and the chapters, sections of chapters, or activities they have written have added greatly to the quality of COMMUNICATION TECHNOLOGY.

Harry M. Shealey
Baltimore County Public Schools
Towson, Maryland

and

David Goetsch
Okaloosa-Walton Community College
Niceville, Florida

TECHNICAL DRAWING SECTION
Chapter 3, Sketching and Illustration
Chapter 4, Technical Drawing Techniques
Chapter 5, Drawings
Chapter 6, Computer-Aided Design and Drafting

Ryan Smith
Milwaukee Area Technical College
Milwaukee, Wisconsin

PRINTING SECTION
Chapter 7 Graphic Design, Image Generation and Assembly
Chapter 8, Desktop and Electronic Publishing
Chapter 9, Image Preparation
Chapter 10, Image Transfer
Chapter 11, Finishing

Ervin A. Dennis
Department of Industrial Technology
University of Northern Iowa
Cedar Falls, Iowa

PHOTOGRAPHY SECTION
Chapter 12, Cameras and Film
Chapter 13, Photographic Techniques
Chapter 14, Darkroom Processes
Chapter 15, Motion Pictures and Animation

Lewis Kauffman
Franklin County Vocational-Technical School
Chambersburg, Pennsylvania

Chapter 21, Recording Systems
All photos in Chapter 21 otherwise uncredited

Douglas Hammer
University of Alaska
Anchorage, Alaska

COMMUNICATION
TECHNOLOGY ACTIVITIES

SECTION ONE

THE NATURE OF COMMUNICATION

Communication, the accurate transfer of information from a sender to a receiver, is a basic human need. It has become a central part of our modern economy, and people now speak of living in the information age. In the information age, more workers are involved in creating, processing, and communicating information than there are in agriculture or manufacturing. Information workers include bankers, attorneys, teachers, computer operators, people in the communication industry, and many others.

Many of the recent advances in the communications field have been brought about by the development and use of computers of all sizes, ranging from very large supercomputers to tiny, single-chip microprocessors. This section includes an overview of the communications process and an introduction to computer concepts. These will lay the foundation for the chapters to follow.

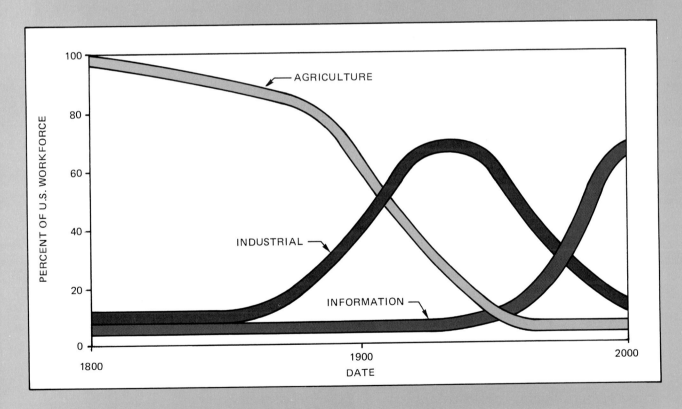

SHIFT FROM AGRICULTURE ERA TO INDUSTRIAL ERA TO INFORMATION AGE

Chapter 1 ■ What is Communication?
Chapter 2 ■ Role of Computers in Communication

What Is Communication?

OBJECTIVES

After completing this chapter you will know that:

- Communication involves having a message received and understood.
- Three kinds of communications are person-to-person communication, animal communication, and machine communication.
- Five purposes of communication systems are to inform, persuade, entertain, educate, and control.
- A system is a combination of elements which function to achieve a desired result.
- Feedback is information which monitors how closely the actual result (the output) of a communication system conforms to the desired result (the input).
- The communication process includes five parts: encoder, transmitter, channel, receiver, and decoder.
- The design of a communication is largely influenced by the type of audience for which the message is intended.
- In designing a communication system, some thought must be given to how the message will be monitored and evaluated after it is received.

KEY TERMS

animal communication	encoding	nonverbal communication
channel	evaluation	person-to-person communication
communication	feedback	propagation
communication process	machine communication	receiver
communication system	mass communication	transmitter
decoding	monitoring	verbal communication

Kinds of Communication

Communication is the accurate transfer of information from a sender to a receiver. The sender and receiver can be people, animals, or machines.

Communication is a two-way street. To have good communication with someone else, we

FIGURE 1–1 To have communication, a message must be received and understood. *(Photo courtesy of 3M)*

FIGURE 1–2 Newspapers are a mass communication medium. *(Courtesy of Knight-Ridder, Inc.)*

must be sure that what we are saying is clearly understood. We must know that the message we are sending has the same meaning to the other person as it does to us.

Just saying something to someone does not mean that we have communicated with that person. Communication means that the message we have transmitted has been received and understood, Figure 1–1.

Person-to-Person Communication

Most communication occurs among people. People use their own senses to communicate. Often, interpersonal communication uses speech and is called **verbal communication**.

People also use **nonverbal communication**.

This means using other senses, like vision, smell, and touch to communicate. For example, people choose clothing that visually communicates a message about themselves. They wear fragrances to make themselves more appealing. Emotions can be communicated through touch or by a glance. Someone's body language can communicate a lot about what that person is feeling.

Person-to-person communication is often enhanced by technological means. Speech is amplified by public address systems. Film and videotape record gestures which can then be seen at a later time, and at another place. By using technology, communication from one person can reach the masses. **Mass communication** systems include radio, television, books, magazines, and newspapers, Figure 1–2.

FIGURE 1–3 This carrier pigeon communication system is for the birds! *(Used by permission, Gannett Co., Inc.)*

Animal Communication

Animals can communicate with people and with themselves. We talk to our pets and they understand us. We can train them to respond to certain commands and to perform tasks, Figure 1–3.

One very interesting case of **animal communication** involved teaching a young gorilla named Washoe to speak using the same sign language used by deaf people. At the end of several years of training, Washoe had a vocabulary of 160 words and was putting them together to make short sentences.

Animals can communicate with other animals. Animals from tiny ants to huge gorillas communicate about gathering food, mating, and defense. Skunks produce odors which are interpreted as messages by other animals. Some animals use body language like grimacing and pointing. Giraffes rub their necks together to show affection, while chimpanzees embrace each other.

Machine Communication

In addition to being able to communicate with people and animals, people are also able to communicate with machines. We push buttons on pocket calculators. These process data and give information in the form of a digital display or printout. We get visual information from machines like televisions. We can read computer output from video display terminals. These are all examples of **machine communication**.

Machines can also communicate with other machines. CAD/CAM is a technology that joins CAD (computer-aided design) with CAM (computer-aided manufacturing). CAD/CAM allows a person to design a part on a computer screen. Then, all the necessary design information (including the size and shape of the part) is communicated directly to a machine tool. The machine tool makes the actual part, Figure 1–4.

Purposes of Communication

A message has a certain purpose. Communication systems are designed with five different goals in mind. We use communication systems to

- inform
- persuade
- entertain
- educate
- control

To inform people, we might use mass media like newspapers, radio, or television, Figure 1–5. News stories give information to the general public. Other examples of using communication systems to inform include announcements about concerts, school or community events, sales, births, deaths, or marriages.

Person-to-person communication is often used to persuade. Salespeople try to get customers to buy products. Negotiators try to persuade companies and unions to agree to salary and benefit packages. Lawyers try to persuade judges and juries. Advertisements use persuasion to sell products, Figure 1–6. Ads are often produced by an agency which employs artists, writers, and public relations people to target an ad to a certain audience.

FIGURE 1–4 In CAD/CAM, computers communicate directly with machine tools. *(Courtesy of Grumman Corporation, 1988)*

FIGURE 1–5 News broadcasts inform the public. *(Courtesy of Knight-Ridder, Inc.)*

FIGURE 1–6 Advertisements are designed to persuade people to buy products or services. *(COCA-COLA is a registered trademark of the Coca-Cola Company. Permission for use granted by the Company)*

Problem Solvers In Communication

**Peggy Tice, Sales Representative
Eastman Kodak Company**

Even as a young girl, Peggy Tice enjoyed graphic design. At the age of 10, she did calligraphy (beautiful handwriting) on signs and posters. When she was a senior in high school, Peggy was looking for an appealing career. She had learned about how photographs were broken down into dots for printing in newspapers. That fascinated her. She brought home a magnifying glass to show her parents how newspaper pictures were made up of dots.

Peggy knew she liked graphics, but didn't want to study pure art. A friend of her father introduced her to the Graphics program at Arizona State University. Peggy enrolled, and four years later received a B.S. degree in Graphic Communications.

Now Peggy works for Kodak as a Copy Preparation Sales Representative. She provides technical assistance and information to the graphics industry. Peggy works with advertising

Peggy Tice

agencies, type houses, printers, camera houses, and small newspapers. She says she visits her accounts "just to see how they're doing." Peggy finds out what their needs are and then tries to help them by giving them the right products. She says, "I'm not a hard-sell type of person. Of course we're trying to sell our products, but it just makes sense to me that if you help the customers solve their problems, you'll reach your sales quotas."

Entertainment is a huge industry. Entertainment is based on communication, Figure 1–7. Movies, books, records and tapes, plays, dances, concerts, and cartoons are all forms of communication used to entertain.

We often use communication to educate people. Teachers communicate verbally with students, Figure 1–8. Newspapers provide education through articles on topics of special interest. Television documentaries educate viewers about new advances in medicine, natural phenomena, foreign cultures, and many other topics.

Communication is also used as a method of control. This can be seen in machine-to-machine communication. Thermostats communicate with furnaces and control them by turning them on and off. Timers communicate with washing machines to control the washing cycle. Photo-

FIGURE 1–7 Televising baseball games provides entertainment to millions of people. *(Courtesy of General Electric)*

FIGURE 1–8 This teacher uses speech and visual materials to educate a student. *(Courtesy of Charles Cherney, Chicago Tribune)*

cells communicate with streetlights to turn them on at dusk and off at dawn.

Sometimes, communication is also used to control people. Propaganda (a point of view held by a government or movement) is an example. The best defense against propaganda is open communication which allows all parties to freely exchange ideas and points of view.

Communication Systems

We are surrounded by many examples of **communication systems**. These systems range from small calculators to large, special purpose computers; from pocket radios to stereo systems; from intercoms to world-wide telephone systems. A system is a combination of elements which function to achieve a desired result. In all communication systems, the desired result is to communicate a message.

Every system has an input (the desired result), a process, and an output (the actual result). In a communication system that involves two people speaking to each other, the desired result (the system input) is what it is the sender wishes to communicate. The process is the technical means you will use to communicate the message. The output is the message that is actually received by the second person, Figure 1–9.

Feedback in Communication Systems

How do you know if the message you sent was received exactly as you intended? Perhaps noise interfered with the message. Maybe the receiver misunderstood what you meant. Generally, communication systems provide a way for the sender to find out whether the actual message received (the output) was the same as the desired message (the input). **Feedback** gives this information.

Feedback is information we get by **monitoring** the output of the communication process. We monitor the output to see if it is what was desired. Feedback can be obtained by asking the

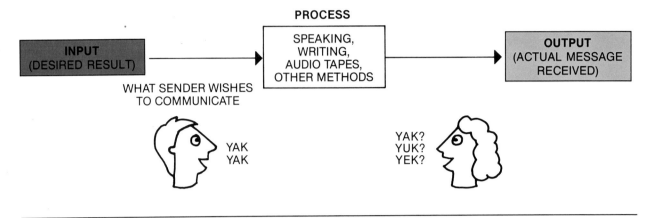

FIGURE 1-9 Person-to-person communication system

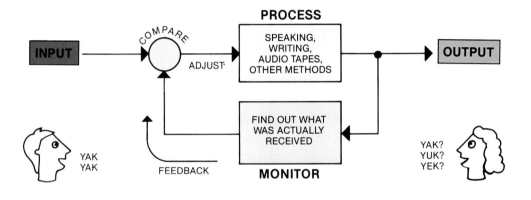

FIGURE 1-10 Monitoring the output provides feedback. The feedback allows us to compare the message actually received to that which was intended. An adjustment is made in the process to make actual results match the desired results.

receiver to repeat the message as received. Or, the receiver can be asked if he or she understood the message. Once feedback is received, the output is compared to the input. If needed, an adjustment is made to the process to get the intended output, Figure 1–10.

Since all communication systems have inputs, process, and outputs, we can model each using the basic system model, Figure 1–11.

The Communication Process

The process is the action part of any system. The process combines resources in order to produce an output. The resources that are combined in

a communication system are the same kinds of resources that are needed by all other technological systems. These include people, information, materials, tools and machines, energy, capital, and time. These resources are added inputs to the system, Figure 1–12.

Every **communication process** includes five parts: a way of **encoding** the message (preparing it for transmission); a means of **transmitting** the message; a **channel** or route which the message takes; a **receiver** which accepts the message; and a way of **decoding** the message (turning it into a form which is understandable to the user). All communication systems—whether the communication involves printing books, transmitting signals to satellites, radio broadcasting, or tele-

COMMUNICATE
DESIRED MESSAGE

INPUT

COMPARE

TELEPHONE
NETWORK

PROCESS

ACTUAL MESSAGE HEARD
BY PERSON ON THE
RECEIVING END

OUTPUT

PERSON RECEIVING
PROVIDES FEEDBACK
TO PERSON SPEAKING

MONITOR
(PROVIDES FEEDBACK)

COMMUNICATION SYSTEM USING TELEPHONES

*(Courtesy of Brodock Press, Inc./Jam
Scherzi photography)*

COMMUNICATE
CURRENT NEWS

INPUT

COMPARE

REPORTERS,
CAMERAPERSONS
TV TRANSMITTER,
ANTENNA

PROCESS

ACTUAL BROADCAST
RECEIVED
BY VIEWERS

OUTPUT

SURVEY, POLLS,
VIEWER RESPONSES

MONITOR
(PROVIDES FEEDBACK)

TV NEWS BROADCASTING SYSTEM

(Courtesy of Goodyear Tire & Rubber Co.)

TO PRINT A
NEWSPAPER OF
HIGH QUALITY

INPUT

COMPARE

PRINTING PRESS,
INK, PREPARED
COPY

PROCESS

ACTUAL
NEWSPAPER

OUTPUT

PRESS OPERATOR
CHECKS QUALITY BY
INSPECTING NEWSPAPER

MONITOR
(PROVIDES FEEDBACK)

NEWSPAPER PRINTING SYSTEM

(Courtesy of Gannett Co., Inc.)

FIGURE 1–11 System models of various communication systems.

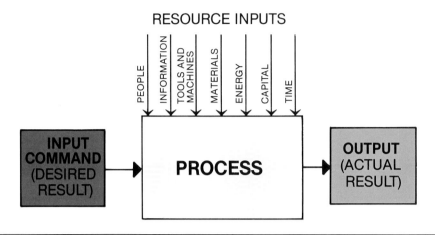

FIGURE 1–12 Seven kinds of resources are combined by the technological process. The process converts the resource inputs into outputs.

vison programming—include these five parts, Figure 1–13.

Encoding Messages

Before a message can be communicated, it must be put into a form that can be transmitted easily. A thought like "I love you" can be whispered to someone. But to communicate that message over a distance, it may take various forms. Suppose you wish to send that sentence over telegraph. You would need to convert the letters into Morse code and send the message as a series of dots and dashes. To process data with a computer, letters must be turned into a digital code like ASCII (American Standard Code for In-

formation Interchange). When we convert a message into a form better able to be transmitted, we encode the message. Encoding changes information from one form into another, Figure 1–14.

A microphone encodes speech (acoustical energy) into electrical energy so that messages can be transmitted by radio. Light energy reflected by objects and people is encoded into electrical energy by video cameras. The television pictures can then be transmitted. Photographs are encoded into halftones (pictures made up of dots) so they can be printed by printing presses. Words and symbols are encoded into digital bits so they can be processed by computers.

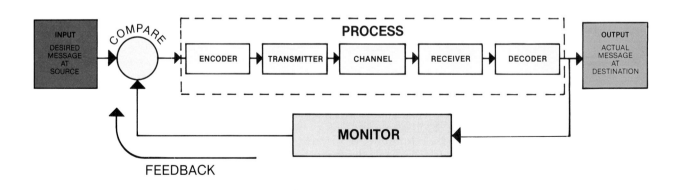

FIGURE 1–13 A complete system model of a communication system

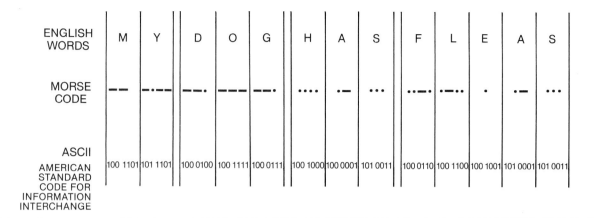

ENGLISH WORDS	M	Y	D	O	G	H	A	S	F	L	E	A	S
MORSE CODE	––	–·––	–··	–––	––·	····	·–	···	··–·	·–··	·	·–	···
ASCII AMERICAN STANDARD CODE FOR INFORMATION INTERCHANGE	100 1101	101 1101	100 0100	100 1111	100 0111	100 1000	100 0001	101 0011	100 0110	100 1100	100 1001	101 0001	101 0011

FIGURE 1–14 Encoding English words into Morse code and ASCII

Transmitters

The transmitter is the technical means used to send a message from the source to the destination. Some common transmitters include radio and televison transmitters; telegraph senders; humans writing with pens, pencils, or typewriters; and cameras. Transmitters convey the message onto the channel, Figure 1–15.

Channels

The channel is the path that the message takes from the transmitter to the receiver. All messages must move through a channel. Some communication processes use air as the channel. Radio transmissions and human speech involve messages transmitted through air. When radio waves leave an antenna, they travel through space at the speed of light (186,000 miles/second). The process of radio waves traveling through space is called **propagation.** Radio waves can be propagated to distant locations as they are reflected by a layer of the earth's atmosphere (between 70 and 200 miles above the earth) called the ionosphere.

FIGURE 1–15 The transmitter and antenna at a ham radio station (*Photos courtesy of American Radio Relay League*)

FIGURE 1–16 Cables installed on the ocean floor serve as the channel for a transatlantic telephone system. *(Courtesy of AT&T)*

Other channels involve wires or fiber-optic cables. Cables carry telephone signals over land and under the ocean, Figure 1–16. Cables also carry video signals from video cameras to video recorders. In photographic systems, film is the channel that conveys information from the camera to the viewer. In systems involving drawing or sketching, paper acts as the channel.

Receivers

The message must be taken from the channel by a receiver so that it can be interpreted by the user. In broadcast radio systems, AM or FM receivers and antennas are used.

Television signals are received by TV sets. These are television receivers and are connected

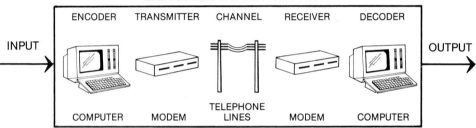

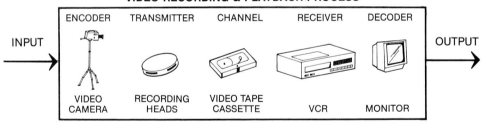

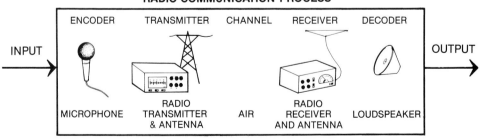

FIGURE 1–17 All communication processes include five components: encoders, transmitters, channels, receivers, and decoders.

to television-receiving antennas. Television sets can be tuned to the frequency of the incoming television signal. Each television station transmits on a different frequency. Tuning the channel selector allows the television set to receive the desired station.

In a photographic system, the film negative is the receiver. When we play a musical instrument, the audience is the receiver. In a Morse code telegraphic system, the receiver is the telegraph sounder.

Decoders

Once the message has been transmitted and is received by the receiver, it must be decoded (put back into a form which is understandable to humans), Figure 1–17. Morse code telegraph signals are decoded by the operator on the receiving end. Electrical signals transmitted by radio are decoded into voice and music by loudspeakers. Television monitors decode electricity. They produce light patterns on video monitors which can be interpreted by people. Printed halftones are decoded by the human brain which visualizes complete pictures and not dots.

Communication by Design

Since messages are designed to reach certain audiences, human psychology drives message design. A photographic advertisement intended to sell products to Japanese school children might not do well with British senior citizens. Likewise, a rap music track for a TV commercial could be appealing to American high-school students, but might leave their parents cold! The content of the communication, the way it is presented, and the choice of the communication medium depend upon the audience for which the message is intended, Figure 1–18.

The design of a communication system needs careful planning and analysis. The target audience and the cost must be considered. Technical aspects relating to the communication medium must also be taken into account. The design of a message is based on the following questions: Why, what, who, when, how, and how much? See Figure 1–19.

FIGURE 1–18 Music that is appealing to one audience may not be appropriate for another. *(Photo by Joseph Schuyler)*

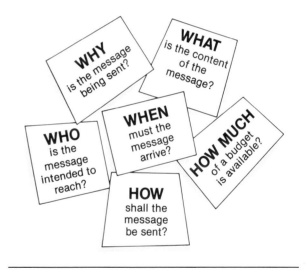

FIGURE 1–19

Why is the message being sent? Human needs or wants establish the need for a communication. As an example, think of a case where there is a life-and-death medical emergency needing immediate attention. Clearly, this need is the reason for a message.

What is the content of the message? In this case, the content of the message might be "This is an emergency. Send an ambulance immediately to 123 South Street."

FIGURE 1–20 For some messages time is critical. Satellite technology helps us to communicate quickly with very remote locations. *(Courtesy of TRW Inc.)*

Who is the message intended to reach? The message is intended to reach the ambulance corps that provides emergency service to your location.

When must the message arrive? The message must be received immediately. A delay could cost a life. In some cases, there is a real sense of urgency about a communication, and time is critical, Figure 1–20. Beside medical emergencies, urgent communications could involve financial transactions, such as the sale or purchase of shares of stock. Another urgent communication might relate to military defense. Some information is not time sensitive. Sending letters to friends is an example. Often it makes little difference whether a letter is received days or even weeks after it was sent.

How shall the message be sent? In the design process, we must recognize that there may be different technical methods that can be used to convey the message. Making the best choice is part of effective design for communication.

Time affects the selection of the communication process. In a medical emergency, we might use a telephone to call an ambulance. In the case where we wanted to let friends know about the incident, we might write letters. If we wished to let the public know about the fine service the ambulance corps offers, we might tape a story to be broadcast at some future time over radio or television.

The communication method we choose affects the message design. For example, if the message is to be communicated by a printing process, thought must be given to carefully choosing the right words to convey the message exactly as intended. We must also be aware of how the message looks on paper. Sizes and styles of lettering (type fonts) must be chosen, along with design elements (like lines or bars across the page). Often, a printed messsage also has some graphics (drawings and pictures). These must be put in the right places to best enhance the message.

If the process chosen is an electronic method, like radio, the design of the message also involves choosing the right words. In radio, however, we are concerned more with the way the message will sound than how it will look when written. Therefore, the quality of the voice used is very important. Television messages rely on high-quality sound and pictures. Clever commentary, attractive actors and actresses, and visual images that convey a mood are all crucial to a successful television production.

Often, the product being advertised will influence the design treatment. For example, a radio commercial ad for natural foods might use some folk music, while a radio commercial ad for an elegant restaurant might use classical music.

How much of a budget is available? Cost is a very important factor in message design. Generally, communications are designed to fit within a given budget. If more money is available, the message can be made to be more appealing. For example, printed messages might be done in color instead of in black and white. Television ads could include computer graphics. Photographs might use papers with special finishes. If we wish to hire well-known actors or actresses to do commercials, we may have a more effective message, but the costs may be too much.

Problem Solvers In Communication

Richard Saul Wurman, Communication Designer

"You've heard about the explosion of information? It's really an explosion of noninformation." So says Richard Saul Wurman, an architect turned graphic designer. Mr. Wurman explains that we have a noninformation overload. He points out that much of the information that bombards us is complex and not easy to understand.

Mr. Wurman sees communication design as a problem-solving activity. To try to make information more accessible to people, Mr. Wurman developed The Understanding Business in 1987. Mr. Wurman explains that the idea behind this business is that "to be understood, things need to be related to things you already know about. For example, if you want to explain what an acre is, you could tell someone that it equals 43,560 square feet, but they won't really understand how big that is. If you tell them that it's the size of a football field without the end zones, then they'll understand."

One of Mr. Wurman's projects is a series of guidebooks called Access Guides. These guides inform people about things to do in cities. They are organized by neighborhood and the listings are "word-maps" which accompany the readers as they walk through the streets. Graphic design principles are used throughout. Restaurants are in red type, parks are in green type. Included is a multicolored map of the places you can easily walk to from where you are.

Mr. Wurman's move from architecture to graphic design is not surprising to those who know him. He believes that architecture is not just building things. Rather, it involves mapping ideas and organizing space. He believes that "it gives precise and meaningful direction to how people move through things." His idea of graphic design does just that. His books organize information in a way that can be best used by the reader.

Richard Saul Wurman

Access Guides inform people about things to do in a city *(Courtesy of Richard Saul Wurman)*

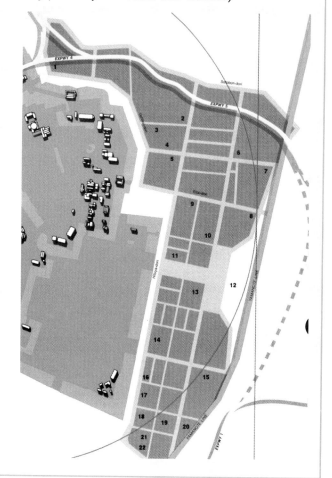

Monitoring and Evaluating the Message

The bottom line for any communication is whether it had its desired effect. Was the communication effective? In designing a communication system, some thought must be given to how the message will be **monitored** and **evaluated** after it is received. The actual message received could be a printed advertisement, a photograph, an image on a TV monitor, or spoken words, Figure 1–21. The output is not always what was intended. TV pictures may be affected by ghost images. Photographs may be too light or too dark.

Generally, the persons for whom the message was intended give feedback about how well the sender's goal was fulfilled. In person-to-person communication systems, feedback can be given by questions asked by the sender designed to find out if the message was understood.

FIGURE 1–22 Newspaper sales volume is a good indicator of the clarity and effectiveness of the message being communicated to the public. *(Photo by Michael Hacker)*

To monitor the quality of a radio or television signal, radio or televison receivers can be installed at the transmitting location and monitored by technicians.

Polls which ask listeners or viewers to respond to questions about a broadcast are another common technique of monitoring and evaluating broadcast quality and effectiveness. Televison networks conduct telephone surveys to get feedback from viewers. Through these surveys, television stations can find out if their programming is appealing to the viewing audience. If the surveys show that viewers are watching other channels, the network may substitute a new show which better matches the likes of the audience.

In the case of products that are printed (like magazines or newspapers) a measure of effectiveness is product sales, Figure 1–22.

Summary

To have effective communication, we must be sure that the message we have transmitted has been received and understood. Just saying something to someone does not mean that we have communicated.

FIGURE 1–21 Facsimile machines enable people to send replicas of documents and graphics around the world. *(Courtesy of Pitney Bowes, Inc.)*

People, animals, and machines can communicate. Person-to-person communication is often enhanced by technical means. Animals can communicate with people and with themselves. Machines communicate with people and people communicate with machines. Machines can also communicate with other machines.

Communication systems are designed with five different purposes in mind. These are to inform, persuade, entertain, educate, and control. Often, machines communicate with other machines to control their processes.

Communication systems like telephone and television broadcast networks surround us. A system is a combination of elements that function to achieve a desired result. In communication systems, the desired result is to communicate a message. A monitoring method is built into communication systems to give feedback about how well the actual results match the desired results.

The communication process has five parts. Encoding is the act of putting the message into a form that can be transmitted easily. The transmitter is the technical means used to send the message onto the channel. The channel is the path through which the message travels. The receiver captures the message so that it can be interpreted by the user. Once the message has been received, it must be decoded back into a form which can be understood by the person, animal, or machine that is the object of the communication.

The design of a communication system is largely determined by the audience for which the message is intended. For example, an advertisement that is designed to sell products to one audience might not be well accepted by another audience.

When designing a communication system, some way must be found to evaluate its effectiveness. Feedback is obtained by monitoring the output. This feedback can be in the form of polls or surveys. It can also be given by technicians who monitor quality. They do this by listening to speech or music, by looking at printed messages, or by viewing TV or radio signals on special test equipment.

REVIEW

1. Communication involves having a message received and understood. Explain why the following are or are not examples of communication.
   A. A bird chirping
   B. A computer controlling a machine tool
   C. The sound of breaking glass
   D. Taking a photograph
   E. Writing a letter which is read by a friend
2. Give two examples each of person-to-person, animal, and machine communication.
3. What are the five purposes of communication systems?
4. How might a communication system be used to persuade?
5. Draw a labeled system diagram of a radio news broadcasting system. Include input, process, output, and feedback elements.
6. If you were publishing a school newspaper, how might you get feedback about its effectiveness from community members?
7. Draw a labeled diagram of the television communication process. Include the components of encoding, transmitting, the channel, receiving, and decoding.
8. Give an example of a message design that has been influenced by the audience it is directed towards.
9. How might a broadcast TV station monitor its television transmissions?

CHAPTER 2

Role of Computers in Communication

OBJECTIVES

After completing this chapter you will know that:

- Computer technology is responsible for causing giant leaps forward in communication technology.
- A computer is a device that processes information according to the rules established by its program (set of instructions), which can be changed by an operator.
- Categories of computers are microprocessors, microcomputers, minicomputers, mainframes, and supercomputers.
- Like other technological systems, computer systems have inputs, processes, and outputs. Feedback is used to modify the process to get the desired result.
- Computer input devices include a keyboard, mouse, magnetic tape, hard disk, floppy disk, optical scanner, and microphone.
- The power of a computer refers to the number of instructions it can handle in a second.
- Computer output devices include video monitors, printers, plotters, and loudspeakers.
- Computer software provides the computer with specific instructions which direct it to do certain tasks.
- Computer applications include on-line communications, word processing, desktop publishing, and computer-aided design.

KEY TERMS

applications program
bits
bytes
central processing unit (CPU)
computer
computer software program
data communications
electronic mail

keyboard
laser printer
mainframe computer
memory
microcomputer
microprocessor
minicomputer

operating system
optical disk
plotter
printer
random access memory (RAM)
read only memory (ROM)
storage

hard copy
icon
input/output (I/O)

modem
mouse

supercomputer
synthesizer

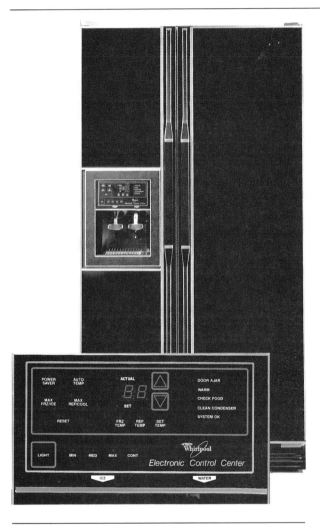

FIGURE 2-2 *(Courtesy of Buick Motor Division)*

FIGURE 2-1 *(Courtesy of Whirlpool Corp.)*

FIGURE 2-3 *(Courtesy of Goldome)*

Introduction

Computers are no longer only found doing difficult or tedious calculations for large companies. Small computers are now used in many home appliances. They provide intelligent displays on refrigerators, and control microwave ovens and washing machines, Figure 2-1. They are found in automobiles where they control the amount of gas and air entering the engine. They are also used to provide dashboard display information, Figure 2-2.

Computer technology is also responsible for causing great leaps forward in communication technology. In the business world, computers are used in automated bank teller machines (ATMs), Figure 2-3. They are contained in the cash registers, and scanning and inventory systems

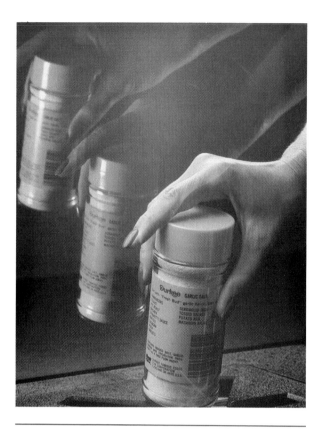

used in major supermarkets, Figure 2–4. In many offices, word processors have replaced typewriters. Computers communicate with machines in factories and command them to operate in specified ways.

Large or small, a **computer** is a device that processes information according to the rules established by its program (set of instructions), which can be changed by an operator. The information processing could include adding, comparing, sorting, or other operations.

Computers in Our World

Today, computers and microcomputers surround us, even though we don't always know that they are there. Computers come in all sizes, from a single chip less than one square inch in size to a whole room full of large cabinets and equipment. They also come in a large variety of capabilities: slow processors to very fast processors; very small memory to very large memory.

FIGURE 2–4 *(Photo Courtesy of NCR Corporation)*

FIGURE 2–5 The MARK I was the first IBM Computer. It first went into operation in 1944. It was about 50 feet long and 8 feet high. It used over ¾ million parts and 500 miles of wire. *(Courtesy of International Business Machines Corporation)*

Bits, Bytes, and Words

Almost all computers in use today are digital electronic computers. In digital computers, information is represented in binary form (ones and zeros). Any number or quantity, no matter how large or how small, can be represented by a binary number using a sequence of ones and zeros. These *binary digits* are called **bits**. A digital 1 represents the presence of a pulse of electricity. A digital 0 represents the absence of that pulse.

Bits are organized into groups of eight to make them easier to deal with. These groups of eight bits are called **bytes**. Each byte can represent one of 256 different characters (numbers, letters, punctuation, or other information). In talking about computers, we often talk about amount of data in terms of bytes, kilobytes (Kbytes = one thousand bytes), and megabytes (Mbytes = one million bytes).

Longer groups of bits are called words. Computer words can contain numbers to be processed (data words), or can be instructions (instruction words) which tell the computer how to process data. Computer words also include other bits which are called parity bits and sign bits. Parity bits are used as a check to see that the correct number of pulses have been processed. Sign bits tell if the data is positive or negative.

PARITY
BIT

| 1 | 1 | 0 | 1 | 0 | 0 | 1 | 1 | 1 | 0 | 1 | 0 | 0 | 1 | 1 | 0 | 1 | 0 | 1 | 1 | 1 | 0 | 0 | 1 |

| 8 BITS (1 BYTE) | BYTE #2 | BYTE #3 |

24-BIT WORD

Most of the computers that we are familiar with are small personal computers. Although they are small, they can be very powerful. There are also very large computers, which are used by government, businesses, and researchers to do large, complex jobs, Figure 2–5. For instance, such computers might record the income tax forms sent in each year by all taxpayers, or help engineers design new airplanes. These large computers have many things in common with the very small computers we find around us in our homes, Figure 2–6. It is the job of a user to select the right size computer for the job at hand.

From Micros to Supercomputers

The rapid development of integrated circuits over the last twenty years has made computers available to everyone. The electronic computer has been in use since the late 1940s. However, it was not until entire computers could be put onto tiny integrated circuit chips that they came into household use.

To better understand the sizes and capabilities of computers it is often useful to put them into categories. These categories are microprocessors, microcomputers, minicomputers, mainframes, and supercomputers.

FIGURE 2–6 Modern lap-top computers are over a million times as fast as the MARK I. *(Courtesy of Hewlett-Packard Company)*

Microprocessors

Microprocessors are integrated circuits that perform control and arithmetic functions. Microprocessors are embedded in appliances like ovens and washing machines. They are then used for single applications, like controlling the time of an on-off cycle.

Microcomputers have microprocessor chips at their heart, Figure 2–7. Microcomputers are the personal computers that are found in homes and offices. They often have a collection of integrated circuits with one microprocessor chip. Microcomputer memories are sometimes very small (one Kbyte), but may reach up to more than one Mbyte.

Microcomputers handle data and instructions 8, 16, or 32 bits at a time (one, two, or four bytes at a time). They are often referred to as eight-bit machines, sixteen-bit machines, or thirty-two bit machines, based on how many bits at a time the microprocessor chip can handle, Figure 2–8.

Minicomputers are somewhat larger than microcomputers, and are often used by several users in a small company, or in a department of a larger company or university. They can handle data 16, 24, 32, or more bits at a time. They often have large disk or tape secondary storage devices attached to them.

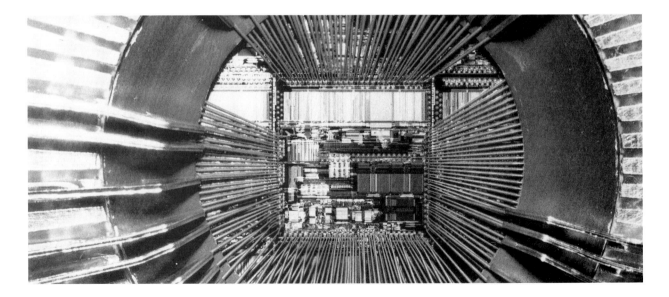

FIGURE 2–7 This chip measures 10 millimeters square. It can operate at 350,000 instructions per second. *(Courtesy of International Business Machines Corporation)*

FIGURE 2–8 The Tandy 4000 microcumputer can handle 32 bits at a time. This high speed allows the user to produce exceptional graphics. *(Courtesy of Radio Shack, a division of Tandy Corporation)*

Mainframe computers are the large computers used by large companies, government agencies, and universities for their administrative work, such as making out the payroll checks each week, keeping personnel records of employees, keeping track of orders, and maintaining a list of all items kept on hand in a warehouse (keeping an inventory). Mainframe computers handle data and instructions 32, 36, 48, and 64 bits at a time. They often have very large secondary storage devices (hard disks and tapes) attached. Mainframes can perform millions of instructions per second, and often have billions of bytes of memory attached to them.

Supercomputers are the fastest and largest computers. They are most often used for research, or for analyzing satellite data, or other very large problems, Figure 2–9. Today's largest supercomputers can perform several billion floating point operations per second $(1.23 \times 2.6 = 3.198$ is a floating point operation). Supercomputers are very expensive to purchase and to operate. They can cost up to $20 million.

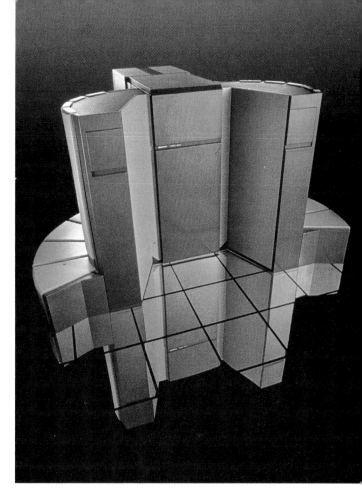

FIGURE 2–9 This Hitachi supercomputer is very fast. It is used for artificial intelligence experiments. *(Photo by Paul Shambroom, courtesy of Cray Research, Inc.)*

Computer Systems

Like other technological systems, computer systems have inputs, processes, and outputs, Figure 2–10. Input devices provide command inputs and information inputs. For example, an operator at a terminal may provide a command input by keying in a short application program. The operator may also provide an information input by keying in data.

A computer's processor acts on the information resources in response to the command input. The processor is thus clearly the process in the system model. The output of the computer system is the processed data. Feedback may be provided by a person, or automatically through hardware or software. Feedback is used to modify the process in order to get the desired result.

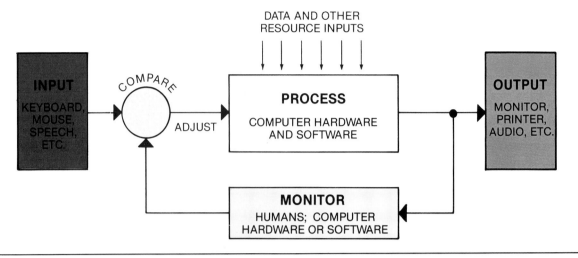

FIGURE 2-10 A computer is a technological system.

Mice in the Computer

An input device that has a body, and a cord that looks like a tail, is called a **mouse**. The computer operator moves the mouse along a flat surface. As the mouse moves, a ball in its base rotates. An electrical signal is sent to the computer, and the cursor is made to move across the screen. Some people find the mouse easier to use than a keyboard.

Certain software packages use pictorial images called **icons** to represent various functions. For example, a wastebasket may appear on the screen as a symbol for deleting a file. A mouse is often used in connection with these software programs to select the proper icon.

(Courtesy of Aldus Corporation, Seattle, WA)

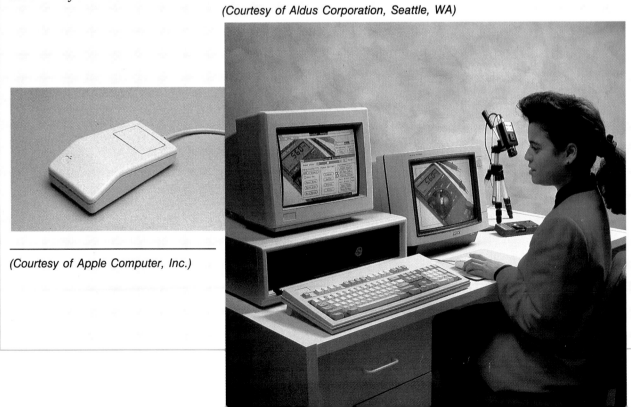

(Courtesy of Apple Computer, Inc.)

Computer Input Devices

In order for the information which the computer processes to be useful, it has to be exchanged with people or other machines outside the computer. This exchange is called **input/output**, or **I/O**. Many techniques of exchanging information with computers have come about because of the many ways in which computers are used.

The form of input (providing information to a computer) that you are probably most familiar with is through a **keyboard**. Using a keyboard which is very similar to a typewriter keyboard, people can issue instructions to (i.e., program) a computer, or provide raw data to it. The data can be processed using a program which was already stored in the computer. Using a keyboard, people can work with a computer on an interactive basis; that is, short instructions can be issued, with immediate results coming from the computer.

Inputs from keyboards and mice have already been described. Other forms of input come from a magnetic tape, hard disk, or floppy disk. These can contain a program or data which the computer will load into its main memory. Using tapes or floppy disks, people who develop new programs for computers can easily sell them, and transfer them to many other people who have compatible computers. Hard disks provide much more storage than floppy disks. Some hard disks contained within personal computers can store 100 megabytes or more of data.

Another form of input is through an optical scanner. This device recognizes the letters and numbers printed on a page, and converts them into a code of bytes that can be understood by the computer. This is one method used to put many pages of typewritten text into a computer's memory for future use. Bar code readers used in supermarkets are another form of optical scanner.

A more recently developed form of input, and one that is still being developed, is speech recognition. Some systems that have microphone inputs use special devices which can recognize some spoken words. These are converted into a series of bytes that are the same as if the words had been entered through a keyboard. This

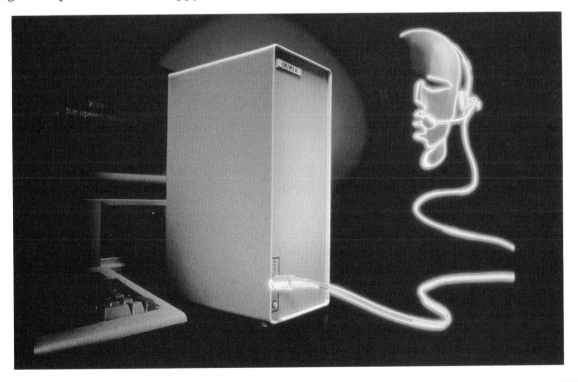

FIGURE 2–11 This experimental voice recognition system has a vocabulary of 10,000 words. *(Courtesy of Kurzweil Applied Intelligence, Inc.)*

technology is still in its developmental stage. The total number of words most devices of this kind can recognize is fairly small (up to several thousand), Figure 2–11.

The Central Processing Unit (CPU)

The **central processing unit**, or **CPU**, is the heart of the computer. It controls the flow of data, the storage of data, and the way in which the computer works on the data. The CPU reads the program (set of instructions) and converts them into actions. These actions might include adding two numbers, comparing two numbers, or storing a number or letter. The CPU includes a control unit and an arithmetic logic unit (ALU), Figure 2–12. The control unit directs the data flow through

the system and controls the sequence of operations. The arithmetic logic unit performs the mathematical calculations on the data.

Most computers use a single processor and do serial processing. Serial processing does one computing job at a time. It is like a single worker building a house all alone.

Very fast computers handle instructions through different paths at the same time to increase the number of instructions handled per second. This is called parallel processing. With parallel processing, computers can handle hundreds of millions of instructions per second. Parallel processing is like a team of people building a house together, Figure 2–13.

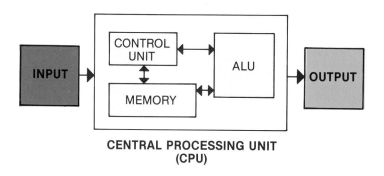

FIGURE 2–12 A process box with an arithmetic logic unit, a control unit, and memory

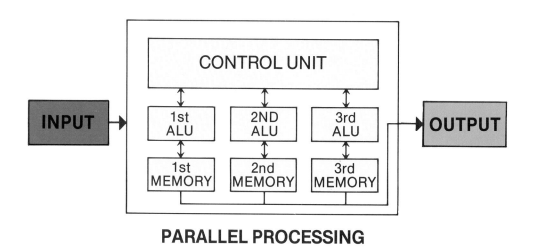

FIGURE 2–13 In a parallel processor, computing tasks are divided among multiple processors.

Computer Processing Power

The definition of a supercomputer is constantly changing, as computers get more and more powerful each year. What was considered to be a supercomputer only twenty years ago is now available in a desktop personal computer.

The power of a computer is largely dependent upon the number of components that can be placed on an integrated circuit chip. In 1960, integrated circuits had 4 or 5 transistors. By the year 2000, it is expected that chips will contain one billion components.

The power of the central processing unit refers to how many instructions it can handle in a second. Personal computers can handle hundreds of thousands of instructions per second. Large computers used in business or research can handle millions of instructions per second (MIPS). The largest supercomputers can handle more than 100 million instructions per second.

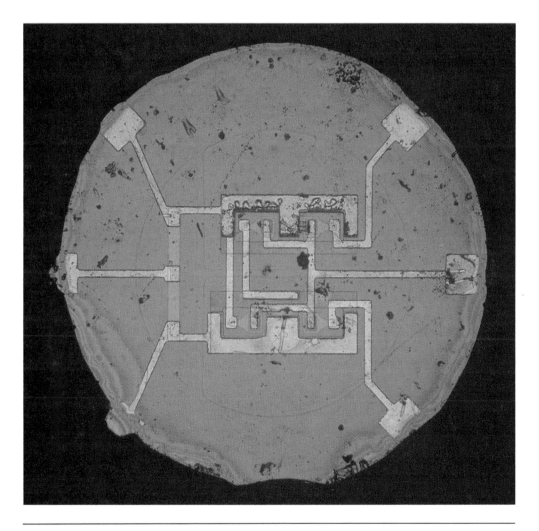

1961 Integrated circuit with four transistors (*Courtesy of National Semiconductor Corporation*)

Computer Memory

The place where the program is stored is called the **memory**. The memory also stores the information which is being worked on at any one time. Within the memory of the computer, information is sent to special registers. Registers are like storage bins. Each register has its own address which identifies it.

Almost all modern computers use integrated circuit memory circuits, sometimes referred to as semiconductor memory. Today, a single chip 3/8″ × 5/8″ can store over 100,000 characters, Figure 2–14. New memory chips are being developed with even higher capacities.

The memory that stores the program and the information that is currently being worked on is called the computer's main memory. When a computer is referred to as a 64 Kbyte computer, it is the size of the main memory that is being described as 64 Kbytes. Modern personal computers have main memory sizes of 8 Kbytes (small lap-top portables) to more than 1 Mbyte (one megabyte, or one thousand Kbytes). Large computers used by businesses and universities have main memories of tens of Mbytes.

Main memory is made up of **read only memory (ROM)** and **random access memory (RAM).** A small part of the main memory is assigned to ROM. ROM is permanent memory that is used to control the operation of the computer when it is first turned on ("booted up"). Computers can only read data from ROM. The computer operator cannot use ROM to store and process new information.

RAM comprises the majority of the main memory. RAM is used to store data and instructions which are used during execution of a program. RAM is referred to as volatile memory because it loses its data when the computer is turned off.

In addition to the main memory, a computer needs to have another, much larger memory so that it can store information for use at a later time. This kind of memory is called **storage**, or secondary storage. It is designed to be very large, so that many different types of information, or large amounts of the same type of information, can be stored.

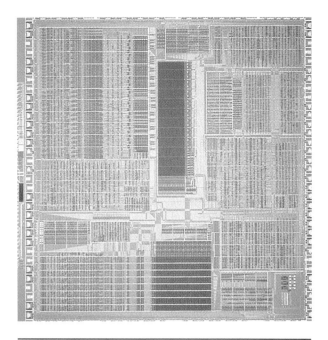

FIGURE 2–14 1985 chip with 132,000 transistors. *(Photomicrograph courtesy of Intergraph Advanced Processor Division)*

Secondary storage includes floppy disks, hard disks, and magnetic tape. In each of these, data is stored magnetically. The surface of the disk or the tape (which is called the medium) is coated with a very thin layer of iron oxide, which is a magnetic material. A very small electromagnet called a head is placed near the tape or disk as it is moving. A voltage applied to the head will magnetize the tiny pieces of iron oxide, thus putting information onto the medium, Figure 2–15. This is called writing to memory. When information is needed from the disk or tape, the electromagnetic head senses the magnetic field stored in the iron oxide coating of the medium, and turns it into electrical impulses. This is called reading from disk or reading from tape, Figure 2–16.

The amount of data that can be stored in secondary storage is almost limitless. If one disk or tape is full, it can be removed and another one can be put in its place.

A typical 5.25″ floppy disk, the secondary storage medium most often used with personal computers, can store up to 360 Kbytes of information, or 360 thousand characters. This is

FIGURE 2-15 This magnetic disk drive spins at 3600 RPM. Recording heads move at about 100 MPH over the surface. *(Courtesy of International Business Machines Corporation)*

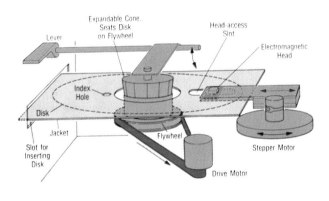

FIGURE 2-16 Typical disk drive mechanism *(Reprinted from USING COMPUTERS IN AN INFORMATION AGE by Brightman and Dimsdale, © 1986 South-Western Publishing Co.)*

roughly equal to 200 pages of typed text. Increasing in popularity is the 3.5″ micro-floppy disk. This disk has a very high storage density. It can store almost 1 Mbyte of data. A small hard disk that is used in many personal computers can hold up to 40 Mbytes of information, or 40 million characters. This is over 20,000 pages of typed text. There are hard disk drives which contain several large disks, with a total storage capacity of several hundred Mbytes. Tapes have similar, very large storage capabilities.

Another technology that holds a great deal of promise for the near future is **optical disk** storage technology. Optical disks can be used to store computer data, just as they are used to store entertainment on audio compact disks and in video disks. The emerging part of this technology is in how to write to the disk, as well as reading from the disk. Optical disks have very large memory capabilities, Figure 2-17. Manufacturers who are developing them talk about billions of bytes, or gigabytes (Gbytes).

Computer Output

Computer output can also be in many forms. Probably the most common is the video monitor, or CRT screen. CRT stands for cathode ray tube. The cathode ray tube is the special tube that converts electronic signals to visual images. CRTs are used in televisions to produce the picture you watch (see Chapter 18), and in electronic test equipment called oscilloscopes, as well as in video monitors. Video monitors can be used to display text (i.e., letters, numbers, and punctuation), graphics (pictures), or a combination of the two. Monitors can be monochrome (black and white, green and black, etc.) or full color.

Another very commonly used computer output device is a **printer**. A printer records the computer's output on a piece of paper which may be mailed to someone else or saved in a filing cabinet. Such a paper record of a computer's output is called a **hard copy**. Printers used in small computer systems are usually dot matrix printers or daisy wheel printers. Dot matrix printers can be used to print letters, numbers, and punctuation, but they may also be used to make drawings. Because they use dots and not continuous lines their letters and pictures are not of very high quality, but are acceptable for many applications, Figure 2-18.

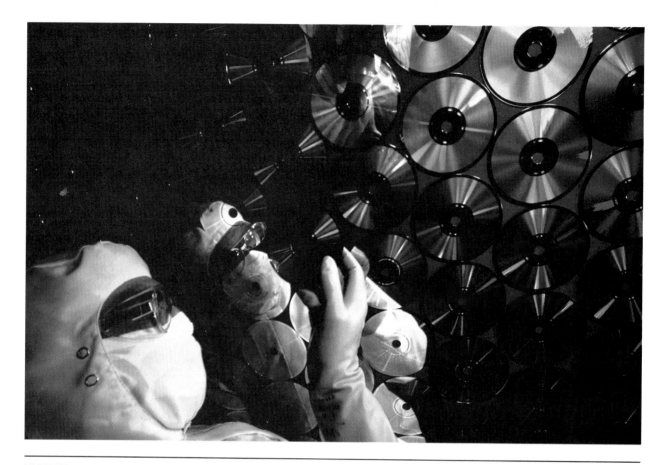

FIGURE 2–17 An 8-inch optical disk can store as much as 15,000 sheets of paper. *(Courtesy of 3M Optical Recording)*

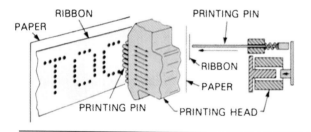

FIGURE 2–18 A dot matrix printer has a print head with pins that are moved by electromagnets to strike a ribbon, making an impression on paper. *(Reprinted from USING COMPUTERS IN AN INFORMATION AGE by Brightman and Dimsdale © 1986 South-Western Publishing Co.)*

Daisy wheel printers give higher quality printing of letters, numbers, and punctuation, but cannot create graphics. Daisy wheel printers are generally more expensive than dot matrix printers, and usually operate at a somewhat slower speed, Figure 2–19.

FIGURE 2–19 A daisy wheel contains many "petals." Each petal has a raised character on it. A daisy wheel printer positions the correct character and strikes it. This pushes the character onto an inked ribbon, which leaves an impression on the paper. *(Reprinted from USING TECHNOLOGY IN AN INFORMATION AGE by Brightman and Dimsdale © 1986 South-Western Publishing Co.)*

FIGURE 2–20 This small desktop laser printer produces very high-quality printed copies. *(Courtesy of Hewlett-Packard Company)*

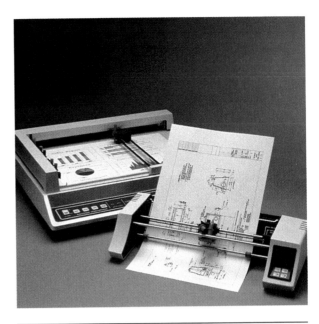

FIGURE 2–21 A plotter may be used to make multicolored drawings. *(Photo courtesy of Houston Instruments, a division of AMETEK)*

For higher quality output, **laser printers** are used. Laser printers can print text and graphics at high speed, Figure 2–20.

Large computers use printers that can print much faster than the dot matrix or daisy wheel printers used with personal computers. These printers can print whole lines at a time (line printers) or whole pages at a time (page printers). One of the most difficult problems with these printers is how to handle the paper they use, because it moves through the printer so fast.

Another output device that is like a printer is called a **plotter**. A plotter uses one or more pens whose position may be controlled by a computer to create drawings on a piece of paper, Figure 2–21. Plotters are often used in systems which are used to create, modify, and store drawings (CAD, or computer-aided drafting systems).

Loudspeakers provide audio output in the form of tones, beeps, music, and voice. Tones and beeps can be used to signal to the operator that the end of a page has been reached, or that the operator has issued an improper command to the computer. Music can be made inside the computer by a device called a **synthesizer**. It can produce a wide range of tones, volume, music, and even percussion (sounds like drums). Speech

can be produced by a voice synthesizer. Modern voice synthesizers can work with a large vocabulary, as they use the rules of pronunciation to generate voice sounds. Voice synthesizers are often used by the telephone company to give the time of day and to give out numbers when you call the information operator.

The output of computers can be sent to other computers, or to terminals located at a distant location. This kind of information exchange is called **data communications**. In order for computers to send data communications, a special device called a **modem** is often used. Modems send computer data over standard telephone lines, using sequences of tones to represent the bits of information to be sent. Using inexpensive modems, up to 1200 bits per second can be sent over telephone lines. At 1200 bits per second, one page of typed print would take about 12 seconds to send. Using more expensive and complex modems, 9600 bits per second can be sent over telephone lines (see Chapter 20).

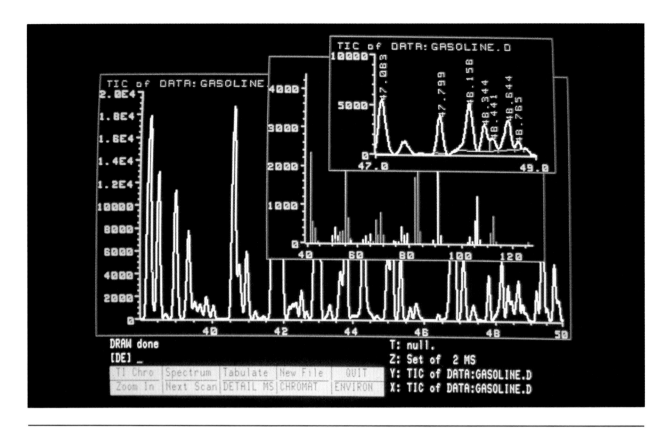

FIGURE 2–22 Special software is required to display information needed to identify and measure chemical substances. *(Courtesy of Hewlett-Packard Company)*

Computer Software

A computer does its work based on a list of instructions, called a **computer software program**. This can be changed at any time. The computer is thus a general purpose tool that can be made to do a given job by giving it the proper instructions. If the software program (set of instructions) is changed, the computer can do a different job. The computer is thus under program control.

Computer software exists at different levels. Three important types of software are **operating systems; applications programs** that are purchased, ready to use; and applications programs written by the user in a programming language.

The computer's operating system is the set of instructions which allows the user to control and access the computer's memories (disk, tape, semiconductor, etc.), printers, and other peripheral devices. The operating system also makes the computer's components available to other software, such as applications programs. Operating systems must be designed with the characteristics of the computer in mind, as well as the likely uses it will have, Figure 2–22. Sometimes more than one operating system is available for a computer. The user must choose which one is needed, based on the kind of tasks the computer will be expected to perform. Examples of operating systems used in personal computers are MS-DOS® and TRS-DOS®.

An applications program gives instructions to the computer to do a certain well-defined task. Applications programs range from computer games, to word processors, to student instruction programs, to income tax preparation programs, to automobile engine control programs and rocket design programs. Applications programs are stored on tape or disk. They may be purchased at a computer store for personal computers, but may also be large and very complex, in the case of mainframe computer programs.

When choosing an applications program, a user must make sure that it will work with (is compatible with) the operating system that is being used on the computer. Many applications programs are available to be used with a number of operating systems; these applications programs can then be used on different computers, or on similar computers using different operating systems.

If an applications program is not available to do a job, a custom-written applications program may have to be written. This may be easy to do, or may be a very large effort, taking many people several years to finish, depending on the problem to be solved. Such custom-written applications programs are written in one of many programming languages available today. Each programming language has unique features which make it useful for writing certain kinds of applications programs.

Some of the common programming languages are BASIC, PASCAL, C, COBOL, FORTRAN, ADA, and LISP. BASIC (**B**eginner's **A**ll-purpose **S**ymbolic **I**nstruction **C**ode) is used by students, business people, and hobbyists. It is perhaps the most widely used programming language today, largely because it is relatively easy to learn and use. BASIC is available for use on most small personal computers as well as large mainframe computers.

Applications of Computers in Communications

On-line Computer Networks

By dialing a telephone number that connects your home computer to a mainframe computer you can have access to a host of information services. For example the Compuserve Information Service gives news, weather, and sports information, as well as banking, travel services, and home shopping, Figure 2–23. General Electric's

FIGURE 2–23 COMPUSERVE® provides access to a very large information network. *(Courtesy of Compuserve Information Systems)*

Computer Viruses

(Courtesy of USA Today Magazine © January, 1989 by the Society for the Advancement of Education)

Just as infections can attack the human body, can be contagious and spread, computer viruses can infect a computer program and spread to other software. A computer virus is a software program that attaches itself to other programs with which it comes in contact. When a virus infects a program, it interferes with the computer's operation by altering or deleting files, or by consuming all the remaining computer memory.

In November, 1988, the ARPAnet (Advanced Research Projects Agency Network) computers were invaded by a virus. Over 6,000 computers in this network, used by the Department of Defense and research institutions, had to shut down. The virus was grown by a computer-science graduate student who thought he was just playing a prank. He didn't realize how much panic he would create, as his virus went out-of-control and threatened to block the use of huge amounts of important and irreplaceable data.

Typically, viruses lodge themselves in the computer's main memory or operating system. Any other program that is executed is then infected. When infected software is shared by other computer users, the virus spreads. This spread could occur when users trade software, use telephones and modems, or communicate through computer networks.

Computer viruses are a real threat to our information-based society. Criminals who wish to steal financial data, or who gain access to computer systems containing classified documents, could paralyze our finance and defense industries. They could threaten our largest corporations. They could even jeopardize the nation's air traffic control system. The Federal Bureau of Investigation now is looking into computer virus cases. Developing a virus and threatening to destroy data is serious business, and not a meaningless joke.

Computer programs are being developed to combat viruses. These immunization programs detect viruses by comparing actual program length to expected program length, or by checking for the presence of certain known viruses. These programs have names like "Disk Defender," "Interferon," and "Vaccine."

In this information age, knowledge is our most precious resource. It must be protected and treated with great respect.

GENIE offers on-line access to a whole encyclo-pedia. The International Technology Education Association has developed Technology Link, which provides communication among technology teachers nationwide. TechNet is a nationwide system that uses a mainframe computer housed at the New York Institute of Technology and provides teacher-to-teacher and student-to-student computer-based communication. For people who travel, the Official Airlines Guide is available on-line. By dialing into their network, you can find out about airline schedules and even book a flight from your home or office.

Electronic Mail

Instead of sending letters through the mail, you can use computers to send electronic messages. In large companies, **electronic mail** systems are becoming a popular way to provide communication among their employees. Memos can be sent to certain people, and copies can be sent to others. The messages are received in ''electronic mailboxes'' that are assigned to individual users.

Word Processing

Word processors offer typists many advantages over typewriters. When you type on a word processor keyboard, the text is displayed on a monitor screen, and is stored in the computer memory. Changes can be made quite easily by simply retyping the text. Only after the typist is satisfied that the text is all correct is it printed on paper.

Word processors have features that typewriters do not have. For example, the whole text can be checked for proper spelling. Paragraphs can be moved from one place to another. The number of columns and lines on each page can be rearranged easily. Most modern offices now use word processors instead of typewriters.

Computer-Aided Design

Computers are changing the way people do technical drawing. With computer-aided design (CAD) systems, drafters can create, change, and store more complex drawings than ever before.

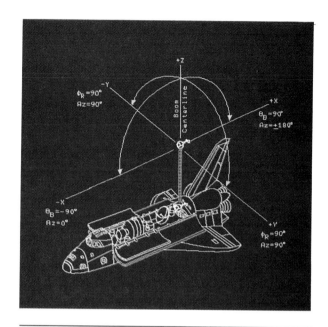

FIGURE 2–24 Computer-aided design is used to show how a satellite reacts to the movement of a space shuttle. *(Courtesy of Hewlett-Packard Company)*

Computer-aided drafting has become one of the most important tools used by engineers, drafters, and designers, Figure 2–24.

Desktop Publishing

Computers are now used to produce whole pages with words, headlines, and illustrations, Figure 2–25. Using special software, a mouse, and a laser printer, desktop publishing lets you arrange copy on a computer screen just as you would like it to appear on paper. The image on a screen is called a WYSIWYG image. WYSIWYG is pronounced whizzy-wig and stands for ''what you see is what you get.''

Computer technology is responsible for causing great leaps forward in communication technology. Computers are all around us. They come in sizes from a single chip to a whole room full of equipment. Almost all computers are digital electronic computers which represent information in binary form (ones and zeros). These binary digits are called bits.

FIGURE 2–25 Some materials generated by desktop publishing methods. *(Courtesy of Aldus Corporation)*

Categories of computers are microprocessors, microcomputers, minicomputers, mainframes, and supercomputers. Microprocessors are found in appliances and are used for single purposes.

Microcomputers include the personal computers found in homes and offices. Minicomputers are often used by several users in a company. Mainframe computers are large computers used by large companies, universities, and government agencies. Supercomputers are very fast and have large memories. They are most often used for research and for analyzing very complex problems.

Like other technological systems, computer systems have inputs, processes, and outputs. Feedback is used to modify the process to get the desired result.

Computer input devices include a keyboard, mouse, magnetic tape, hard disk, floppy disk, optical scanner, and microphone. Hard disks used in personal computers can hold 40 megabytes or more of data. Floppy disks can hold up to a megabyte of data.

The central processing unit (CPU) reads the computer program. It controls the flow and storage of data, and the way in which the computer works on the data.

The power of a computer refers to the number of instructions it can handle in a second. Computer power depends upon the number of components that can be placed on an integrated circuit chip.

Computer memory is the place where the computer program and the information being processed are stored. Main memory refers to read only memory (ROM) and random access memory (RAM). A computer also needs secondary storage provided by hard and floppy disks, tapes, and optical disks.

Computer output devices include video monitors, printers, plotters, and loudspeakers.

Computer software provides the computer with specific instructions which direct it to do certain tasks. Since computer programs can be changed to suit a particular application, a computer can be thought of as a general purpose tool.

Computer applications include on-line communication, word processing, desktop publishing, and computer-aided design.

REVIEW

1. Explain how computers have caused giant leaps forward in communication technology.
2. What are the five major categories of computers?
3. Draw a system diagram of a computer system. Explain the input, process, output, and feedback portions of the system.
4. Name five input devices used in computer systems.
5. What is the function of the central processing unit (CPU)?
6. Why is it that the power of computers is increasing, although the size is decreasing?
7. Explain the difference between ROM and RAM.
8. What types of output devices do the computers in your technology lab use?
9. What is the difference between operating system software, and applications software?
10. What applications have you personally used computers for?

SECTION ACTIVITIES

 A DAY IN MY LIFE

OVERVIEW

This activity gives you an opportunity to describe a typical day in your life and how modern communication technology affects the way you do things. In this activity, you will explore a large number of technologies (especially communication technology) that impact your daily activities.

For example, if you live in the suburb of a modern city, you might be awakened by a digital clock radio. While having breakfast, you would check your telephone answering machine to see if you missed any calls from the previous night. One of your parents might drive you to school. While in the car you would listen to the car radio for news, weather, and traffic information, while your parent uses the car telephone to make a business call. While on the phone, your parent is paged on a pocket pager, and asks the company operator to be connected to the company's office in another city over a satellite network.

This illustrates some of the communication technology around you everyday. What other technological products do you use or come in contact with on an everyday basis?

MATERIALS AND SUPPLIES

To complete this activity, you need a notebook and pencil to keep track of your daily activity.

PROCEDURE

The following procedure will assist you in the preparation of your paper:

1. Get a pencil and paper, and outline each of your activities for the day.
2. Be sure your day's activity is in the proper order.
3. Expand your notes into a short statement about each activity. Explain how technology affects each activity. For example, what technologies affect your morning wash-up routine? Your meals? How you are transported to school? How you spend your free time?
4. Place an asterisk (*) next to all activities that are impacted by communication or communication technology.
5. If available, use a word processing program to prepare your final copy.
6. Prepare one paragraph in which you anticipate new developments in communications technology that will provide additional impacts on your daily life.

FINDINGS AND APPLICATIONS

In this activity you found that a large number of technologies currently impact each of us as we go through our daily activities. You are probably also impressed with the quality and complexity of some of these technologies that we take for granted.

ASSIGNMENT

1. Prepare a paper describing a new communication technology that will soon impact on our lives. How will the technology affect us?
2. Observe communication activities around you for one day. Classify the communication that you observe under the following categories:

   - Human to human
   - Human to animal
   - Animal to human
   - Human to machine
   - Machine to human
   - Machine to machine

   Prepare this information in an easy-to-interpret form and submit it to your instructor.

 COMMUNICATE YOUR MESSAGE

OVERVIEW

This activity expects you to convey a message to someone else, and from that person get some feedback to indicate completion of the communication process. A typical activity may include inviting someone to the school dance using a creative method, and then receiving that person's response. The message may be communicated using a skywriter, a local ''rent-a-clown,'' with a song, in a letter, over the school announcement system, by modem, during half-time at the football game, or any one of numerous methods. The choice is up to you. Each message will ask a particular question of an individual. Responses will be given in similar methods, but must be received to complete the communication loop. You may choose any communication process which involves developing a message, sending the message, and having someone receive the message and send a response to be received by you.

As you prepare a message, be sure to provide enough information so that the receiver gets the correct message. You will then be able to anticipate the response that your communication activity evokes.

MATERIALS AND SUPPLIES

Numerous technologies may be used to transmit the initial message. It will be your task to select which technology you will use.

PROCEDURE

To prepare and communicate an appropriate message, the following steps must be included:

1. Select a person to receive your message.
2. Develop the message to be transmitted. Be sure sufficient information is contained.
3. Develop whatever assistance is needed to send your selected message.
4. Transmit the message.
5. Evaluate the response to determine whether or not the correct message was received.

FINDINGS AND APPLICATIONS

In this activity you found that communications is *not* a simple process consisting only of sending information. Communication requires the receipt and interpretation of the message, and some feedback to determine whether the correct message was received. You also may have discovered a number of different methods of communication, and were able to analyze each to determine which one would best meet your needs.

ASSIGNMENT

1. Develop a message that can be transmitted to the rest of your class using only hand signals and facial expressions, but no audio or voice. Deliver that message. What feedback do you receive?
2. Design a chart to illustrate the process of encoding, transmitting, receiving, and decoding a message. Indicate what type of channel is used. This chart would be titled ''Communication Cycle.''
3. Describe briefly a communication method used between human and animals.
4. Describe the communication process between humans and machines. Illustrate one example.
5. Describe the communication process between machine and machine. Illustrate one example.

SECTION TWO

TECHNICAL AND COMPUTER-AIDED DRAWING

Drawings on cave walls are the earliest known form of nonverbal communication between people. Drawings are still used to communicate complex ideas, but special paper and computers are now used to create and store the drawings. Whether drawn on paper or with the aid of a computer, technical drawings form a communication system.

In order to have complex architectural, industrial, or other ideas understood exactly, rules and procedures for creating the drawings have been developed. These rules and procedures, and the tools used for creating the drawings, are the subject of this section.

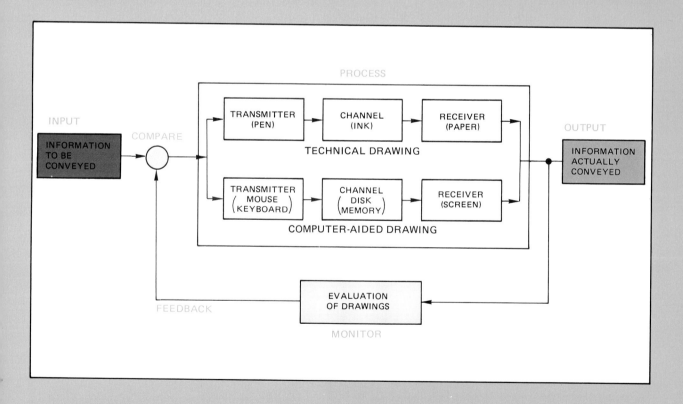

Sketching, Drawing, and Illustration

KEY TERMS

American National Standards Institute (ANSI)	irregular curves	scale
blueprint	lines of sight	sketches
charts	metric measure	technical drawing
compass	object	technical illustration
diazo print	orthographic projection	templates
dividers	parallel projection	title block
freehand drawing	perspective projection	U.S. customary measure
graph	piercing points	vanishing point
guides	projection plane	view point

Drawing to Communicate

Drawing is one of the oldest and most basic forms of communication. Even before there were spoken languages, cave dwellers communicated by drawing simple figures on the walls of caves, Figure 3–1. As the world became more and more complex and as people began to construct and manufacture, the need for drawings increased. As a result, technical drawing technology

FIGURE 3–1 Drawings have been a main form of person-to-person communication throughout history. *(Photo courtesy of Westinghouse)*

evolved as a way to ensure uniform drawing communication.

In our technological world, the ability to read and interpret drawings is needed in many occupations and is important to all of us. Perhaps you have experienced trying to follow the instructions for putting together a new toy or connecting a stereo system. Such instructions usually include a drawing that you must be able to read and interpret.

Drawings can be used to convey simple and complex messages in a clear and precise manner. Information can be given and ideas expressed using drawings and graphics that would take thousands of abstract words to express.

A drawing is a graphic representation or picture of an idea in the mind of the person creating the drawing, Figure 3–2. Object details such as size, shape, and locational relationships can be clearly described with drawings. A drawing is a communication tool.

Drawing Categories

Generally, drawing is divided into three areas:

1. Sketching and freehand drawing
2. Technical drawing
3. Technical illustration.

Sketches are freehand drawings used to formulate, express, and record information and ideas. They vary widely in quality based on the intended use and the person doing the sketching.

The ability to sketch well is a basic communication skill in today's world. A picture is worth a "million" words today. The sketched destination map made by a truck driver solving a routing problem and the incision sketches on a skull made by the brain surgeon are uses of the same problem-solving and communication techniques.

A quality working sketch has all the information found on a technical drawing. The sketch

FIGURE 3–2 The process of communication using technical drawing involves transforming the visualization into a graphic form. *(Courtesy of Simpson Industries, Inc.)*

FIGURE 3–3 Sketches are freehand drawings used to formulate, express, and record information. *(Courtesy of Faber Castell)*

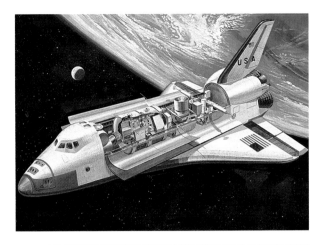

FIGURE 3–4 Technical illustration is a blend of freehand and technical drawing as seen in this illustration of the space shuttle. *(Courtesy of NASA)*

may be used by craftspeople to make the first samples of prototypes, Figure 3–3.

Freehand drawing is often done by artists and used to express aesthetic ideas. The person drawing with an artistic goal may not be concerned with perfect accuracy in shape, size, and relationships. Because the drawing is not to be used as a guide to construction, details are often not developed.

Technical drawing is used to convey information accurately and completely. The craftsperson or machine operator should be able to create the object exactly as it was visualized by the drafter. For example, a drawing of a machine part should have all of the information needed to produce the part without having to ask for more instructions or look up more information. A set of house plan drawings should have all of the information needed by the various tradespeople to build the house. Such drawings are tools for communicating the ideas of the designer as well as specifications to the people who will make the design.

Technical illustration blends technical and freehand drawing, Figure 3–4. The goal of the technical illustrator is to present a picturelike drawing of the design which shows how it would appear in the real world. Many technical illustrations look like photographs rather than drawings. Technical illustrations are used in technical journals, magazines, instruction packages that

come with consumer products, and as part of sets of architectural and industrial drawings.

A Universal Communication Form

Drawing was one of the first communication technologies. It is still the closest thing to a universal language form. A technical drawing created by a designer, engineer, or drafter in India can be read by a machine operator in France or Australia. An integrated electronic circuit diagram using standardized schematic symbols designed and drawn in the United States, can be used in China to make the microchip for a computer to be put together in Brazil.

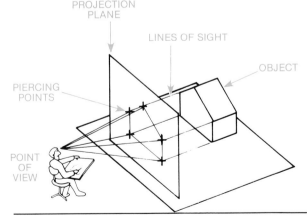

FIGURE 3–5 The basic visualization elements are viewpoint, projection plane, piercing points, lines of sight, and object.

Drawing Presentations

When we visualize, we create a picture of an object or idea in our mind. We see this mental picture as if we were standing at some point and looking at the object. The elements we use to build our visualization are the **object, view point, projection plane, lines of sight,** and **piercing points** on the projection plane, Figure 3–5.

Converting what we see in our mind into a technical drawing can be done in two ways: in **perspective projection** or **parallel projection**, Figure 3–6.

Perspective Projection

Notice in Figure 3–6 that the perspective method imitates the way the human eye actually sees an object.

Perspective projection drawings appear more natural to the viewer than parallel projection drawings do, because they are drawn as they would appear naturally to the eye of a viewer. The best way to understand perspective drawing is to stand on a railroad track and look at a point on the tracks off in the distance. You will notice that the tracks seem to merge at what is called a **vanishing point**. Perspective drawing reflects this. Parallel projection does not. In parallel projection, the railroad tracks would stay parallel forever. The two most widely used types of perspective drawings are one-point and two-point perspective, Figure 3–7.

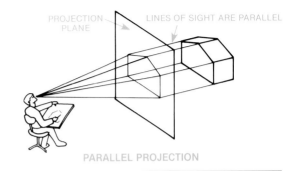

PROJECTION PLANE · LINES OF SIGHT ARE PARALLEL

PARALLEL PROJECTION

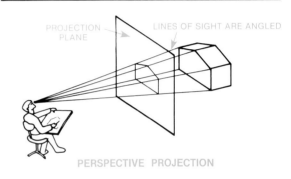

PROJECTION PLANE · LINES OF SIGHT ARE ANGLED

PERSPECTIVE PROJECTION

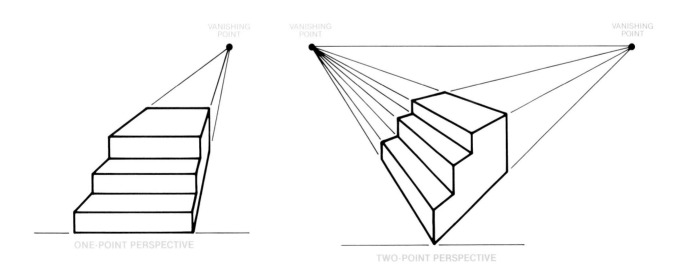

VANISHING POINT · VANISHING POINT · VANISHING POINT

ONE-POINT PERSPECTIVE

TWO-POINT PERSPECTIVE

FIGURE 3–7 One-point and two-point perspective projection

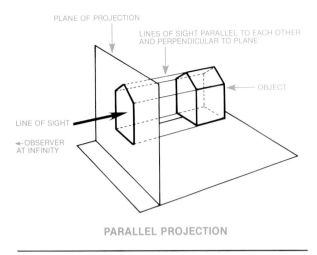

FIGURE 3–8 Parallel projection drawing

Parallel Projection

Parallel projection drawings are drawn with the lines of sight parallel to each other, Figure 3–8. The most common type of parallel projection drawing is called **orthographic projection**. In orthographic projection, the lines of sight are parallel to each other and are also perpendicular to the plane of presentation. Orthographic projections are the most important form of technical drawing used to communicate throughout the world today. The orthographic projection technique is used to create multiview drawings which show several views of an object (see Chapter 5). Because multiview drawings show the surfaces of an object in full or scale shape and size, all readers see objects the same way.

Drawing Materials and Equipment

Many tools and special materials are used to produce modern, high-quality technical drawings.

Materials To Draw On

There are three types of commercial materials used for drawing: paper, cloth, and plastic film. Drawing paper varies greatly in quality. Papers having high rag or long fiber content are considered the highest quality. Papers are bought in either sheet or roll form.

Tracing paper is used to produce drawings to make prints or copies. The most widely used tracing paper is called vellum. A cloth material called linen was used for many years for all quality drawings. Cloth drawing materials are rarely used anymore for technical drawings.

Modern plastic, or polyester film, is the best drawing material used today. One type of plastic film that is widely used is called ''Mylar'', which is a manufacturer's tradename. Drawing film is made with a matte finish on one surface to allow graphite, ink, or other drawing medium to adhere well. Plastic film is very sturdy. It will not distort or be damaged by water, and makes excellent prints. Most high-quality drawing work today is made on plastic film.

Drawing sheets come in standardized dimensions, Figure 3–9.

Size	INCHES Dimensions	
A	8½ x 11	9 x 12
B	11 x 17	12 x 18
C	17 x 22	18 x 24
D	22 x 34	24 x 36
E	34 x 44	36 x 48

Size	MILLIMETERS Dimensions
A–4	210 x 297
A–3	297 x 420
A–2	420 x 594
A–1	594 x 841
A–0	841 x 1189

FIGURE 3–9 Drawing sheet size standards

Pencils and Pens

The purpose of the marking instrument is to mark or inscribe an image on drawing materials.

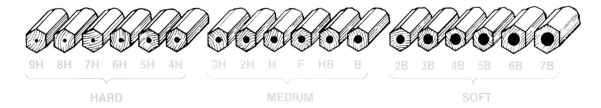

9H	8H	7H	6H	5H	4H		3H	2H	H	F	HB	B		2B	3B	4B	5B	6B	7B

HARD   MEDIUM   SOFT

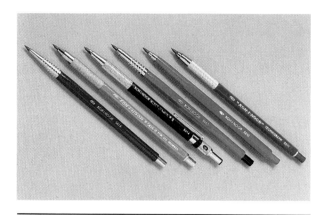

FIGURE 3–10 (A) Drafting pencils (B) Cased pencils and lead grades *(Courtesy of Koh-I-Noor Rapidograph, Inc.)*

Pencils. A pencil works by transferring particles from its point to the surface of the drawing material. For technical drawings, mechanical pencils or lead holders are used instead of wooden pencils, Figure 3–10.

Pencil leads are classified and graded from 9H very hard to 7B very soft. For general line work and lettering, 2H to F leads are used.

Ink. High-quality drawings are made using ink. Ink gives opaque, high-contrast lines which make excellent prints, and equally excellent photographic images and negatives. Computer plotters use ink pens and markers to draw with because the pen tip is not used up in the drawing process.

Pens. Technical pens are basically fountain pens, Figure 3–11. They have a reservoir of ink that flows through a tube to the drawing material surface. The ink is held in the pen by a needle valve that releases ink when the tip of the pen is in contact with the drawing surface. The principle of capillary action makes the ink flow from the pen.

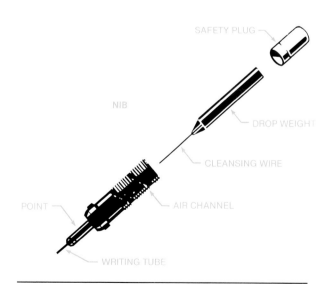

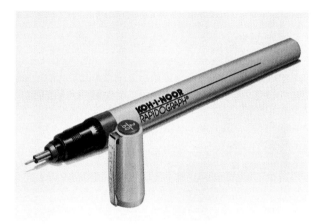

SAFETY PLUG

NIB

DROP WEIGHT

CLEANSING WIRE

POINT

AIR CHANNEL

WRITING TUBE

FIGURE 3–11 (A) Technical pen (B) Technical pen construction *(Courtesy of Koh-I-Noor Rapidograph, Inc.)*

Pen points are designed to draw varying width lines. Point sizes range from 0.005″/.13mm (very narrow) to 0.079″/2.0mm (wide).

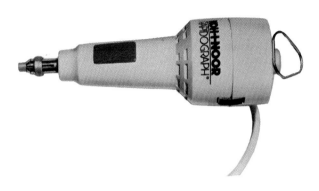

FIGURE 3–12 Erasing machines are designed to erase pencil and ink on paper or film. *(Courtesy of Koh-I-Noor Rapidograph, Inc.)*

Erasers, Erasing Shields, Cleaners, and Brushes

In the process of drawing there are often lines or parts of lines that must be removed or changed. When erasing, the desired result is to remove only the pencil or ink particles and not damage the drawing material surface.

Erasers come in many shapes and sizes depending on the use, Figure 3–12. Erasers have an abrasive material in a binder. Generally, the abrasive is pumice and the binder a plastic material.

Pencil erasers are commonly colored: red — very abrasive, white — medium abrasive, and pink — minimum abrasive. Ink erasers are often gray and have large amounts of pumice. Vinyl and kneaded erasers, which have no pumice abrasive, are used for plastic film, paper, and vellum. Today, handheld electric erasers are widely used.

Shields and Removers. Erasing shields are used to erase small areas while protecting or masking other areas, Figure 3–13. The many different-shaped holes in the metal or plastic shield allow very accurate shielding of areas beyond the area to be erased.

Cleaning Pads and Brushes. To avoid smudges cleaning pads containing finely ground gum erasing materials are available. When the pads

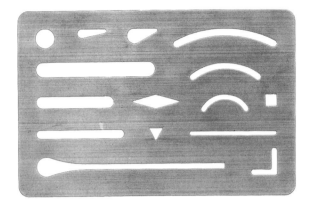

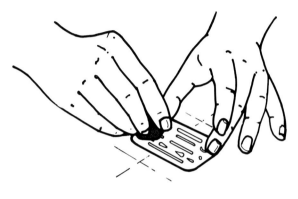

FIGURE 3–13 (A) Erasing shield (B) Erasing shields are designed to shield areas not to be erased. *(Courtesy of Koh-I-Noor Rapidograph, Inc.)*

are shaken over the paper the particles fall through the pad fabric onto the paper surface. When instruments are placed on the surface they tend to "float" on the surface, thus smudging does not occur.

Soft, long-bristle dusting brushes are made for cleaning drawings, Figure 3–14. Cleaning unwanted erasure materials and other particles from the drawing surface with a soft brush reduces the chance of smudging lines.

FIGURE 3–14 Cleaning brush *(Courtesy of Koh-I-Noor Rapidograph, Inc.)*

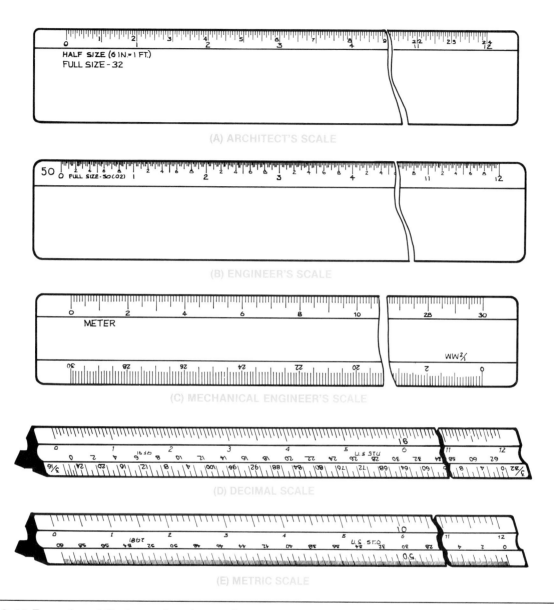

FIGURE 3–15 Examples of Scales and scale notations.

Scales

The **scale** is similar to a plastic or wooden rule, Figure 3–15. It is the standard instrument of reference for the engineer, architect, drafter, tool maker, or other technician. A scale provides a straight edge which can be used to measure objects in proportion to their true size (to scale). For example, scaling techniques allow the drafter to represent a large object, like a house, to scale, on a sheet of drawing paper.

Scales are graduated in either **American National Standards Institute (ANSI) U.S. customary** (feet and inches) or **metric** (centimeters and millimeters) standard graduations.

There are three basic types of scales: architect's scales, mechanical engineer's scales, and civil engineer's scales. There are also metric scales for engineering and architectural uses.

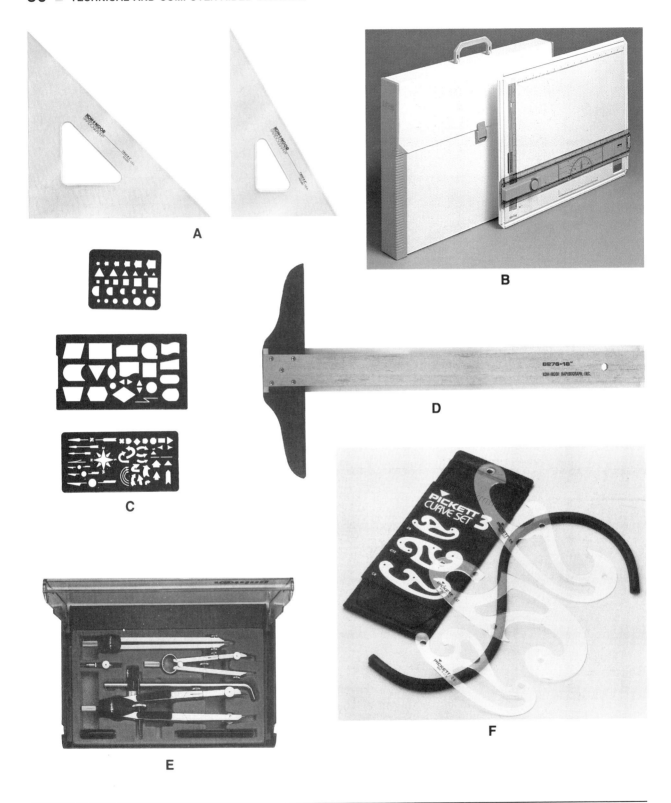

FIGURE 3–16 (A) Triangles **(B)** Drawing boards and parallel edge **(C)** Templates **(D)** T square **(E)** Dividers and compasses **(F)** Curves *(A-E Courtesy of Koh-I-Noor Rapidograph, Inc.) (F Courtesy of Chartpak)*

Drawing Instruments and Aids

Instruments are the specialized tools that aid in the technical drawing process, Figure 3–16.

Drawing Boards. Drawing boards provide a surface for mounting paper or film, a straight edge for guiding a square, a smooth surface for drawing, and a surface to engage the centering tip of a compass or divider.

T Squares and Parallel Edge. T squares are designed to give a true right angle straight edge from the edge of the drawing board. Because the head of the T square is kept at a right angle to the blade, it will always give a straight edge for drawing parallel lines on the drawing paper. A parallel edge has the same function as the T square, but it uses cables to keep a stable angle. The bar is superior for large drawing surfaces. It is used most often in architectural work.

Triangles. Fixed triangles are made in two standard forms: 30–60–90 and 45–45–90. There are also adjustable triangles and drafting machines with adjustable arms. Triangles are designed to be used with a T square, parallel bar, or straight edge to provide a surface for drawing angled lines. When used in various combinations, triangles can make lines at 15 degree increments. Today there are adjustable triangles that can create any angle.

Compasses. Compasses are used to draw circles and arcs. Compasses have one leg or end designed to pierce the paper or film and hold a firm center point for drawing an arc or circle. The other leg may be made to hold a pen or pencil point.

Dividers. Dividers are made to transfer points or divide distances into equal parts. Dividers have been used for centuries to measure and calculate

Duplicating Drawings

Drawings, like other messages, must often be communicated to a large number of people. The three basic methods for making copies of drawings today are diazo, electrostatic, and photographic reproduction.

Diazo prints are made using a direct printing process. The original drawing must be drawn on translucent materials such as Mylar or vellum. Chemically treated paper under the original drawing is exposed to ultraviolet light. The light is blocked from passing through the areas on the original drawing where opaque lines are present. The print material is then exposed to ammonia vapor which creates a chemical reaction. This reaction causes light areas to darken and thus a print of the original drawing appears. Diazo printing is a somewhat inexpensive process. Diazo prints are sometimes called **blueprints**. This is a carry over from an earlier process that is rarely used anymore. ''Diazo print'' should now be used instead of ''blueprint.''

Electrostatic printers work in the same manner as regular office photocopiers. The original does not need to be translucent because the process is based on the use of reflected light.

Diazo whiteprinter *(Courtesy of Dietzen)*

Photographic copying involves making prints or photographs from a photographic negative of the original drawing. This process provides very high-quality copies of the original drawing. Coupled with the advantages of photographic enlarging and reduction techniques, and the advances in modern photography, this process is widely used with today's highly advanced printed circuit board technology.

distances in marine navigation. Dividers look like compasses, but have two piercing points rather than one piercing point and one drawing point.

Guides and Templates. Guides and **templates** are used to letter and draw special shapes repeatedly. Common templates and guides are made to draw circles, ellipses, hexagons, squares, and specialized symbols. Templates are widely used in architectural and engineering drawing.

Curves. There are many forms and sizes of **irregular curves**. Curves are used to draw lines with changing radii. A common form of curve guide is called a French curve. French curves have constantly changing radii. Flexible curves can be formed into special curved shapes.

Measurement and Drawing

The development and use of a standard measurement system has been a basic technological goal of all societies. The measurement system of a society sets the standard for all trade and becomes a major part of its value system. Fairness and equity are often concepts expressed in the measurement system of a society; ''Go the extra *mile*'' or ''Worth its *weight* in gold'' are examples.

Measurement Standards

Standards for measurement are closely controlled in the United States by the American National Standards Institute (ANSI). Measure-

QUANTITY	UNIT OF MEASURE
LENGTH	YARD, FOOT, INCH, METER
MASS/WEIGHT	POUND, OUNCE, GRAM, KILOGRAM
TEMPERATURE	FAHRENHEIT, CELSIUS, KELVIN
TIME	SECOND, HOUR, MINUTE
ELECTRIC CURRENT	AMPERE
LUMINESCENCE	LUMEN, CANDELA
AMOUNT	MOLE

FIGURE 3–17 There are seven basic areas of customary and metric measurement.

ment standards are internationally controlled by the International Standards Organization (ISO). Of the seven basic quantities measured in science and industry, length is the central unit used in drawing communication, Figure 3–17. In today's integrated world society, the International Metric System is the standard for worldwide technical drawing and communication.

Lettering

When drawing, there is often a need to communicate more information than the drawing can convey. This added information is expressed in letters and numerals. Letters and numerals are symbols that can have meaning alone or in groups.

While there are many mechanical, electronic and computerized lettering devices, hand lettering is still widely used. Designers, engineers, and architects hand letter to record information on drawings or sketches.

ANSI recommends the use of single-stroke gothic-style lettering, Figure 3–18A. This style of lettering is easier to read and draw than other styles. Gothic lettering also copies and prints clearly. When drawings are going to be microfilmed or photocopied, a special style of lettering called microfont is recommended to give added clarity to selected letters, Figure 3–18B.

In most engineering applications only uppercase lettering is used. However in civil engineering and cartography lowercase lettering is often used. Architects can stylize their lettering to meet their own standards of style and clarity. Many designers and drafters are able to letter as fast as they can write.

Lettering Aids

There are many mechanical and electronic lettering aids. These range from templates for drawing guidelines, to electronic image processors. Most templates use the straight edge of the T square or drafting machine as the parallel reference surface to make uniform letters.

Specialized typewriters can mount directly on the drawing board. There are machines that print on tape which is then affixed to the draw-

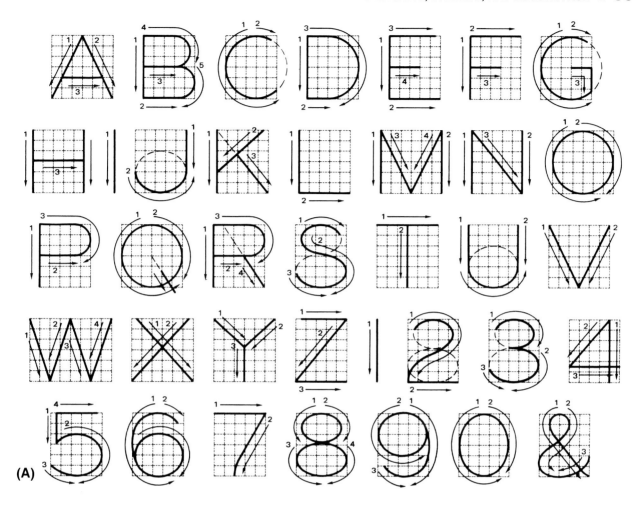

(A)

ABCDEFGHIJKLMNO

PQRSTUVWXYZ

(B)   1234567890

FIGURE 3–18 (A) Single stroke gothic lettering and (B) Microfont lettering *(Drawing A reprinted from DRAFTING FOR TRADES & INDUSTRY by John A. Nelson, © 1986 Delmar Publishers Inc., Drawing B reprinted from MECHANICAL DRAFTING by Madsen, Shumaker, and Stewart, © 1986 Delmar Publishers Inc.)*

ing. Transfer sheets of various size letters, numbers, and symbols are available for many drawing and graphic uses.

Computers and Lettering

Modern electronics has had a great impact on lettering. A computer can accurately and pre-cisely produce letters and numbers in a wide variety of fonts and sizes. A font is a complete series of letters of one design style. Today, the drafter, engineer, or architect types the text and then directs the computer to adjust, align, and insert the lettering on the drawing where needed. Use of a standard typing keyboard is a basic skill for the technical drawing professional.

THE QUESTION IS. . .

WHAT ARE THE CATEGORIES *JEOPARDY!* HAS USED MOST FREQUENTLY SINCE ITS RETURN TO TV IN SEPTEMBER 1984?

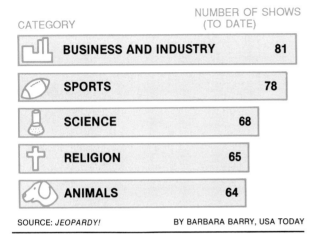

CATEGORY	NUMBER OF SHOWS (TO DATE)
BUSINESS AND INDUSTRY	81
SPORTS	78
SCIENCE	68
RELIGION	65
ANIMALS	64

SOURCE: *JEOPARDY!*   BY BARBARA BARRY, USA TODAY

HOME SALES

NEW HOMES
(seasonally adjusted annual rate)

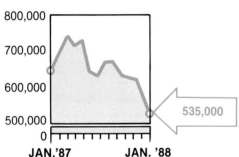

535,000

SOURCE: COMMERCE DEPARTMENT
BY MARTY BAUMANN, USA TODAY

FIGURE 3–19 Graphs are graphic shorthand for communicating information. *(Copyright 1988, USA TODAY. Reprinted with permission)*

Charts and Graphs

Often the information to be presented is not about an object, but about a concept, relationship, comparison, process, or technique. When this is the case, **graphs** and **charts** can be used. Graphs and charts are a type of graphic shorthand, Figure 3–19.

Graphs and charts are widely used forms of graphic communication. They are used to

- Show trends,
- Present data analysis and comparison,
- Display mathematical relationships (e.g., formulas), and
- Show the behavior of systems.

Many specialized materials and aids can help in the development of charts and graphs. These include grid papers, templates, and overlay film. When the type of graph needed to present the idea has been selected, available commercial materials should be researched.

Types of Charts and Graphs

Graphs can be classified by the method used to present data. Generally charts and graphics can be categorized as either technical or semitechnical.

Technical Application Graphs. Technical application graphics are mainly used to communicate between individuals with specialized technical knowledge and background. Types of technical graphs include

- Rectangular coordinate
- Semilogarithmic coordinate
- Logarithmic coordinate
- Trilinear coordinate
- Polar coordinate.

Semitechnical Charts and Graphs. Semitechnical charts and graphs are widely used in business and industry today to illustrate a comparison, show a trend, and/or give information. Widely used types of semitechnical graphs and charts include

- Bar charts
- Surface charts
- Pie charts
- Volume charts
- Distribution charts
- Flow charts
- Organization charts

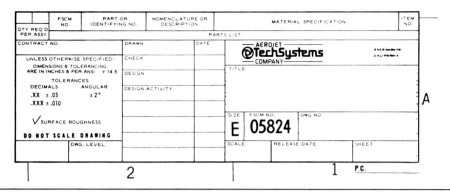

FIGURE 3-20 Sample title block *(Courtesy of Aeroject Techsystems Co.)*

Borders and Title Blocks

Borders are placed around drawings to mark the presentation area. The main purpose of the **title block** on a drawing is to serve as a reference key for readers of the drawing, Figure 3-20. The title block gives information to the reader about the specifications the drafter used in developing the drawing.

Computer-developed Graphs

Computer image generation makes use of a grid system. This system lends itself to the easy development of any of the rectangular coordinate system type graphs. A number of modern business and technical computer programs have

FIGURE 3-21 Many computer programs are available to change data into graphs automatically. *(Courtesy of Compaq Computer Corp.)*

specialized graph generation programs, Figure 3-21. These programs let the operator select from a predetermined series of graph types to make bar, pie, volume, and other types of graphs.

Summary

Drawing has been a universal system of person-to-person communication throughout history. Modern technology has accelerated the need for drawings.

Sketches are freehand drawings used to formulate, express, and record information and ideas. Freehand drawing is often used to express aesthetic ideas. Technical drawing is used to convey information accurately and completely to those producing objects, structures, or systems. Technical illustration blends technical and freehand drawing.

Converting an image of what we perceive in our mind to a technical drawing can be done using perspective projection or parallel projection. Perspective projections imitate the way the human eye actually sees an object. Object lines are projected to a vanishing point. Orthographic projections are a common type of parallel projection. Traditional drawings are done in pencil, or pen and ink on paper, cloth, or plastic film. Other tools for drawing include erasers, erasing shields, cleaners, and brushes.

Scales are similar to plastic or wooden rules. They provide straight edges which can be used to draw objects in proportion to their true size (to scale). Other drawing instruments include drawing boards, T squares, parallel edges, triangles, compasses, dividers, guides, templates, and irregular curves.

Drawings can be duplicated by Diazo printers, electrostatic printers, and by photographic methods.

Drawing standards and measurements are controlled by the American National Standards Institute (ANSI) and by the International Standards Organization (ISO).

Lettering is done by hand, mechanical, electronic, and computerized methods. Lettering must be easy to read, and easily photocopied or otherwise duplicated. Lettering aids such as templates and lettering machines are commonly used. Much of today's drawing and lettering is done by computer.

Charts and graphs are special types of drawings used to show trends, present data, display mathematical relationships, and show the behavior of systems. Many charts and graphs are now easily generated by computers and specialized software.

REVIEW

1. Why is technical drawing considered a universal language?
2. What are the three general types of drawing?
3. What element makes technical drawing different from freehand drawing?
4. Why is a title block an important part of a technical drawing?
5. In the development of a drawing, if the lines of sight are not parallel from the object to the view point what type of drawing is it?
6. What is parallel projection drawing?
7. What is the most important form of technical drawing?
8. Identify two types of sheet materials used in making drawings.
9. What does ANSI mean?
10. Why are charts and graphs important forms of technical communication?
11. What is the major difference in erasers?
12. Describe the advantage of using a cleaning pad when drawing.
13. What is the purpose of a drawing scale?
14. What is the scientific principle that causes the ink to flow from the tip of a technical pen when it is in contact with the surface of the drawing materials?

CHAPTER 4

Technical Drawing Techniques

OBJECTIVES

After completing this chapter you will know that:

- Technical drawing, as a communication medium, includes two phases: the design phase and the actual drawing phase.
- Multiview drawings are developed using orthographic techniques.
- Orthographic projection drawings usually present the top, front, and right-side views of an object.
- Sectional and auxiliary view drawing techniques are used to present specialized views of objects.
- Specific American National Standards Institute (ANSI) standards and conventions are followed in the development and dimensioning of technical drawings.
- Special pictorial drawing techniques are used to develop isometric, oblique, and perspective drawings.
- Pattern development techniques are used to draw surface patterns.

KEY TERMS

aligned dimensions
alphabet of lines
auxiliary view
cutting plane
development drawings
dimensioning

full section view
half section view
hidden lines
isometric axis
isometric drawing
isometric lines
multiview drawing

oblique drawing
orthographic projection
perspective drawing
scale notation
sectional views
unidirectional dimensions

Drawing Techniques

Technical drawing is a specialized graphic communication process. There are two basic parts

to this communication process: the design phase and the actual technical drawing phase.

The design phase of the process involves collecting information, refining the design idea, and finding out what is needed to present the idea. The second phase, technical drawing development, deals with the actual process of drawing, including on it any specifications needed to make the object. A technical drawing is often a medium for communicating a design to the people who will make it. For example, you and a friend decide to build a doghouse. You agree that you will design it and your friend will build it. In designing a doghouse, the following questions must be answered:

1. What will the doghouse look like?
2. How big will it be (length, width, height)?
3. What will it be made of?
4. What color(s) will it be painted?

Answering these and many other questions is part of the design process. Having answered the questions, how do you communicate the design to your friends? Verbal instructions are risky because they can be misinterpreted. However, if you make a technical drawing, your friend will be able to see exactly what you want.

Multiview Drawings

Viewing Planes and Projection

Drawings present pictures and information on two-dimensional surfaces such as paper or film; or a monitor screen in a CAD system. All presentations or drawings are based on a relationship between the object, projection plane, and viewer.

A **multiview drawing** is one that presents the object in two or more of the six normal views: front, top, right side, left side, back, and bottom, Figure 4–1. The number of views presented on a drawing and which views are selected are determined by the need for detailed information to produce the object. The views selected should be those that together show all of the information needed to make the object.

Orthographic Projection

In the mid-1790s, Gaspard Monge developed a system of technical drawing based on the

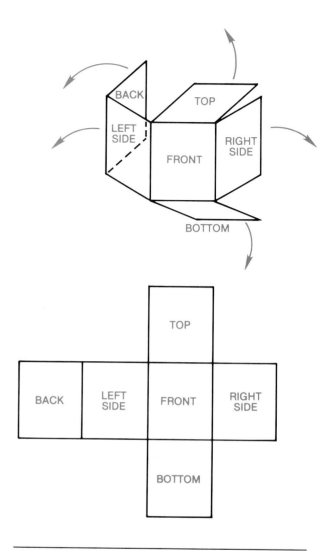

FIGURE 4–1 The six normal views developed for multiview drawings include the front, top, right-side, left-side, bottom, and back views.

arrangement of specific projection planes on a drawing presentation. Monge's system involved the presentation of horizontal, frontal, and profile viewing planes in a right angle or perpendicular relationship. The term for this relationship is **orthographic projection**. As Monge's system was widely accepted as the principal method for developing multiview drawings, it became known as orthographic projection.

Orthographic projection drawings are made as if a transparent glass or plastic plate were positioned parallel to a major surface of the object.

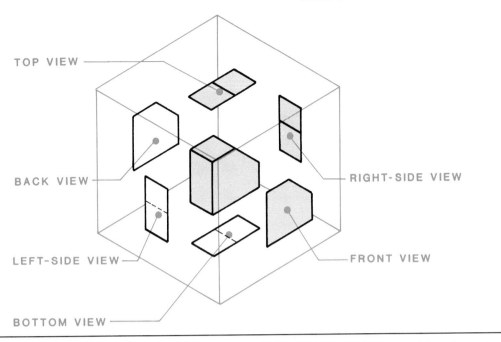

FIGURE 4–2 The views of an object are developed as if they were being projected onto the six sides of a transparent cube around the object. *(Reprinted from DRAFTING IN A COMPUTER AGE by Wallach and Chowenhill, © 1989 Delmar Publishers Inc.)*

If such planes were placed around the whole object, a transparent cube would be formed. The six surfaces of the cube form the primary projection planes for the development of orthographic projection drawings, Figure 4–2. An orthographic drawing shows two-dimensional views of various sides of the object. There is no depth perception in an orthographic view. The imaginary line of sight from the viewer's eye to the surface of the object is perpendicular to the surface.

Orthographic drawings normally show three principal views of an object—the front, top, and right-side views.

Figure 4–3 shows the two main systems of orthographic projection: first-angle and third-angle projection. Drafters and designers in the United States use third-angle projection. In countries that follow the metric system of measurement, first-angle projection is used. Notice in Figure 4–3 that the difference comes in the placement of views on the page.

In orthographic projection, the three principal views (the front, top, and right side) always share common surfaces. The surfaces may appear as a surface edge in one view and full surface

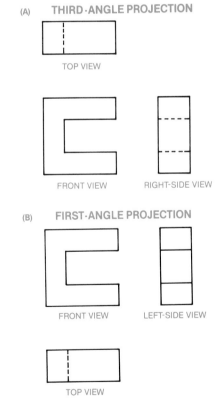

FIGURE 4–3 The two principal types of orthographic projection are (A) Third-angle projection; (B) First-angle projection.

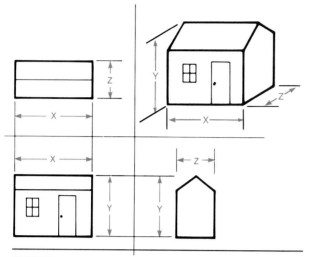

FIGURE 4–4 The views presented in a multiview orthographic projection share common surface dimensions.

in the other. The size of these common surfaces and edges are shown as common dimensions in drawing views, Figure 4–4.

Hidden Surfaces and Features

Hidden surfaces and features are qualities of an object that may not be visible in one of the presentation views because they are under or behind a solid object surface, Figure 4–5. To show the existence of these features, uniformly broken or dashed lines are used. These lines are called **hidden lines**.

Sectional Views

Sectional views are developed to expose important interior details of an object. These views are made as if a saw or knife were used to cut into an object and remove a part. The imaginary line along which the section is removed is called the **cutting plane** line. This line is drawn on the principal view of the object where the imaginary cutting plane shows as an edge, Figure 4–6.

The two basic types of sectional views used in drawing are **full sections** and **half sections**. Full and half sections are shown in Figure 4–7. Notice that with a full section, the cutting plane cuts fully through the object. With a half section it cuts halfway through.

Auxiliary Views

There are objects that are shaped so that one or more of their surfaces will appear angled or oblique in all principal projection planes. The true shape and size of these surfaces can only be developed using **auxiliary view** drawing techniques.

To develop auxiliary views, projection lines are extended from the edge view of the oblique surface to a projection plane perpendicular to the edge view of the surface. Then, using the true surface dimensions from one of the two other views, the appropriate surface dimensions are

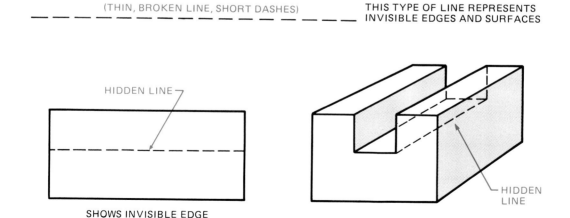

FIGURE 4–5 Hidden lines are used to identify object features that are hidden by the surfaces parallel to the projection plane in the view presented. *(Reprinted from BASIC BLUEPRINT READING AND SKETCHING, 5E by Olivo, Payne, and Olivo, © 1988 Delmar Publishers Inc.)*

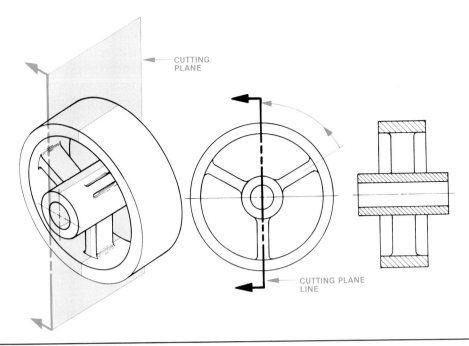

FIGURE 4-6 Sectional views are drawn as though a portion of the object were removed along a cutting plane line to expose details. *(Reprinted from DRAFTING IN A COMPUTER AGE by Wallach and Chowenhill, © 1989 Delmar Publishers Inc.)*

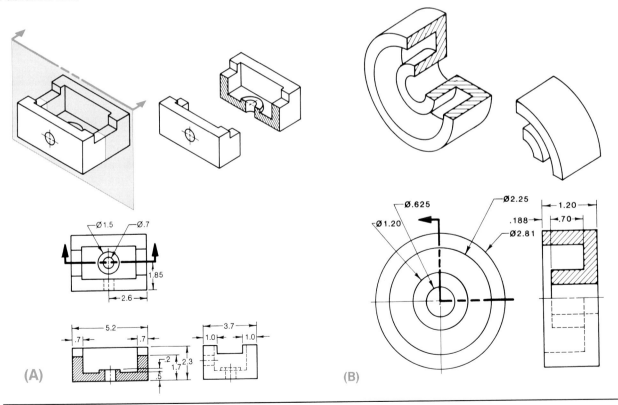

FIGURE 4-7 Examples of (A) full sections and (B) half section. *(Reprinted from DRAFTING IN A COMPUTER AGE by Wallach and Chowenhill, © 1989 Delmar Publishers Inc.)*

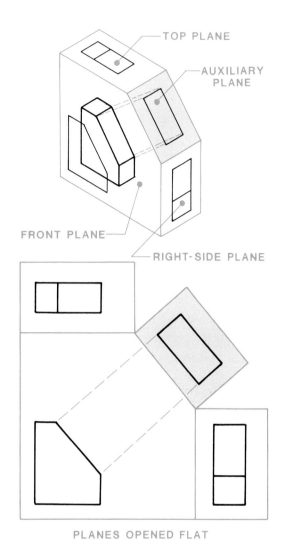

PLANES OPENED FLAT

FIGURE 4–8 When a surface or feature is not parallel to the projection plane it will not appear true to size and shape in the normal projection views. An auxiliary view is needed. (Reprinted from DRAFTING IN A COMPUTER AGE by Wallach and Chowenhill, © 1989 Delmar Publishers Inc.)

plotted on the projection lines. The oblique surface of the auxiliary view developed parallel to the auxiliary projection plane will be shown in true shape and size, Figure 4–8.

Dimensioning

Dimensioning is the process used to describe the size of an object and locations of features on the object. In dimensioning, symbols, number values, and notations are included on a drawing to describe the object in great detail. Because of the symbolic nature of dimension notation on a drawing, one must closely follow standards to ensure clear communication. ANSI standards are very specific regarding the use of symbols, placement of information, and type of information needed.

There are two classifications of dimensions: size dimensions and location dimensions. Size dimensions are placed directly on the object and give the size of each geometric shape. Location dimensions establish the relationship of features or geometric shapes in the object. By following the standards for developing size and location dimensions readers are able to accurately understand the total drawing.

The two major systems used for dimensioning drawings are

■ **Unidirectional** — all dimensioning reads from the bottom of the drawing, Figure 4–9A.
■ **Aligned** — all dimensioning reads from the bottom and right side of the drawing, Figure 4-9B.

Dimensioning Practices

The basic purpose of dimensioning is to communicate size and locational information about an object. Line types and practices used for dimensioning are (Figure 4–10)

■ Extension lines — thin light lines that begin 1/16″ from the object and end 1/8″ beyond the dimension line of the last notation.
■ Dimension lines — thin dark lines that extend from extension line to extension line, beginning at least 3/8″ from the object surface or edge and spaced at intervals of 1/4″.
■ Arrowheads — used at the end of dimension line to indicate the area described by the dimension notation. Arrowheads are 1/8″ to 3/16″ long and about one third as wide.

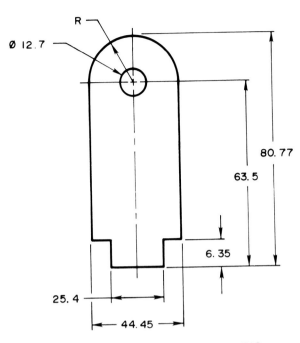

(A) UNIDIRECTIONAL DIMENSIONING

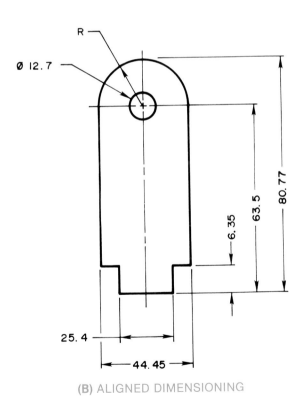

(B) ALIGNED DIMENSIONING

FIGURE 4–9 The two main systems for dimensioning drawings are (A) Unidirectional (B) Aligned. *(Reprinted from MECHANICAL DRAFTING by Madsen, Shumaker, and Stewart, © 1986 Delmar Publishers Inc.)*

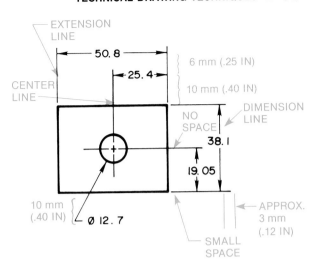

FIGURE 4–10 Standards are followed for drawing dimension lines and symbols, and for their positioning on the drawing, to ensure clear communication. *(Reprinted from MECHANICAL DRAFTING by Madsen, Shumaker, and Stewart, © 1986 Delmar Publishers Inc.)*

▪ Leaders — thin lines used to indicate the specific object surface or feature to which the dimension or note applies. Leaders should align with the intersection of the center lines of a radius or hole size notation.

Line Conventions and Standards

Line conventions and standards are used so that drawings are readable and copyable. Lines are symbols, just as the numbers and letters on a drawing. American National Standards Institute (ANSI) line and lettering standards should be followed.

When a real object or a mental picture is made into a drawing on paper or film we use lines to symbolize features of the object. The symbol that means the quantity one unit is the numeral 1. The symbol for you as a person is your name. In technical drawing, lines are used as symbols to identify surfaces, edges, projection planes, and more. Often the listing of line types used in drawing is called the **alphabet of lines**.

The use of standard symbols in drawing is controlled by ANSI. ANSI has set a standard for lettering and lines which provide a guide for all technical drawings. This standard is called

American National Standard Line Conventions and Lettering — ANSI–14.2M–1979.

Measurement and Scale Notation

One of the first decisions to make in the technical drawing process is what measurement system will be used to develop the drawing. It must also be decided if the drawing will give dimensions in both the metric and customary measurement systems or in only one system. This is important because conversion between the metric and customary systems is hard and time consuming. Modern computer-aided drafting

Making a Multiview Drawing

Before beginning a multiview drawing, take some time to plan the process. Making a technical drawing is a step-by-step process that is much simpler to do when the steps are taken one at a time and in the proper order.

Step 1

Visualize the object you are about to draw. Make sure you have a good mental picture before starting.

Step 2

Decide which view should be the front view. Remember the following rules:

a. The front view should show the most detail.

b. The front view should show the most basic shape in profile.

c. The front view should be drawn with the widest or heaviest part on the bottom so it appears stable.

d. The front view should be selected so that the other views have as few hidden lines as possible.

Step 3

Decide how many views you should draw to completely describe the object.

Step 4

Center the views in the work area.

Step 5

Lay out all views using light lines. Leave at least one inch between views.

Step 6

Check all views for correctness and accuracy. Then darken all views using the correct line thicknesses.

Step 7

Add dimensions and dimension lines.

Step 8

Recheck all work.

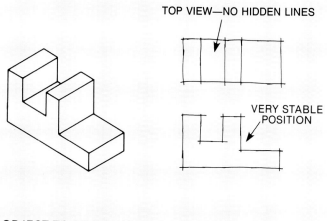

OBJECT TO BE DRAWN

BEST POSITION - FRONT VIEW

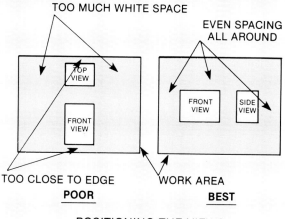

POSITIONING THE VIEWS WITHIN THE WORK AREA

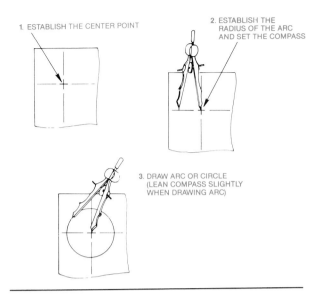

FIGURE 4–11 Drawing circles and ellipses involves (1) setting a center point, (2) setting a compass to desired radius, and (3) drawing the arc or circle.

(CAD) programs, however, give immediate conversion functions and dual labeling.

The second decision is the scale of the drawing. **Scale notation** is the proportional relationship of the actual size of the object to the size of the drawing presentation (see Figure 3–15, page 49). In architectural work, drawings are usually completed in 1/8″ = 1′-0″ or 1/4″ =

1′-0″ scale. A 1/8″ long line on a drawing prepared at a scale of 1/8″ = 1′-0″ scale is defined as one foot or 12 inches long on the actual object.

Circles and Curves

When drawing circles, arcs, or curves the first step in the process is finding the center point and radius of the arc. The center point is then marked with intersecting center lines. This point can then be used to draw the arc or circle with a compass or proper template, Figure 4–11.

Pictorial Drawings

A pictorial drawing is one that shows more than one side or surface of an object in a single view of the object. It shows the width, height, and depth of an object in an approximation of how the human eye sees it. The word pictorial comes from the word picture. Pictorial drawings appear at times to be photographs, particularly when they are developed into technical illustrations.

The three main types of pictorial drawings are isometric, oblique, and perspective, Figure 4–12.

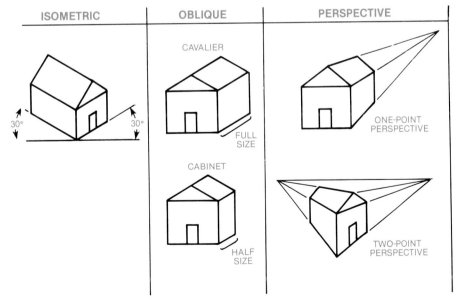

FIGURE 4–12 Types of pictorial drawings: (A) Isometric (B) Oblique (C) Perspective

Isometric drawings are developed by projecting the lines of sight parallel and perpendicular to the projection plane.

Oblique drawings are developed by projecting the lines of sight parallel, but oblique to the plane of projection. The term oblique means at an angle less than 90 degrees.

Perspective drawings are developed by finding the relationship of the view point, picture plane, ground line, horizon line, and object. The lines of sight are projected in relation to these elements to form a picture of the object that appears most natural to the viewer.

Isometric Drawings

The word isometric comes from the root word "iso," meaning "equal," and "metric," meaning "measure." Isometric drawings are often used by illustrators to develop technical illustrations.

Isometric drawings are developed along the X, Y, and Z axes to show width, height, and depth. The relationship of these axes is 30 degrees to the horizontal base lines, and 120 degrees to each other. These three lines form the **isometric lines** or **axes** which are used as the base reference lines for all measurements and line developments, Figure 4–13. Measurements are plotted directly on these lines in isometric drawings.

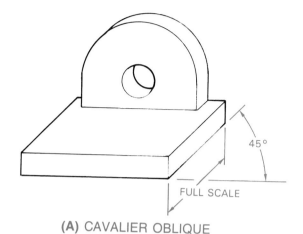

(A) CAVALIER OBLIQUE

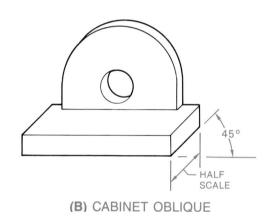

(B) CABINET OBLIQUE

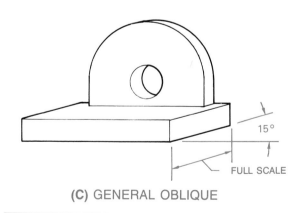

(C) GENERAL OBLIQUE

FIGURE 4–14 Types of oblique technical drawings: (A) Cavalier (B) Cabinet (C) General. *(Reprinted from MECHANICAL DRAFTING by Madsen, Shumaker, and Stewart, © 1986 Delmar Publishers Inc.)*

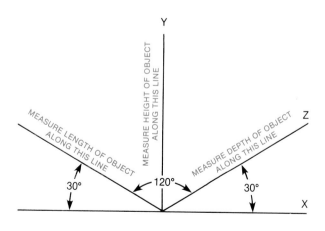

FIGURE 4–13 Isometric axes and angles

The intersection of lines on the X, Y, and Z axes form planes called isometric planes. All drawing measurement is referenced to positions on isometric lines and planes, which are the same measurements used to develop each of the projected views in an orthographic drawing.

Oblique Drawings

The term "oblique" means other than horizontal or perpendicular. Because oblique drawings present the least picture-like view of an object they are the least used of the pictorial drawing types.

Oblique drawings are developed with the front view parallel to the front projection plane. The same procedure used to develop the front view in a multiview drawing is used with oblique drawings. The other views are developed projected at an angle from the front view. The three types of oblique drawings are cavalier oblique, cabinet oblique, and general oblique, Figure 4–14.

Cavalier oblique drawings are developed with the X and Y axes perpendicular to each other. The X axis is parallel to the horizontal baseline. The Z axis is projected at an angle (usually 45 degrees) to the X and Y axes. Object measurements are plotted full size on the planes of the oblique drawing.

Cabinet oblique drawings are developed with dimensions along the Z axis cut in half. Carpenters and cabinet makers are the major users of this type of drawing because it is easily developed, and will show cabinet detail in true shape and size in the front view.

General oblique drawings are developed as cabinet oblique drawings except the angle of projection for the Z axis is less than 45 degrees and the depth projection is changed from half to full scale. Although any angle is acceptable, the angle of projection is usually 30, 45, or 60 degrees.

Perspective Drawings

Perspective drawings are the most picture-like of all the types of technical drawings. They are often developed by engineers and architects to present a design idea to people untrained in reading technical drawings. Perspective drawings are often used by illustrators to develop technical illustrations.

Perspective drawings differ from other types of technical drawings with regard to the way in which the lines of sight are projected to the projection plane. In perspective drawings, the lines of sight are not parallel. They converge from the object to imaginary vanishing points.

The three types of perspective drawings are one-point, two-point, and three-point perspective, Figure 4–15. The term "point" refers to

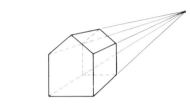

A. ONE-POINT PERSPECTIVE

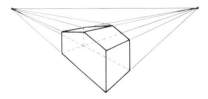

B. TWO-POINT PERSPECTIVE

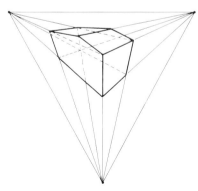

C. THREE-POINT PERSPECTIVE

FIGURE 4–15 Types of perspective technical drawings: (A) One-point perspective (B) Two-point perspective (C) Three-point perspective

vanishing points that are used to make perspective drawings. A two-point perspective has two vanishing points.

One-point perspective drawings are often termed parallel perspectives because the front surface of the object drawn is parallel to the picture plane. The front surface is drawn full size on the X and Y axes. Projectors extend on the Z axis to a single vanishing point on the horizon line.

In two-point perspective drawings, the surfaces of the object are at an angle to the picture planes in both the X and Y axes. The projectors in the Z axis are parallel.

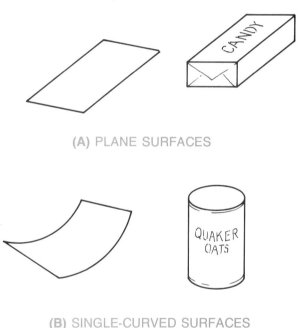

(A) PLANE SURFACES

(B) SINGLE-CURVED SURFACES

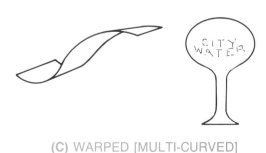

(C) WARPED [MULTI-CURVED]

FIGURE 4–16 Surface features of objects are classified as (A) Plane, (B) Single-curved, and (C) Warped (multicurved).

Three-point perspective drawings are the most picturelike of the types of perspective drawings and appear most natural to the viewer. The X, Y, and Z axis lines each radiate from separate vanishing points. Three-point perspectives are time consuming to develop. They are mainly used to develop specialized visual presentations and effects.

Development Drawings

Development drawings show how an object would look if it were cut along its edges and stretched out flat. Patterns which are drawn on sheet metal, then cut out and folded into shapes, are examples of development drawings.

These three-dimensional qualities must be considered in making the object from sheet materials.

The surface features of all objects may be classified as plane, single-curved, or warped (multicurved), Figure 4–16. Only plane and single-curved surfaces can be formed using pattern methods from sheet materials. Warped surfaces are formed using pressing or molding methods.

When we visualize and draw objects we usually focus on their surface qualities. For most objects, that is an incomplete description of their nature and form. Most nonsolid objects are formed from a solid sheet of material that has three-dimensional qualities: length, width, and thickness.

Summary

Technical drawing, as a communication medium, includes a design phase and an actual drawing phase.

Multiview drawings present the object in two or more views. The orthographic projection technique is used to create multiview drawings. An orthographic drawing normally shows three principle views of an object; the front, top, and right-side views. Features of an object that may not be visible in one of the views are represented by dashed (hidden) lines.

To expose important interior details of an object, sectional views are used. These can be full-sections or half-sections. Surfaces that are at an angle to the projection plane are drawn using auxiliary views.

Standard techniques are followed in the development and dimensioning of drawings. Dimensioning is the process used to describe the object's size and location of features. When dimensioning an object, extension lines, arrowheads, and leaders are used. Line conventions dictate the line types to be used on drawings. The American National Standards Institute (ANSI) standards should be followed.

A pictorial drawing is one that shows more than one side or surface of an object in a single view. Three types of pictorial drawings are isometric, oblique, and perspective. Development drawings show how an object would look if it were cut along its edges and stretched out flat.

REVIEW

1. Why are the various types of lines used in technical drawing called the alphabet of lines?
2. Why are ANSI standards important in technical drawing?
3. If the scale notation on a drawing is stated as 1/4″ = 1′ 6″, how long is the drawing of a 3′ wall section?
4. Diagram the relationship of the object, viewer, and projection plane in a multiview drawing.
5. Explain the term orthographic as it applies to multiview projection drawing development.
6. Identify the principal views presented in an orthographic multiview drawing.
7. Outline the drawing development sequence used to develop multiview drawings.
8. How do drafters represent a hidden feature on a drawing?
9. Describe the purpose of sectional views in multiview drawings.
10. When must an auxiliary view of an object be developed?
11. Sketch a multiview drawing of a sphere.
12. What is the advantage of unidirectional dimensioning on large drawings?
13. Identify the three principal types of pictorial drawings.
14. Sketch a cube in oblique and isometric presentation form.
15. Explain the concept of proportion as it is applied to sketching.

CHAPTER 5

Drawing for Production

OBJECTIVES

After completing this chapter you will know that:

- Working drawings are the central communication element in product design and development.
- Special drawing symbols and techniques are used to communicate product tolerance, fit, and finish information.
- Architectural drawing involves the development of solutions to problems using specialized communication processes and techniques.
- Beyond drawings, other elements in the development of an architectural design plan include permits, codes, schematic diagrams, specifications, and cost analyses.
- CADD technology adds new dimensions to the architectural and engineering design process.
- Topographical drawing and cartography are very important specialized forms of graphic communication.
- CADD technology has added important new dimensions to topographical drawing and cartography.

KEY TERMS

architectural design
architectural drawing
architectural plan drawings
assembly drawing
bearings
bench marks
bubble diagram
building permit laws
cartography
chart
computer-aided design
  and drafting (CADD)

conceptual design
electrical plan drawings
elevations
field notes
finish
fit
floor plans
foundation plans
framing plans
map
mechanical plan drawings

plot
production drawings
schedules
site plan drawings
specifications
structural plan drawings
tolerance
topographical drawing
transit
working drawing
zoning laws

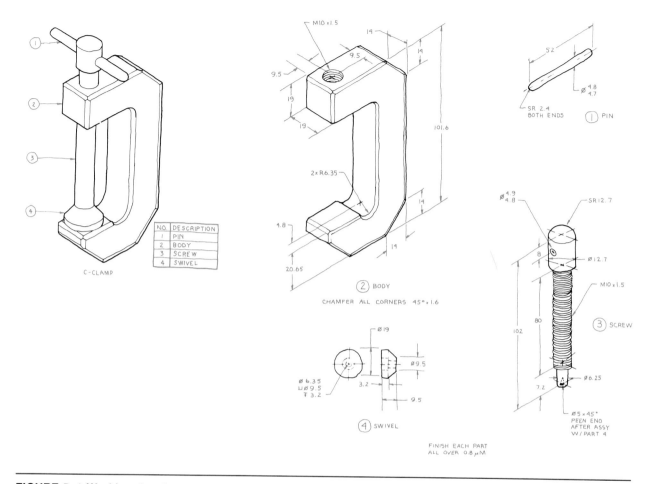

FIGURE 5–1 Working drawings are the vehicle for communicating the design solution for the product to those who will produce it. *(Reprinted from MECHANICAL DRAFTING by Madsen, Shumaker, and Stewart, © 1986 Delmar Publishers Inc.)*

Industrial Drawings

Working Drawings

In modern business and industry an engineer, designer, or drafter is part of a team. That team might include managers, accountants, craftspeople, machine operators, and purchasing agents. To design an object first involves a clear understanding of the total problem to be solved. Designers and engineers propose design solutions, usually as sketches and illustrations. Once the general design is accepted, detailed design work is done and technical drawings are developed. Next, design groups are assigned to develop specifications and drawings for special components. Parts as small as knobs for a radio or the hinge screws for eyeglasses need individual detailed technical drawings.

Drawings that direct the construction, manufacture, or assembly of object components are called **working drawings**. Working drawings are first made as full size or scale layout sketches showing the relationship of the parts in a component, Figure 5–1. After a review of the layout drawings by the engineering staff, a drafter or detailer will make a formal technical drawing. Working drawings serve as the basis for production and assembly drawings.

Production drawings give directions to craftspeople, such as machinists, on how to make the product. These drawings contain detailed engineering specification data in lettered notes

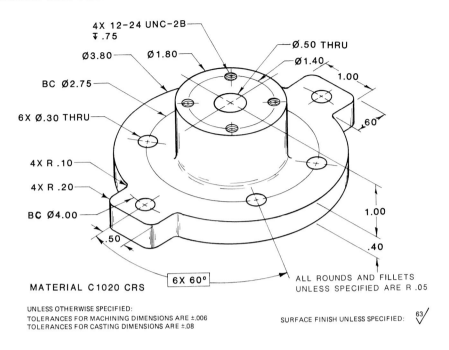

FIGURE 5–2 Sample production drawing (Reprinted from DRAFTING IN A COMPUTER AGE by Wallach and Chowenhill, © 1989 Delmar Publishers Inc.)

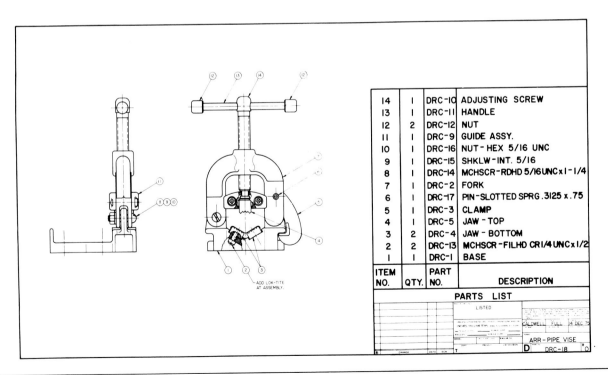

FIGURE 5–3 Sample assembly drawing (Reprinted from MECHANICAL DRAFTING by Madsen, Shumaker, and Stewart, © 1986 Delmar Publishers Inc.)

and tables as part of the drawing, Figure 5–2.

Assembly drawings for components give information to craftspeople who assemble parts into components, which are then assembled into products. The assembly drawings are working drawings for the assemblers, just as production drawings are working drawings for the machine operator making a part. Assembly drawings often use balloons and leaders together with parts lists to identify parts and give detail specifications, Figure 5–3.

Threads and Fasteners

Plans for the development of complex objects usually include information about special details such as the use of fasteners, cutting of threads, and other details needed for construction or assembly, Figure 5–4.

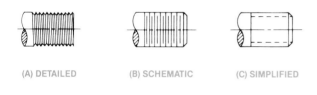

(A) DETAILED (B) SCHEMATIC (C) SIMPLIFIED

FIGURE 5–4 Methods for drawing threaded fasteners: (A) Detailed (B) Schematic (C) Simplified *(Reprinted from MECHANICAL DRAFTING by Madsen, Shumaker, and Stewart, © 1986 Delmar Publishers Inc.)*

Tolerances, Fits, Finishes, and Specifications

A **tolerance** is the amount of variation allowed between the exact size of the object as noted on the drawing and the size of the object after it has been made. It is impossible to make an object or a series of objects without some very small size differences.

When parts are made at different locations and by different craftspeople, it is important that the degree of tolerance be considered in the design and specified in the drawings. Only when the variation in size is within specified limits can interchangeable parts be developed. When making working drawings the designer, engineer, and drafters must specify the tolerances of surfaces and parts to ensure interchangeable parts, Figure 5–5. Tolerances may be noted in terms of a ± value from the specific size stated.

Fits. A **fit** refers to the difference in size between mating parts. The three types of fits are clearance, allowance, and interference.

Finishes. **Finish** refers to surface characteristics, such as smoothness, that are to be given on the completed object. Notations showing the type of finish are important elements in the specifications for working drawings.

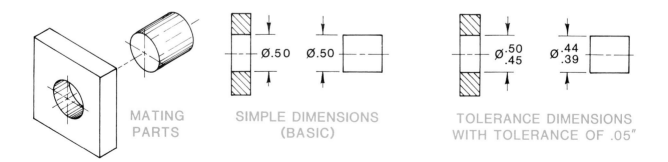

MATING PARTS SIMPLE DIMENSIONS (BASIC) TOLERANCE DIMENSIONS WITH TOLERANCE OF .05″

FIGURE 5–5 Working drawings for mating parts must specify the allowance between the parts. Standard allowances are sometimes specified as classes of fit. *(Reprinted from DRAFTING IN A COMPUTER AGE by Wallach and Chowenhill, © 1989 Delmar Publishers Inc.)*

(A) PATENT DRAWING

(B) WELDING DRAWING

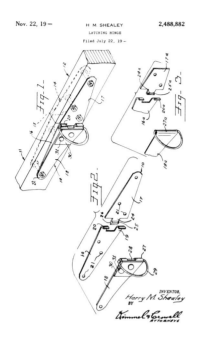

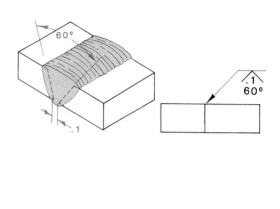

FIGURE 5–6 Sample special drawings: (A) patent drawing (B) welding drawing *(Drawing B reprinted from DRAFTING IN A COMPUTER AGE by Wallach and Chowenhill, © 1989 Delmar Publishers Inc.)*

Specialized Drawing Areas

As new areas of technology have developed, specialized forms of drawing have also developed to communicate special information. The use of specialized forms of drawing allow more information to be presented. It is assumed that the user of the drawing has an advanced understanding of the technology area; therefore, special presentation techniques, symbols, and notations can be used. Examples of these specialized areas include the following:

- patent drawing (Figure 5–6A)
- structural drawing
- aeronautical drawing
- marine drawing
- electronic drawing
- architectural drawing
- welding drawing (Figure 5–6B)
- piping drawing
- mechanical drawing.

Architectural Drafting

Architectural design began with the need to develop shelters for protection from the environment. **Architectural drawing** began when the design ideas were first recorded as graphics for craftspeople to use as construction guides. Architecture has developed into a specialized form of art and technology.

Design

The first step in the architectural design and drawing technology process is to identify the problem. The architect must find out what function(s) are to be provided for in the proposed structure. Usually, a written contract is drawn up and signed by the architect and client specifying the scope of the architect's activities in a project, the budget for a project, and the schedule for completion of the various items in the contract.

(A) BUBBLE SCHEMATIC

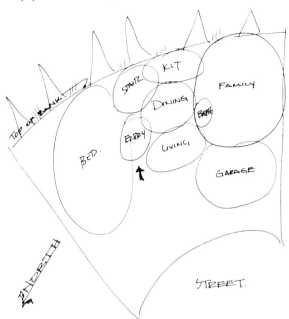

A basic design principle is "form follows function." A complete study of the use of the structure and the needs of the user must be made to determine what functions the structure is to serve. An analysis of the setting for a new building should be done to find what factors, such as other buildings and site topography, might influence a structure's design. After all factors are considered, a schematic drawing of the design is developed.

This schematic drawing is first developed as a **bubble diagram**. The bubble diagram is a rough sketch showing the relative relationships of the structure's parts as interrelated circles, Figure 5–7A. After analyzing the relationships in the bubble diagram, a more formal **conceptual design** sketch is made, Figure 5–7B. The conceptual design is drawn using sketching and some

(B) CONCEPTUAL DESIGN DRAWING

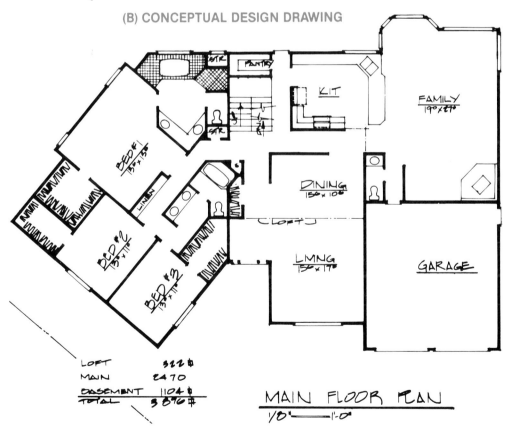

FIGURE 5–7 A bubble schematic (A) is used to graphically define the relationships of the elements of the architectural design. The bubble schematic is refined and formally drawn as a conceptual design drawing (B). *(Reprinted from ARCHITECTURAL DRAFTING AND DESIGN by Jefferis and Madsen, © 1986 Delmar Publishers Inc.)*

drawing techniques to give an accurate proportional presentation of the proposed design.

The conceptual design drawings are analyzed and evaluated in terms of the client's needs and the contract specifications. After needed changes are made in the conceptual drawings, the last phase of the design process, design development, begins with the refinement of sketches. Sketches are refined throughout this phase, pictorial sketches are developed, and at times models are made to present to the client. As sketches are approved by the client, detailed working drawings and specifications are prepared for the construction of the project.

Planning and Codes

When architects are planning projects they must be sure the project will satisfy the regulations required by land use and building design laws. Laws differ throughout the nation. Land use laws are often termed zoning ordinances or building codes. **Zoning laws** are set by communities to ensure that buildings are uniformly and correctly positioned in the communities and on building lots. The laws often specify the type of building that may be built in an area or on a lot, the areas of the lot that may not be occupied by the building, and even the minimum or maximum size of the buildings that may be built on a lot.

In addition to the zoning laws of most communities, states and financing banks require that buildings be designed to meet national or regional construction standards. These standards are called codes. Codes ensure the safe construction and occupancy of a building, and the protection of the environment from damage caused by the construction of the building.

There are national and local standard codes for the areas of structures and framing, electrical, plumbing, heating and ventilation, and fire safety. Codes for owner-occupied residential buildings often differ from the codes for commercial buildings.

Building permit laws ensure that designers and builders follow code regulations as they design and construct buildings. The permit process usually requires an inspection and approval of the building process at three points:

■ Drawing and specification development.

■ Construction stages — site preparation, foundation development, framing, electrical, and plumbing.

■ Construction completion — occupancy.

Architectural Working Drawings

A completed set of drawings developed by an architect or building designer will have the following parts:

■ **Site plan drawings**—drawings showing the project in relation to the building plot, surrounding community, roads, and compass orientation, Figure 5–8.

■ **Architectural plan drawings**—drawings showing the orientation of spaces within a project and their specific location measurements. The drawing sets will also include furniture and equipment layouts. These drawings are coded in the title block as "A" drawings.

■ **Electrical plan drawings**—drawings based on architectural floor plans that show the location of electrical features such as control panels, switches, lights, etc. These drawings are coded in the title block as "E" drawings.

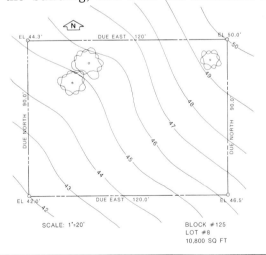

SCALE: 1"=20'

BLOCK #125
LOT #8
10,800 SQ FT

FIGURE 5–8 Sample site plan. A site plan shows the various heights of the ground surrounding the building as contour lines. Other site features on site drawings are easements, utilities, and sediment control. *(Reprinted from DRAFTING IN A COMPUTER AGE by Wallach and Chowenhill, © 1989 Delmar Publishers Inc.)*

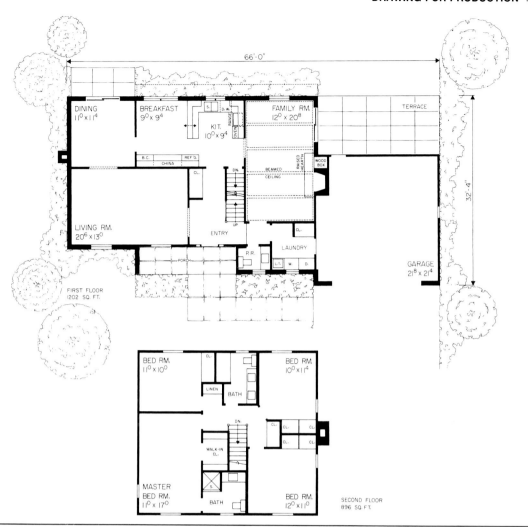

FIGURE 5–9 A floor plan *(Courtesy of Home Planners, Inc.)*

- **Mechanical plan drawings**—drawings based on architectural floor plans that show the location of mechanical features such as ducts, plumbing fixtures, vents, heaters, etc. These drawings are coded in the title block as "M" drawings.
- **Structural plan drawings**—drawings based on architectural floor plans that show features such as framing, footings, roof structures, etc. Structural drawings are used primarily in commercial construction drawings. These drawings are coded in the title block as "S" drawings.
- **Schedules and special features**—drawings are often included that contain composite schedules or listings of the materials used in the construction of a building. Schedules for doors, painting, windows, and electrical fixtures contain important information for builders as they purchase construction materials. These drawings will be accompanied by a set of written specifications that further communicate the design.

Floor Plans and Elevations

A **floor plan** is a drawing showing spaces or rooms drawn as if the viewer were seeing them with the ceiling removed, Figure 5–9. An **elevation** drawing is drawn to show the location of

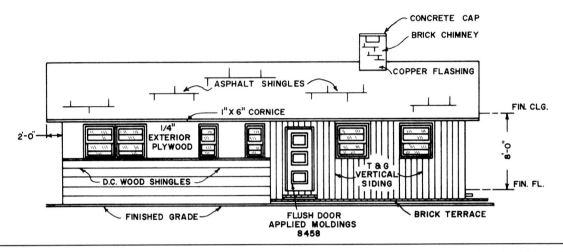

FIGURE 5-10 Architectural drawings often show special wall features by including elevation drawings of wall areas. *(Reprinted from ARCHITECTURAL DRAFTING AND DESIGN by Jefferis and Madsen, © 1986 Delmar Publishers Inc.)*

elements in relation to the ground or floor level of a room or space, Figure 5–10. Floor plans and elevations are usually developed in scales of $\frac{1}{4}'' = 1'0''$ or $\frac{1}{8}'' = 1'0''$, although other scales may be used.

Foundation and Structural Plans

A foundation is the part of a building that distributes the weight of the structure to the ground. The foundation is built at or below ground level in most cases. Foundations are usually made of masonry materials, although wood, steel, and other materials are also used.

A **foundation plan** or structural drawing describes the construction details and relationship of the parts including footings, walls, columns or piers, sills, shielding, girders, and anchors.

Architects must think about the scope of the project, local codes, and the topography of the lot when designing a building foundation. Detail sectional drawings are often included in construction drawings to fully describe the construction of foundation and structural features, Figure 5–11.

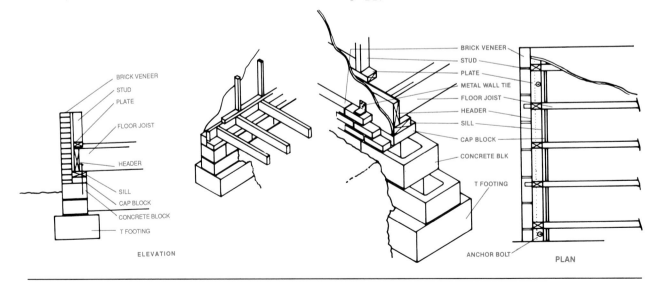

FIGURE 5-11 Structural drawings are drawn to present special details about the construction of the structural and foundation elements of the structure.

Framing Plans

Framing plans describe in detail the best method for building a project. Consideration must be given to the following:

- The type of structural design to use—bearing wall or skeleton frame.
- The type of loads the structure is to serve—live or dead loads.
- The type of materials to use—wood, steel, masonry based on structural design and load.

Specialized plans are developed for floor, wall, and roof framing features. Detail drawings and sectional drawings are often developed to give added detail to framing drawings.

Electrical and Mechanical Plans

Using structural features given on the architectural floor plans as a base, electrical and mechanical plan drawings are made, Figure 5–12. These drawings show the location of electrical, heating, ventilating, plumbing, security, and other special systems. Special symbols and techniques are used to draw mechanical, plumbing, and electrical features. The location drawings and specifications for special features are developed by engineers and technicians with expertise in those areas.

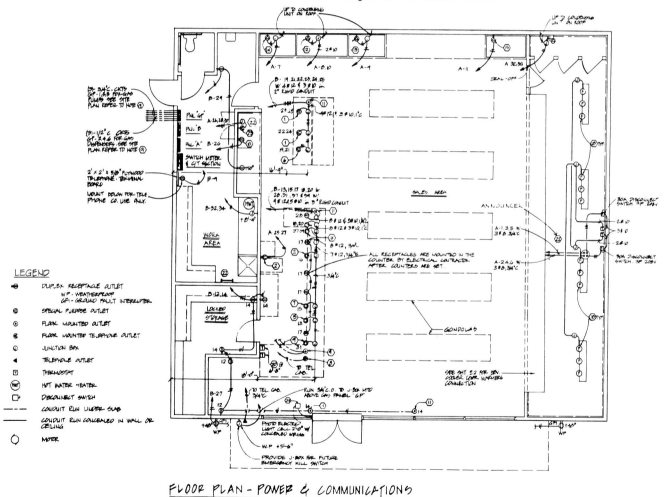

FIGURE 5–12 Floor plan with power and communications for a 7-ELEVEN convenience store. *(Courtesy of The Southland Corporation, Washington Division)*

FIGURE 5–13 CADD programs consistently create high-quality architectural drawings. *(Courtesy of Hewlett-Packard)*

Specifications

Specifications for projects are used to clarify and simplify the basic directions for building a project. They may be used by builders for contract bid information or to select materials. Specifications give the fine detail in the construction documents and ensure all parties working on a project receive the same information.

Computer Uses in Architectural Drafting

In less than ten years the development of **computer-aided design and drafting (CADD)** technology has greatly changed architectural design and drawing. Architectural CADD programs are performing any and all tasks, not just repetitious ones. The following are some of the activities a CADD system can do:

- Draw high-quality lines, letters, and symbols, Figure 5–13.
- Recreate drawings to alter or change features, or provide alternate presentations.
- Keep records of duplicated or typical features recorded on drawings.
- Perform standard engineering computations related to architectural, structural, or mechanical features.

- Record the number of every building part as it is drawn to develop schedules and obtain cost analysis information.

The development of CADD technology has given architects new chances to study and test ideas for project design. This can only lead to new levels of creativity in architectural design.

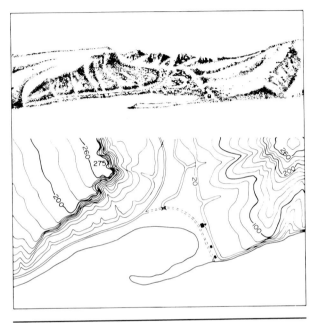

FIGURE 5–14 Topographical drawings show contour lines to indicate changes in land height. This topographical map has contour lines at 20-foot intervals and a corresponding land relief pictorial. *(Courtesy of the U.S. Dept. of Interior, Geological Survey)*

Topographical Drawings

Topographical drawings present the physical features of a section of the earth's surface using standard symbols and notations. These drawings give information about roads, population distribution, navigation, and mineral resources locations. Topographical drawings use symbols called contour lines as well as notations to show changes in the height of surface features, Figure 5–14.

Cartography is the process of making maps and charts using topographical data. Throughout history the development of accurate maps has been one of the most important technological goals.

Mapping Standards and Terms

A **map** is a drawing of a large section of the earth's surface, Figure 5–15. Maps that are drawn of small sections of land are called plat or **plot** drawings. **Charts** are maps of water surface areas used for ship navigation.

Surveying and Data Collection

Surveying is the process of measuring and recording the various features of the earth's surface using the principles of geometry and trigonometry. A surveyor uses a **bench mark** or known point to establish a starting point for work. The notes the surveyor takes are called **field notes**. They are used to develop a plot drawing.

Using a **transit**, which is leveled and aligned to compass headings, and a distance-measuring instrument, the surveyor plots lines, Figure 5–16. These are straight lines of specific length and at

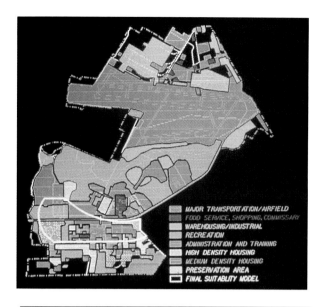

FIGURE 5–15 This map was computer generated for land use and resource management applications. *(Courtesy of Intergraph Corporation)*

Computers In Cartography

The adaptation of computer technology to topographical drawing and cartography has been dramatic. The process of surveying involves making and recording specific measurements as referenced to axes of a three-coordinate system. A computer can easily process and interconnect two-axes data to develop contour and profile maps. It can integrate third axis data to make pictorial topographical drawings and maps.

Very accurate maps can be made by making use of images generated by cameras carried by orbiting satellites. Since these satellites are high above the earth, their electronic cameras can provide pictures of large land masses. Contours can be easily and clearly seen. In some cases, existing maps have been shown to be incorrect as they did not conform to satellite images.

Computers are used to integrate the vast amount of data produced by these satellite cameras. The computers digitize the photographic image, and through enhancement techniques, improve the quality of the picture. Better maps of land masses and geologic forms can be produced.

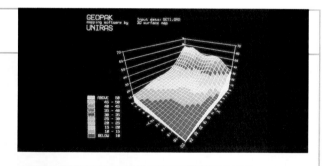

A computer can use three-axes data to make pictorial topographical drawings and maps. *(Courtesy of McDonnel Douglas Manufacturing and Engineering Systems Company)*

Computers, CADD, and satellite data and pictures provide accurate maps and charts. *(Courtesy of NASA)*

specific angles in the horizontal and vertical planes. The angles of the lines are recorded as **bearings**. Bearings are compass-referenced angles recording minutes and seconds measurements. The corner points of lines are usually marked with stakes.

FIGURE 5–16 Surveying methods are used to collect data on the topography of the proposed building site. *(Reprinted from CONSTRUCTION TECHNOLOGY by Mark Huth, © 1989 Delmar Publishers Inc.)*

Summary

For engineers, technicians, craftspeople, architects, and technical drafters, technical design, planning, and drawing are one process. The common element in the process that allows each individual to communicate clearly and concisely with the others is the technical drawing.

The development of more and more complex technologies has made necessary the development of specialized areas of technical drawing. Today, there are special drawing standards and conventions for such areas as welding, piping, aviation, marine design, and electronics. Each of the specialized drawing areas reflects the need for special forms of communication.

Architectural design and drawing, like other specialized technical drawing areas, keeps evolving. It has been a central technology throughout history because it deals with one of the most important human needs, shelter. The future of architectural design and drawing will change greatly as CADD technology affects the way architects, engineers, and drafters develop and test design solutions.

The greatest changes in technical drawing are in the areas of topographical drawing and cartography. The impact of CADD technology has enabled the development of maps and charts never before possible. Regardless of whether technical drawings are made by hand or with a CADD system, they still serve the same purpose. They communicate a design or the solution to a problem.

REVIEW

1. What are working drawings?
2. What is the purpose of a production drawing?
3. Why are assembly drawings important to craftspeople?
4. What does the term fit mean?
5. If the tolerance notation shown on a diameter measurement of a shaft was $+0.005''$, and the shaft was $0.050''$ in diameter, how large could the shaft be made and meet specifications?
6. What is the purpose of the contract between the architect and the client?
7. What is the purpose of a bubble sketch?
8. Why are building codes important to clients and architects?
9. What is the purpose of building permit laws?
10. Why is it important to plan the location of furniture and equipment when designing a structure?
11. A wall is designed to be $8'0''$ high, and it is presented on a $1/4'' = 1'0''$ scale drawing. What will be the actual height of the wall drawn on the elevation drawing?
12. Why is the design of a building foundation and structural elements so important to the architectural design of a structure?
13. If the construction superintendent was looking for information about the installation of heating equipment for a building, where in the building plan would this information be found?
14. What is the difference between a map and a chart?
15. If the contour lines of a site plan form an irregular circular pattern with the notation $4'$, $5'$, $8'$, and $10'$, what do the contour line drawing and notation indicate?

Computer-Aided Design and Drafting

KEY TERMS

computer-aided drafting (CAD)
computer-aided design and drafting (CADD)
computer-aided manufacturing (CAM)
computer-integrated manufacturing (CIM)
computer-numerically controlled (CNC)
coordinate geometry
cathode ray tube (CRT)
display terminal

dot matrix printer
geometric modeling
hard image
input device
jaggies/stair steps
mass properties
menu

modeling
output device
plotter
soft image
solid modeling
surface modeling
wireframe model

Introduction

Computer technology has greatly enhanced people's ability to express graphic images. In the field of design and drafting, computers have become an important tool. CAD stands for **computer-aided design** or **computer-aided drafting** (or drawing); CADD stands for **computer-aided design and drafting**.

FIGURE 6–1 Hard images were the only way to share ideas for thousands of years. CAD allows drafters to create soft images, which are flexible and easily modified. *(Courtesy of Computervision, a division of Prime Computer, Inc.)*

For thousands of years, the visible expression of a thought or idea was a **hard image**. A hard image might have been as simple as a sketch or as complex as the set of architectural drawings for a 50-story building. Hard images are fixed. If changes are needed they can be time consuming and difficult to make.

Computers offer the drafter the ability to create **soft images**, Figure 6–1. A soft image is one that can be electronically manipulated and displayed. Soft images are in the computer system as groups of interdependent reference points that can be modified. The image created by the operator is flexible because a change made to one point can create a sequence of related changes in other reference points. This changes the configuration of the soft image held by the computer.

Computer Graphics

Computers can be programmed to give every point in a drawing a value based on a X, Y, and Z coordinate system. The location of any point or series of points is referenced to the origin (or 0, 0, 0 point) of the X, Y, and Z axes, Figure 6–2. Circles, ellipses, lines, cubes, cylinders, and other shapes can then be expressed as a sequence of points related to the established coordinate system. It can be seen from this that the computer graphic development process is made up of applied mathematical principles and concepts.

Coordinate Systems

Generally, CADD systems locate and draw points, lines, and shapes using **coordinate geometry.** Coordinate geometry uses relative or absolute geometry. It involves the use of the cartesian coordinate system to locate the turning points for lines in relation to the X, Y, and Z axes system. Some systems use polar geometry which specifies positions by line length and angle from the origin, Figure 6–3.

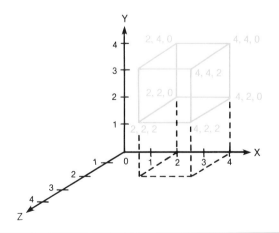

FIGURE 6–2 CADD systems using a coordinate geometry graphic development system use an X, Y, and Z set of established axes.

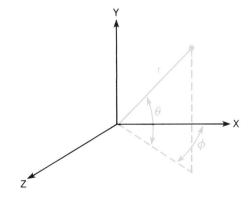

FIGURE 6–3 In polar geometry, a point's position is specified by its distance from the origin (r), its elevation angle (θ), and azimuth angle (Φ). (θ is the Greek letter "theta" and Φ is the greek letter "phi.")

FIGURE 6–4 The central element in a CAD system is the human operator. *(Courtesy of Intergraph Corporation)*

CAD Systems

A CAD system is a computer system specially assembled to perform computer-aided design or drafting applications (see Chapter 2).

It has three main parts: hardware (input and output devices), software, and human operators. The operator must be skilled in the development of technical drawings, as well as in the use of the CAD hardware and software. The capabilities of even the best CAD hardware and software are limited by the skill of the operator, Figure 6–4.

CAD Computer Processors

Computers used in CAD systems range from desktop personal computers to mainframe computers. Small engineering companies use personal computer workstations for CAD, while larger companies use minicomputers that several workers can share. Large aerospace companies use mainframe computers, and sometimes supercomputers for CAD.

The computer must be capable of handling many mathematical computations quickly to move objects or create new objects using their (X, Y, Z) coordinates. When personal computers are used, they are often equipped with a separate math coprocessor for added computation speed.

CAD Input Devices

Input devices allow the operator to communicate with the computer. In many cases operators use multiple input devices for CAD operations. To use a CAD system, a skilled operator must learn to operate input devices such as keyboards, number pads, and tablets. Some of the most widely used CAD input devices include a tablet, digitizer, (Figure 6–5) scanner, light pen, mouse, joy stick, thumbwheel, function board, and keyboard.

CAD Output Devices

Output devices for CAD computers are used by the computer system to display processed information in soft or hard form. Devices used for soft output are display terminals. Plotters and printers are used for hard output.

In a CAD system, drawings and other data appear on a video **display terminal**. This device is sometimes referred to as just the terminal and the screen itself is called the display. The terminal is the device that looks like a television monitor.

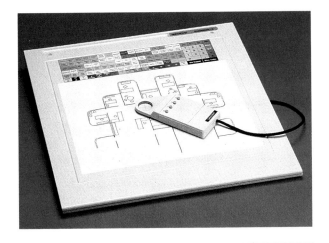

FIGURE 6–5 This digitizer includes a graphics tablet and a mouse. *(Courtesy of Calcomp)*

FIGURE 6-6 Color monitors enhance communication by making the drawing displayed easier to understand. *(Courtesy of International Business Machines)*

Terminals come in monochrome (single color) and color models. A good color model can display over 250 different colors. The use of color on a CAD display can enhance communication by making the drawing displayed easier to read and understand, Figure 6-6. Displays used in

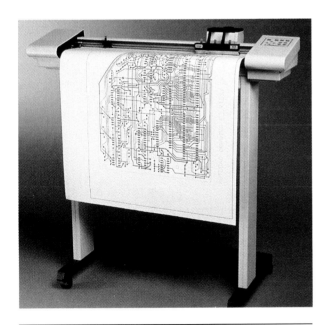

FIGURE 6-7 Plotters can draw very accurate technical drawings with smooth contour lines. *(Courtesy of Houston Instruments Division, AMETEK)*

CAD systems usually have very high resolution (the ability to resolve, or show, small pieces of a picture) so objects can be displayed accurately and in great detail.

The basic technology in most terminals is the **cathode ray tube** or **CRT**. This is why some people call display terminals CRTs. Other displays are flat, such as a liquid crystal display (LCD) or gas plasma display.

A **plotter** is an output device that converts the electronic signals from the computer into a hard image or technical drawing, Figure 6-7. There are several types of plotters. The most widely used is the electromechanical pen. The plotter creates the drawing image by moving a print head containing a pen across a stationary paper, moving the paper in relation to the stationary print device, or moving both the paper and print device at the same time. Often, plotters use different color pens to produce multicolor drawings.

Dot matrix printers are sometimes used as output devices for first-draft CAD drawings. (See Chapter 2 for a description of how dot matrix printers work.) Dot matrix printers have very poor resolution and are unacceptable for all but the simplest CAD drawings. Curved lines and lines drawn at angles will be drawn with **jaggies** or **stair steps**, Figure 6-8. **Laser printers** are sometimes used for smaller, single-color CAD drawings.

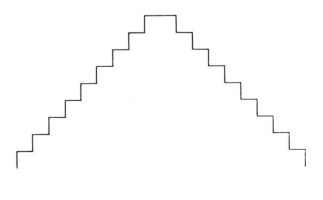

FIGURE 6-8 Curved lines and lines drawn at angles will be plotted with stair steps by a printer.

Software

CAD systems use math calculations intensively, taxing the computer's processor and memory. Every point, line, or surface on an object is identified in the computer memory in terms of its mathematical value. Until the 1970s, only very large mainframe computers or the slightly smaller minicomputers were able to do the many calculations needed to develop technical drawings. With ever increasing growth in computer hardware and program technology, today's microcomputer is able to do CAD operations that a minicomputer could not do only a few years ago.

There are many CAD software developers and publishers providing application programs. Some large companies have developed their own special software system, or contracted with a CAD publisher to modify an existing commercial system to meet their unique needs. For most users today, existing commercial CAD software can give all the CAD abilities needed.

In addition to the computational portion of software needed to create computer drawings, most CAD packages also include a data base. The data base program is used to store and quickly retrieve information used often, such as screw types and sizes, wall thicknesses, and cable sizes.

CAD and CADD

There is sometimes confusion over what is meant by CAD. Does it mean computer-aided design or does it mean computer-aided drafting? Some people use the term for design, some for drafting, and some for both. The term first meant computer-aided design. This is because the first systems designed were used as aids to the design process. Then systems were developed to aid in the drafting process. Now there are systems that can do both. Such systems are known as **computer-aided design and drafting** or **CADD** systems. Most modern systems fall into this category. Systems that are mainly concerned with such activities as solid modeling, finite element analysis, and surface modeling are known as computer-aided engineering or CAE systems.

FIGURE 6–9 CADD systems can create a 3-D soft image so precisely that it appears to be the real object. *(Courtesy of McDonnell Douglas Manufacturing and Engineering System Company)*

A CADD system allows the computer to extend the design and drawing process beyond just the development of graphic images in soft and hard forms. A CADD system can be programmed to consider outside influences which may affect the design of an object. For example, a CADD system can be programmed to consider the heat expansion rate of the steel beams in a bridge, and draw the bridge in the true scale size on a 90° day. A designer using a CADD system can see the design develop just as if the real object were being constructed, Figure 6–9.

System Functions and Menus

While some standards are evolving, CAD systems made by different manufacturers have a variety of features and are used for different tasks. These different uses demand that different systems offer operators several ways of entering and managing data.

Drawing Functions

Every CAD-system publisher and manufacturer designs his or her system with unique functions and command patterns, Figure 6–10. Therefore, to learn to use a specific system the

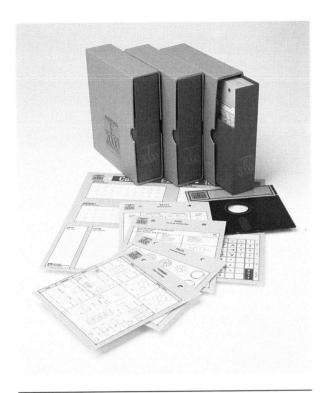

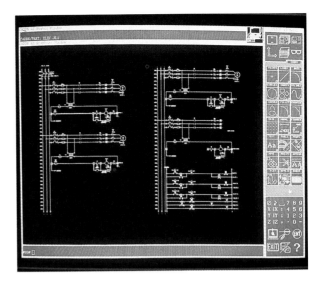

FIGURE 6–11 Menus which allow a computer operator to select specific commands are displayed on the terminal screen. *(Courtesy of Computervision, a division of Prime Computer, Inc.)*

FIGURE 6–10 Today there are a number of major manufacturers of CADD systems, specialized auxiliary software, and hardware packages. *(Courtesy of Carrier Corporation)*

user must obtain the manuals and documents produced by the publisher for training system operators. But, there is a standard core of universal CAD commands and functions that are found in almost every system.

CAD Menus

Programs are designed with **menus** which allow the operator to select groups and subgroups of CAD program functions. The menu allows the operator to identify the group of drawings or system operation functions needed, and direct the computer to perform those functions. Menus are built-in tools that allow operators to interact easily with the program. Often menus will be displayed on the terminal screen when special commands are entered into the computer, although some systems continually display a menu on the monitor screen, Figure 6–11. Many CAD systems use special keys on the keyboard, or locations on input devices such as a graphics tablet to input menu commands.

3-D Drawing, Modeling, and Design Analysis

Basic CADD systems can be easily used to develop two-dimensional (2–D) pictorial drawings. The nature of the image development process allows the operator to

- develop oblique drawings using a grid,
- develop isometric drawings, and
- develop one-point perspective drawings using construction lines.

Advanced CADD systems offer the ability to develop true three-dimensional (3–D) drawings. A 3–D CADD system moves beyond just X, Y, and Z axis plotting. The 3–D system lets the operator rotate the drawing about each axis singularly or in combination with each other. The 3–D computer development requires many more mathematical calculations by the computer than does 2–D drawing. Only microcomputers with fast processors and large memory capacities, minicomputers, and mainframes can process the quantity of data necessary to develop 3–D drawings.

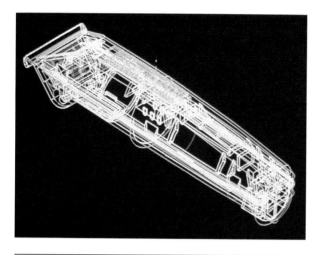

FIGURE 6-12 3-D CADD drawings are developed as wireframe drawings, such as this screen display of the Oster cordless, rechargeable hair trimmer. *(Courtesy of Oster Corporation)*

Wireframe Models

3–D CADD drawings are developed as multiplane or surface views of an object. The process is much the same as drawing an isometric drawing. The process results in a **wireframe model** of the object, Figure 6–12. Using a designated point of view and axis angle orientation, CADD programs will delete or edit the hidden lines or surfaces in the object by determining which are visible. The editing process leaves a true pictorial view displayed on the screen or in hard copy.

While 3–D CADD programs may display only pictorially correct presentations, they keep all input information about the object in memory as interrelated mathematical data. Using this complete object information base, the computer can recalculate and reform the object whenever the point of view or axis orientations are changed. It is possible for the operator to turn the object around to view the back. This is done by reversing the positive and negative mathematical relationship of the X, Y, or Z axis.

Modeling

The concept of 3–D CADD **modeling** is based on the principles of relative mathematics. As the CADD operator inputs information about the

object, the computer system directs the computer to assign specific numeric values. Each data input is related to the next through a mathematical relationship.

The concept of creating 3–D CADD models has moved ahead in recent years with advances in computer and programming technology. Today, there are three forms of computer modeling in use.

Geometric modeling uses the development of a wireframe image as the basis for further computer data input referencing and presentation development.

Surface modeling uses the data input base of the geometric model wireframe as the boundary for identifying object surfaces. All the points on the surface, within the boundaries of the wireframe construction, are assigned values. These values become part of the data base. Surface modeling displays present objects that appear to be 3–D solids.

Solid modeling requires very high levels of data input and calculation. It involves assigning a value to every point within the surface boundaries of an object. The mathematical calculations for this level of modeling are very complex and require powerful computers and programs. Solid models react to change just as if they were actual objects. A change in any point will create changes in every other point.

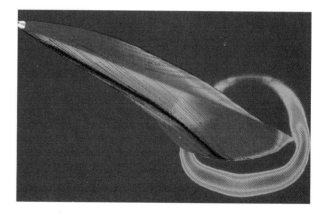

FIGURE 6-13 CADD modeling systems allow the designer to test designs before they are actually produced. For example, the colors on this wing body indicate pressure levels. *(Courtesy of NASA)*

Architectural CADD

Sverdrup Corporation, St. Louis, Missouri, is a leading provider of professional services for development, design, construction, and operation of capital facilities. Projects range from office buildings to stadiums and convention centers; from airports to the space shuttle launch complex; from state-of-the-art breweries and testing facilities to award-winning bridges and tunnels throughout the world.

Sverdrup has discovered that CADD is a valuable tool not only for design but for coordinating, controlling, and communicating information for the duration of a project.

Sverdrup's Intergraph CADD system has the ability to work three-dimensionally, which contributed to an ingenious design solution for one of the nation's largest, most respected teaching and research hospitals.

In 1986, Barnes Hospital in St. Louis had three related problems. First, a busy boulevard separated the 1.7 million-square-foot health care complex from its major parking facility. Second, visitors and patients crossing the roadway and entering the hospital's ground floor did not directly reach the main lobby and reception area. Third, any sort of construction linking the hospital and underground parking garage had to meet stringent design standards, since it would be built partially in St. Louis' famous Forest Park.

"CADD helped us develop an enclosed pedestrian bridge that answers all three concerns," says Sverdrup architect Mark Gustus. "A new bank of elevators and escalators within the garage takes newcomers up to the bridge level, which then leads them across the street directly to Barnes' central reception area. The bridge itself is an intricate truss of steel tubes clad in transparent glass, which provides a pleasant view for both hospital visitors and Forest Park visitors."

"CADD enables us to try more imaginative, complex designs because it works in three dimensions all at once," says Gustus. "Ordinary paper drawings aren't much help, and manmade models are time-consuming. CADD's gaming capabilities, along with its ability to rotate and animate an image, made it an ideal medium for a project like this."

(Courtesy of Louise Wiedermann, CADD System Manager for Sverdrup Corporation. Sverdrup is engaged in the Development Design, Construction, and Operation of Capital Facilities.)

Sverdrup solved several problems for St. Louis' Barnes Hospital by developing an enclosed pedestrian bridge with Intergraph CADD.

An even more complex extension of solid modeling CADD technology is the development of programming technology that provides **mass properties**. Mass properties extend the CADD systems to consider factors of volume, weight, area, etc. The assignment of mass qualities allows the operator to ''tell'' the computer to consider if the object is made of concrete, steel, or rubber as it considers modifications to the original data input.

Solid modeling allows the designer to truly analyze the performance of an object in soft image form. Today, solid modeling is used to test designs and find problems before constructing the object, but with the same reliability as if the object itself was being tested, Figure 6–13.

CAM, CNC, and CIM

While the output from a CAD system is usually converted into drawings, it can be used by industry as the input to a manufacturing system. This process is termed **CAM**, which means **computer-aided manufacturing**. In CAD/CAM systems, CAD systems communicate directly with manufacturing systems.

The CAD operator sets the physical size and shape of the object as digital reference points in the computer. Harware and software have been developed to transfer the information from the CAD system to one or a group of **computer numerically controlled** machines **(CNC)**. CNC machines, such as lathes, milling machines, and drill presses, are then programmed to use the CAD information to manufacture the object, Figure 6–14.

Today, it is possible to go from the CAD soft image of the object directly to the manufacture of some objects with little or no human involvement. The electronics industry uses this technology to create the highly complex integrated circuit devices so widely used today.

Communications technology, including factory-wide local area networks (LANs) (see Chapter 20) are connecting all parts of the manufacturing system together. New computer software is being developed to take advantage of these integrated systems. This concept is

FIGURE 6–14 Computer numerically controlled machines can be programmed to perform a variety of tasks. *(Courtesy of Cincinnati Milacron, Inc.)*

known as **computer-integrated manufacturing** or **CIM**. CIM includes management functions such as automatic inventory stocking levels, machine scheduling, and personnel assignments. It represents the latest in the machine communication technology used in manufacturing plants.

Summary

Computer technology has enhanced people's ability to express graphic images. In the fields of design and drafting, computers have become an important tool. CAD stands for computer-aided design or computer-aided drafting. CADD stands for computer-aided design and drafting. Computers offer drafters the ability to create soft images that can be modified electronically before being drawn, or converted to hard copy.

Computers use coordinate geometry to represent images. In coordinate geometry, points on an object are described relative to the origin. Lines are expressed as a series of points or by a mathematical equation. Rotation of an object to show views from different angles is accomplished by mathematical equations.

CAD systems are made up of hardware, software, and operators. The hardware includes: computer processors that manipulate the geometric equations; input devices such as keyboards, a mouse, and digitizers; and output devices such as display screens, plotters, and printers.

The software contains the geometric equations needed to describe objects, and the interface for people to use them. Simple operator commands that make the system easy to use are translated into complex equations by the software. Many CAD software packages also contain data bases for retrieving standard information or for creating parts lists.

Modern computer-aided design systems are being combined with computer drafting systems, creating computer-aided design and drafting (CADD) systems. CADD systems allow designers to test the properties of complex structures such as bridges and airplanes before actually building them.

Three-dimensional drawings are developed as wireframe models. The CAD program determines which lines are hidden because of viewing angle, and deletes them on the display terminal. Three types of modeling are geometric modeling, surface modeling, and solid modeling.

The output of a CAD system can be used to drive automated machinery in manufacturing systems. This is called CAD/CAM (computer-aided manufacturing). In some systems, an operator can design a part on a CAD system, and the system will make the part directly with little or no intervention by people. Machines that operate under computer control are sometimes called computer numerically controlled, CNC. When computers are used to automatically schedule, stock, and control a manufacturing environment, it is called computer-integrated manufacturing, CIM.

REVIEW

1. What does the term ''soft image'' mean?
2. What mathematical concept is central to CAD system functioning?
3. Explain the concept of modeling using CADD technology.
4. What is the purpose of the CAD menu?
5. In CAD systems, the location of any point or series of points is referenced to what position on the X, Y, and Z axes?
6. What types of output devices are used to create hard images?
7. Why is solid modeling an important extension of CADD modeling technology?
8. Why are the computer processors used for CAD usually much faster and larger than those used for simple office tasks, such as word processing?
9. What is the purpose of the software in the CAD system?
10. Why does the development of surface models require the development of geometric wireframe models?
11. Why is solid modeling technology a very important development in CADD technology?
12. What does the term CRT mean?
13. What is the purpose of the menu in the CADD system?
14. Why is the operator the most important element in a CADD system?

SECTION ACTIVITIES

DESIGN A GRAPH OR CHART

OVERVIEW

Many times as we report information we need to use a graph or chart to show comparisons between different quantities or amounts. In this activity you will record information in graphic form using the equipment available to you in your laboratory. You will compare different types of graphs and charts, and select the one that best shows the information required.

As you work through this assignment, it will be necessary to anticipate the best format for representing the data as well as the amount of space and calibration for your particular graph. A quick sketch may keep you from having to redo your activity.

MATERIALS AND SUPPLIES

Equipment and materials used for this activity depend on the amount of material available to you in your lab. The easiest approach and perhaps the best results can be obtained by using a computer to develop the graph, shade the areas, and prepare the text. Another approach would be to use basic drafting instruments including straight edges, shading film, etc.

PROCEDURE

1. In this activity you will develop a graph to illustrate and report particular data. The first step is to decide what data you wish to collect and present. An example

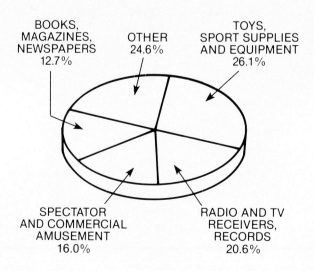

PERSONAL CONSUMPTION
EXPENDITURES FOR RECREATION

BOOKS, MAGAZINES, NEWSPAPERS 12.7%

OTHER 24.6%

TOYS, SPORT SUPPLIES AND EQUIPMENT 26.1%

SPECTATOR AND COMMERCIAL AMUSEMENT 16.0%

RADIO AND TV RECEIVERS, RECORDS 20.6%

PIE CHART

might include a graph of the test scores from the last test given in class. In graphing this information you first want to determine what comparisons you want to report. This may be a comparison of the number of each grade received (how many 90s, how many 85s, etc.). The Y or vertical axis would indicate the score and the X-axis would indicate the number of people that attained a specific score. Or, you may wish to compare the number of students buying hot lunches with those bringing their lunch. Perhaps you could grow a plant and graph the amount of growth over the course of time. You could graphically present data on your basketball team's performance. Brainstorm with a friend to determine what data you will collect and present.

2. Select a graph or chart style that will allow you to best demonstrate what you wish to show. If you choose to show percentages, you may want to select a pie chart (see page 94).

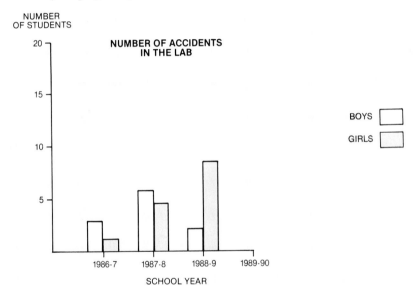

GRAPH

3. Remember that the information on the graph or chart should be complete enough so that no explanation other than the graph or chart and notes is necessary.

4. Prepare a rough sketch of the information graph or chart you wish to construct. For example, the bar graph above shows the number of accidents in the lab for boys and girls over the course of three years.

5. If you are using a computer, begin by accessing the software and adding the necessary information lines, borders, text, and shading.

6. Make a hard copy of your graph or chart. The best quality for a hard copy may be from a laser printer; however, many dot matrix printers do an excellent job of printing the graphics.

7. If you do not have access to a computer, then you must begin by using basic drafting procedures. Tape down your paper. Begin to carefully draw a graph

or chart that will represent your intended information. Care should be taken to present the material as clearly as possible using good drafting techniques including line quality and lettering.

8. Try using a different type of graph or chart to present your data. Which graphic form works better for you? Why?

FINDINGS AND APPLICATIONS

In this activity you may have found that the computer was able to generate a better quality graphic display for the amount of time spent than would be done using traditional drafting methods. You also found that in many cases a simple graph is easier to understand than a lot of written material. As you developed your graphic, you may have noticed graphs or charts being used in a number of places. You may also find that a good graph or chart in a report for another class is a good way to improve your grade.

ASSIGNMENT

1. Develop a graph or chart that displays the amount of time students spend studying technology education outside of class in the form of homework. To do this you will need to survey the class and then select a suitable graphic format.
2. Collect samples of three different kinds of graphs or charts. Write a brief paragraph explaining the information shown on each graph or chart.
3. Prepare a graph or chart showing the percent of students in your school who take part in technology education courses.
4. Make a graph or chart to show how many of the technology education students belong to the local technology education club.

 DESIGN AND SKETCHING

OVERVIEW

In this activity you review the techniques of sketching as you improve your skills. Once you have completed your first sketch, you have the chance to use this sketching technique as you design solutions to other problems.

Sketching requires that you form in your own mind the direction and solution to the various problems. In these exercises you are expected to review basic sketching procedures as you produce the first drawing. As you progress through this activity, you will be required to do additional designing and sketching.

MATERIALS AND SUPPLIES

To complete this activity you will need the following:

■ Pad of plain or graph paper
■ Pencil
■ Computer with drafting software

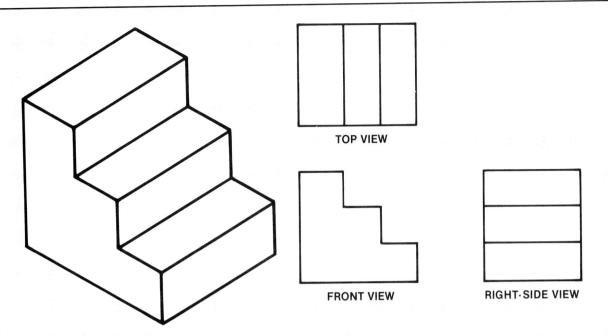

TOP VIEW

FRONT VIEW

RIGHT-SIDE VIEW

PROCEDURES

To complete this activity you will need to perform the following tasks:

1. Redraw the isometric figure shown in orthographic projection. Show the front, top, and right side in their correct relationship to each other. Sketch in the dimensions. Pay close attention to proportion. Also think about the quality of lines and which lines are to appear darker or wider. While the intent of sketching is to allow us to quickly put down our thoughts on paper, we need to remember that neatness and accuracy make it much easier to convey our message.

2. If you have access to a computer to perform the sketching and design function, you will find it easy to do basic corrections and revisions as well as to develop quality sketches and drawings.

3. Using the equipment and materials available, select one project from the following list of design problems. Sketch designs to show your solution to the problem. This list includes:

   a. A catapult that will launch a marshmallow over a 1-foot wall. The catapult will probably be powered with a rubber band or some other elastic material to give it the thrust to flip the marshmallow over the 1-foot barrier. Your task is to determine the shape of the catapult, and what kind of catapult could hold a marshmallow long enough to get it into the air so that it would fly across the 1-foot wall. You might want to use a balsawood frame or some other construction material.

   b. A 2-foot long truss that will hold 40 pounds of weight. Trusses have always been able to increase the strength and load carrying ability of the materials. As you design this truss, try to get maximum support with the least amount of materials.

c. A container that will protect an egg when dropped from a height of 10 to 15 feet.

d. A dragster that has good aerodynamic design.

Your designs will be evaluated based on their clarity, ability to logically solve the problem, creativity, and the quality of the sketch.

FINDING AND APPLICATIONS

As you did this activity, you found that you had to try two or three possible solutions and select from the best of each for your final design. You noted that practice seems to make it easier to sketch. You are aware that almost everything that we use has been drawn at one time or another and probably began with a sketch. This sketching process is used by all types of people in a wide variety of jobs.

ASSIGNMENT

1. List five uses of sketching from your experiences and observations.
2. Complete the required sketches and design problems. Give them to your instructor for evaluation.
3. Identify another design problem that you would like to work on. Develop the sketches and solution.

 CARTESIAN COORDINATES

OVERVIEW

The cartesian coordinate system is used in computer-aided drafting (CAD) and in manufacturing. This system provides an exact method for representing points on a graph. Most of you have used two-dimensional cartesian coordinates in mathematics classes as you used the X- and Y-axis to graph lines. In this activity we will review the cartesian coordinate system using three basic input modes common to CAD — absolute, relative, and polar. Using these input modes we will draw a simple dot-to-dot type sketch explaining what happens in each.

Following the first experience, you will be asked to develop your own cartesian coordinate sketch or drawing. As you design your activity, you will need to predict and plan various points on your graph. Before you begin the demonstration activity, try to predict what the shape will be when you finish, before you have drawn any lines.

MATERIALS AND SUPPLIES

To complete this activity you will need the following:

■ Graph paper   ■ Pencil   ■ Straight edge

PROCEDURE

1. Begin the procedure by following these rules:
   a. We use three input modes — absolute, relative, and polar. Absolute means in relation to the graph paper. Any position entered as absolute is exactly in the position, and is always related to the origin or 0,0.

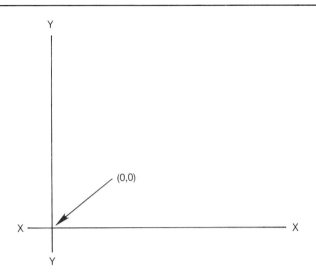

b. Relative means as a position is entered. It is relative to the last position.

c. Polar gives the distance and angle of the vector or line.

2. On a regular sheet of paper begin by drawing two lines, one about an inch from the left side of the page and parallel with that side, and the other about an inch from the bottom of the page and parallel with the bottom. Label the vertical line as Y and the horizontal line as X. Where they cross will represent 0,0. When drawing in the cartesian coordinate system, 0,0 is used as the base position and the place from which we begin counting to determine other locations. In addition, from this 0,0 position any movement from left to right is considered a positive direction, and any movement upward is considered a positive direction.

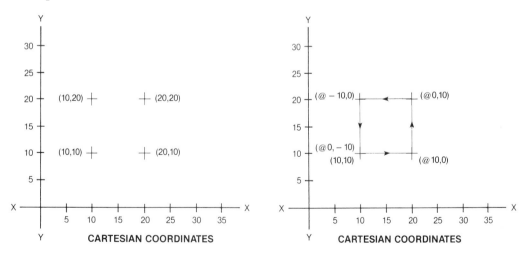

3. Find the point 10,10 (absolute). To find this point go 10 grid spaces to the right of 0,0 (that's positive) on the X-axis and 10 grid spaces up on the Y-axis, or vertically. Locate this point as shown. Your drawing should look like a square that is 10 units wide and 10 units high.

4. Draw the same shape using the relative mode. Begin at point 10,10 (absolute). Enter the distance relative 10,0 (@10,0). The @ symbol is used to replace the word relative and means relative when it is used in this context. Relative means that relative to this last position, we go 10 on the X-axis and 0 on the Y-axis. From this point go @0,10. From this point go @ – 10,0. From this point go @0, – 10. Again, does your drawing look like a square? Is it 10 units wide by 10 units high? If so, it is done correctly.

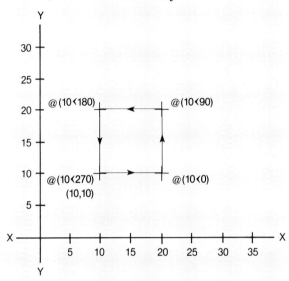

CARTESIAN COORDINATES

5. Draw the same shape using the polar mode. Begin at the point 10,10 (absolute). Enter the distance relative to that point 10 units at an angle of 0° (this is written @10 < 0). 0° in this software system is about 3 o-clock or positive X. Relative to this position go 10 units at an angle of 90°. This is written @10 < 90. Relative to this position go 10 units at an angle of 180° (@10 < 180). Relative to this position go 10 units at an angle of 270° (@ < 270). Did this make the same square shape? Which one was easiest? Many drawings will use all three. Most use an absolute position to begin the drawing and future lines are drawn in either relative or polar. Let's draw a shape and see what it looks like when we get done.

Remember, the A in the circle (@) means relative and the open caret (<) means angle. On a sheet of graph paper find where the point 0,0 is. This point is where the X- and Y-axis intersect. Ready? Now begin at point 5,5 (that means over 5 on the X-axis and up 5 on the Y-axis). This starts our object. The line now goes to the right to point 10,5. This should move the line 5 more points to the right and 0 points up or down. From this point go relative, which is the @, to 0,10. This means that we move from our last point (relative), up 10 points on the Y-axis. The 0 indicated no movement on the X-axis. From this point move to a point @ 5, and the open caret 180° (@ < 180). This is read: relative to the last point go 5 units in the direction of 180°. Finally let's go @0,-10.

Note: All the degree angles are read with 0° being the direction of 3 o'clock or horizontal from left to right. If you want a 90° reading, then we would go perpendicular to that line or toward 12 o'clock so that the included angle would be 90°. If we went 180° we would go to 9 o'clock which would be a horizontal line going from right to left. If we want to go straight down, we would go 270° or -90°. Either would give us a straight down line.

The shape you were drawing using the coordinates was a rectangle. How did you do? At this point, enter the following coordinate points for seven different shapes. See if you identify each of the shapes.

1. 7,9; @0,1; @–3,0; @0,–5; @3,0; @0,2; @–2,0
2. 8,8; @3,0; @0,–3, @–3,0; @0,3
3. 12,8; @3 < 0; @3 < 270; @3 < 180; @3 < 90
4. 19,10; 19,5; @–3,0; @0,3; @3,0
5. 25,10; @5 < –9; @3 < 180; @0,1.5
6. 26,8; @0,–3; @3,0; @0,3; 26,8
7. 30,10; 30,5; 33,5; 33,8; @–3,0

FINDINGS AND APPLICATIONS

If you entered the points and distances correctly, you were able to find the correct message. In computer-aided drafting as well as other manufacturing processes, cartesian coordinates are used to tell the computer what to do. While the symbols for the prompts may be different with other CAD software or other applications, the process stays constant.

ASSIGNMENT

1. Complete the practice exercise. Give it to your instructor.
2. Write your initials in block letters using the cartesian coordinates.
3. Draw a silhouette or shape of a rocket using cartesian coordinates.

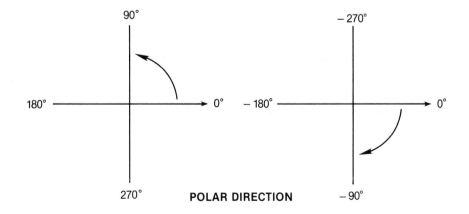

POLAR DIRECTION

COMPUTER-AIDED DRAFTING (CAD) AND THE PLOTTER

OVERVIEW

Many changes are taking place in the drafting field. Many of today's drafting firms use or have plans to use CAD as a tool for drafting. In this activity you will look at a basic introduction to CAD and the process of developing a hard copy of your drawing. You will review some of the fundamental functions used by various CAD systems and make a drawing using CAD.

As you proceed with this activity, think of places where CAD could be used to enhance the drawing capabilities. Think about some ways that you may be able to use CAD in your own experience. In which ways do you think CAD might be better than the traditional paper and pencil drafting techniques? Some things to consider might be speed, accuracy, neatness, line quality, lettering, and ease of drawing storage.

MATERIALS AND SUPPLIES

To complete this activity you will need the following:

- CAD station: Monitor
        Central processing unit
        Keyboard
        Input device (i.e., digitizer pad, mouse)
- Disk
- Output device (dot matrix printer or plotter)
- A copy of the software (may already be in your system)

PROCEDURE

Because computer systems and software programs vary, it may be necessary to perform these activities slightly differently, however, the basic functions should remain about the same.

1. Boot the system. This refers to turning the computer on and bringing up the CAD software program.
2. Access the main menu of the software. From this menu you will begin a new drawing and be asked to enter a new name for this drawing.

Note: At this point some systems require that the CAPS lock be on so that all letters are capitalized.

---

3. Once you have begun the new drawing, you should be able to add lines and other shapes to the screen. In some cases, it is necessary to set the parameters of the screen; i.e., screen size. This is done by following the prompts in the software program.
4. Begin by making an orthographic projection of a simple object. Remember this generally requires two or three views of the object. The views must line up and relate to each other.

5. To draw a line, you will select LINE from the menu. You will be asked to digitize a start and end point. To move the cursor (the cross on the screen) you may use the arrows on the keyboard or the other input device such as a graphics tablet or mouse.

6. Draw the items, the orthographic views, as directed by your instructor. Use the prompts on the software program.

7. When you have made the drawing, enter the text for the title of the drawing, your name, and title block, draw an appropriate border around the object.

8. Save your drawing on computer disk. Following the prompts on the screen you may save the drawing using your own disk by inserting the disk at the proper time in the SAVE sequence.

9. At this point you should be able to make a plot or print of your drawing following the steps.

   a. First, be sure the plotter or printer is turned on and paper is correctly installed. Check, in the case of the plotter, that pens are installed and have been tested to see if the ink in each is flowing correctly.

   b. Proceed with the plotting!

   c. After making the plot put the caps back on the pens, turn the plotter off, exit from the software program on the computer, and turn the computer off. Remove any disks as needed. Make sure the area is cleaned up and the cover is replaced on the computer if there is one.

►**CAUTION:** Do not put the cover on the computer if the computer is left running.

FINDINGS AND APPLICATIONS

In this activity you found that the computer does some things very well. The line quality and lettering are perfect. The arrows and arrowheads are perfect. This is all refreshing if you have tried it using a pencil and paper. You also found that you were able to make corrections and revisions without completely redrawing. Many of today's major industries use only CAD as the process for producing plans and preparing materials for manufacturing. Some of the benefits include speed, accuracy, quality, storage capabilities, the ability to change the drawing easily, and the ability to communicate drawings from one location to another using a modem.

ASSIGNMENT

1. Make an orthographic drawing using CAD. Then make a plotter print, or printer print of this drawing.
2. Using the computer and software available, make a second drawing. Look into the other capabilities of the software package. Such functions as GROUP, ARRAY, and COPY provide a great increase in speed and capabilities of the CAD software.
3. Look in newspapers, magazines, or books for other applications of CAD. Write a one-page paper explaining the information you were able to find.
4. Visit a drafting firm that uses CAD, or visit the drafting program in your own school where CAD is used. List some of the other advantages you were able to discover.

SECTION THREE
GRAPHIC REPRODUCTION

In 1450, Johannes Gutenberg invented movable type and the printing press, paving the way for mass production and the distribution of printed works. The printing industry now uses many different technologies, each selected to match the needs of the job to be done. In addition to the printing itself, there are many other activities, including creating the images to be printed, platemaking, and finishing processes, that must be completed to deliver a finished product.

While many different printing processes are described in this section, each can be thought of as part of a communication system. A printed paper or other object conveys the information to the reader.

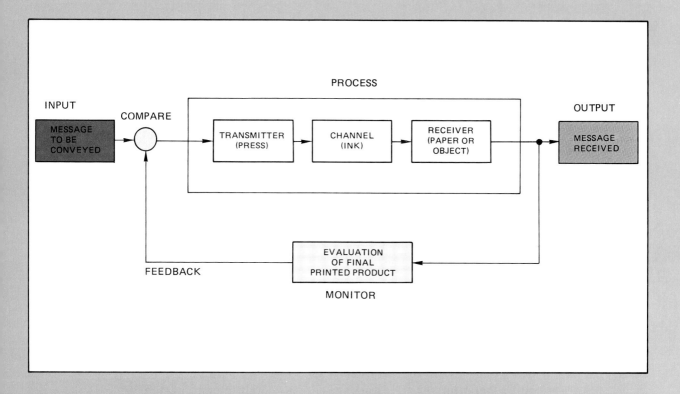

Graphic Design, Image Generation and Assembly

OBJECTIVES

After reading this chapter, you will know that:

- The design process presented in Chapter 7 is related to the design questions in Chapter 1.
- The three steps in the process of designing a printed job are making thumbnail sketches, creating a rough layout, and producing a comprehensive layout.
- The purpose of copy markup is to specify type size, style, line length and line spacing.
- Good typography requires that typeset pages be of the same size, that columns of type do not begin with a widow or end with an orphan, and that no fewer than two letters may be left on the preceding line.
- A paste-up, or mechanical, is a typeset page that has been combined with other job elements, like artwork, into a form suitable for exposure by a process camera.
- The rough layout is the plan which controls the appearance of the paste-up.

KEY TERMS

camera-ready copy	job	spot color
color process printing	justified	substrate
color separation	leading	thumbnail sketch
compositor	line art	type
comprehensive layout	mechanical	typesetting
copy	orphan	type size
copy markup	pagination	type style
galley	phototypesetting	typography
heads	pica	widow
image assembly	points	WYSIWYG
image generation	rough layout	

FIGURE 7–1 A variety of formats are available to the designer. *(Courtesy of International Thomson Organization, Ltd.)*

Client Needs Assessment

Gathering Design Information

Several questions were introduced in Chapter 1 that must be answered before the actual design of a printed job can begin. What does the customer want? What is the purpose of the printed piece? How will it be used? Who is the audience and what is their reading level? What are the limits of time and money? The designer must have an exact idea of the customer's goal before beginning to design the printed material that will accomplish that goal.

Once these questions have been answered, a method of printing will become apparent to the designer. Offset lithography, gravure, screen process, or relief—each method has certain advantages and limitations. An effectively designed printed piece will utilize the advantages of the method while working within its limitations.

The designer, in discussion with the customer, will determine the format of the printed piece (called a **job**). Format refers to the shape, size, and style of a printed job. There are a wide variety of formats available. One format is flat work, which is designed to be printed on one side only. Others are manuals, advertising material, brochures, books, newspapers, newsletters, forms, and folders. Figure 7–1 shows a variety of format samples. How the printed material is to be used will often suggest the best format. In turn, the format and end use of the printed job may determine the type of printing process to be used.

For instance, the book you are reading follows a book format that was designed to be 448 pages long. Each page was to have two columns of reading material called text or printed words. The design format called for each page to have one illustration, diagram, or photograph. Each chapter was to begin with chapter objectives. You may have noticed some similarities among the formats of the chapters. Key terms were to be identified in boldface type (that is the term for the extra heavy lettering you see on most pages of this text). These and other design considerations were decided upon before any of the pages were written.

The Use of Color

The audience, the customer's needs, and budget considerations influence the use of color. Generally, the more specialized and technical the subject and the more the reader already knows about the subject, the less color is used. Less color is needed to attract and hold the reader's attention. The authors are using colors in this introductory text to help make the information more interesting. Figure 7–2 shows several technical manuals.

Adding color affects the printing cost and therefore increases the budget. The customer and designer have two choices. They can use spot

FIGURE 7–2 Technical manuals need less color to hold a reader's attention than an introductory textbook. *(Courtesy of Nodaway Valley Co.)*

STAGES OF PRINTING PRODUCTION

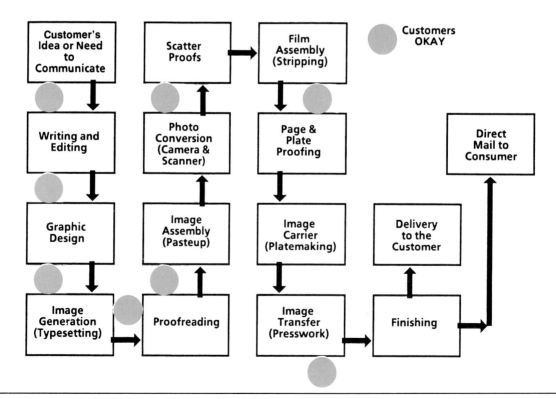

FIGURE 7–3 Spot color is used to show the flow of work from one production stage to another.

color or process color. **Spot color** is the use of one or two colors that highlight important parts of the text. For example, spot color is used to highlight the major ideas, called **heads**, which are printed in cyan (blue) ink for this book. Spot color can also be used to highlight important parts of the illustrations. Figure 7–3 shows the use of spot color to illustrate the work flow from one stage to the next.

Color process printing involves printing photographs and illustrations in full color. The method separates all the colors of the rainbow into three distinct colors of ink and black. Those four inks, when printed in various combinations, can reproduce all the colors of the rainbow, Figure 7–4. The three process ink colors are magenta (red), cyan (blue), and yellow. The process of separating the colors of the rainbow into the three process colors and black is called **color separation**. It is explained in more detail in Chapter 9. Extra production time and materials are required to print a photograph in color rather than in just black and white.

FIGURE 7–5 Paper is available in a variety of colors, thicknesses, and types. *(Paper sculpture by Leo Monahan, reprinted with permission of Hammermill Papers)*

Substrate Selection

Substrate is a fancy term for the material that will accept the ink as the job is printed. The most common substrate is paper. However, almost any type of material can be printed upon with the appropriate printing method. The end use of the printed material determines the substrate to be used. The final product may be a plastic wrapper to hold a loaf of bread, a ceramic coffee cup, a glass soft-drink bottle, a clock face, a T-shirt, or any number of other items.

If the substrate is to be paper, the designer has many decisions to make. Papers differ by type, color, thickness and surface. Figure 7–5 shows a variety of paper types and colors. The needs of the customer, the end use of the product, and the budget are all considerations in choosing the appropriate paper. An expensive paper would be inappropriate for a flyer used to advertise a social club's pancake breakfast and placed on the windshields of cars in a parking lot. Use of an expensive paper would be appropriate for the menu of a fancy restaurant. The printing method and use of process color must also be considered in the choice of paper.

FIGURE 7–4 The Kodak color spectrum *(Reprinted courtesy of Eastman Kodak Company)*

For the highest quality of four-color reproduction, a coated paper must be used to hold the printed ink. Coated paper is paper that has a special clay coating on its surface and a very smooth paper surface. This coating keeps the ink on the surface rather than allowing it to be absorbed into the fibers of the paper. Pictures printed on coated paper show more detail and the colors look richer and more vibrant. Coated papers are more expensive than uncoated papers due to the additional processing required. Compare the quality of the color pictures in this text with that of the color pictures in your Sunday newspaper or *USA Today*, and you can see the effect different paper has on color reproduction.

Written Text

The purpose of a printed job is to communicate a message. Part of this message is communicated through written text, sometimes referred to as **copy**. For example, the local bank may want its logo printed on coffee mugs that will be given to each customer opening a new account. In this instance, the copy (the logo) already exists and no additional copy must be written. A textbook, on the other hand, requires a great deal of written material. The writers produce a manuscript. The manuscript is typewritten and double spaced with uniform space at the top, bottom, and both sides. What is written will influence, and in many instances will dictate, the design of the document.

Early Forms of Writing

The earliest writing known to us is the cuneiform writing that was done by people who lived in the Middle East about 4000 B.C. This kind of writing was done by drawing wedge-shaped symbols on cave walls and trees. Later, the Egyptians developed their own method of writing called hieroglyphics. Hieroglyphics involved pressing a tool into a soft clay tablet leaving a series of pictorial impressions. Each symbol had a different meaning.

Both cuneiform and hieroglyphic writing required people to know hundreds of symbols. This limited people's ability to read and write. In about 1500 B.C., scribes in Syria developed an alphabet that used symbols to represent the sounds which make up words and syllables. The alphabet used only about 30 characters. Most people had no difficulty memorizing it. The alphabet was one of the most important early means through which knowledge spread.

The Rosetta stone is a tablet found in 1799 at Rosetta, a town in Egypt. It provided the key to understanding ancient Egyptian writing because it included a Greek translation of cuneiform and hieroglyphic writings. *(Reproduced by courtesy of the Trustees of the British Museum.)*

The Design Process

Once the manuscript is written, more specific design considerations can be resolved. Designs for the cover and the text layout must be developed. Artwork and photographs that will help to clarify the ideas presented in the manuscript need to be developed.

Before artwork can be drawn or photographs taken, a visual representation of the work must exist so the artist and photographer know what to produce. Print designers may design the same project several times before producing a design acceptable to the customer. There are three steps in the process of designing a printed job. These include making thumbnail sketches, a rough layout, and a comprehensive layout.

Thumbnail Sketches

Thumbnail sketches are small, very quickly drawn pencil sketches of various designs that may be appropriate for the printed job. Thumbnail sketches are similar in character to doodles. The artist tries out several ideas to see which will best fulfill the communication needs of the job and remain within the general format that was decided upon earlier. While smaller than the actual job, thumbnails are proportional to it in size. They show different arrangements of type and illustrations or photos. Only large-sized type is sketched in. Smaller type is represented by pencil lines. Figure 7–6 shows two thumbnail sketches developed for a textbook cover.

FIGURE 7–6 Two of the thumbnail sketches developed for a textbook cover.

Rough Layout

The next stage in the design process is the production of the **rough layout**. The rough layout begins as a thumbnail sketch and is developed to a full-size sketch of the work. Sketches representing artwork or photographs are rather detailed. They show more clearly what the artist wants to communicate. As in the thumbnail sketches, only the larger type, called display or headline type, is drawn. However, this time it is drawn in its correct size, style, and position. Smaller type is represented by lines drawn to represent lines of type and is drawn to show correct column width. Page margins follow the general format established earlier. Figure 7–7 shows a rough of a text cover.

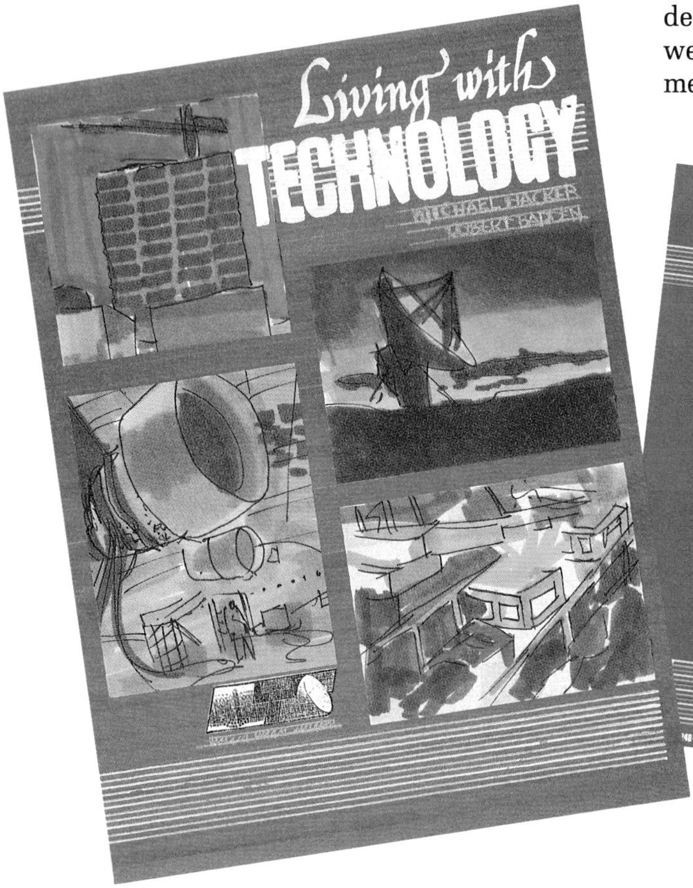

Comprehensive Layout

The **comprehensive layout** (sometimes called the comp) is an artistic representation of exactly what the final product will look like. Figure 7–8 shows a comprehensive of the same design shown as a rough in Figure 7–7. Colors are used that closely match the ink to be used.

Copies of the actual photos to be used may be in place. The comprehensive is produced using the same paper color, thickness, and style as the final job.

The designer must allow enough space in the layout to accommodate the type and artwork. If there is not enough room for the type and art, then one of three things must be done. (1) Copy can be cut out to fit existing space. (2) The design can be reworked to accommodate the extra copy. (3) The type can be set with less leading or in a smaller type size, thereby requiring less space. Because these steps are so closely related, the designing agency will often design the job as well as set the type and produce the paste-up or mechanicals.

FIGURE 7–7 The rough layout of the cover.

FIGURE 7–8 Note the differences between a rough layout and a comprehensive layout.

Pantone® Matching System (PMS)

The PANTONE® MATCHING SYSTEM is a color matching system that defines color pencils, color markers, color paper, and color inks. Each color is identified by a number. When a customer or designer specifies that number, each person involved knows exactly what color is being used. This is important when the customer, designer, and printer are in different buildings, in different cities, or even in different countries. By relying on the PANTONE® Matching System, it is not necessary to send ink samples to each person for every job. There are a number of ink suppliers that will provide Pantone inks. The figure shows the various media available to the artist.

THE PANTONE® * MATCHING SYSTEM is a comprehensive color communication system for the graphic arts. It consists of seven publications which are designed as tools for color communication between the specifier and the producer. The PANTONE MATCHING SYSTEM is based on a palette of 747 PANTONE Colors plus four process colors; each identified by a unique name or number to ensure accurate color communication. Graphic art materials such as markers, papers and acetate overlays in PANTONE Colors are also available in support of the system. In addition, approved inks in PANTONE Colors are available from over 1,000 licensed printing ink manufacturers worldwide.

*Pantone, Inc.'s check-standard trademark for color reproduction and color reproduction materials. Process color reproduction may not match PANTONE® -identified solid color standards. Refer to current PANTONE Color Publications for the accurate color.

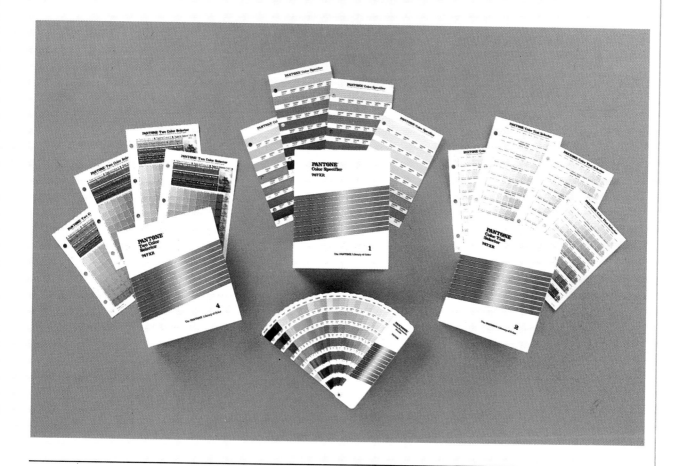

The PANTONE MATCHING SYSTEM allows for a uniform system of color matching. *(Courtesy of PANTONE® Inc.)*

The Design Model

Even though the comprehensive layout is produced, it is usually the rough layout that will be used by artists, illustrators, and photographers. The rough layout serves as the design model from which the final camera-ready art for the job will be produced.

The actual artwork for the job will be produced oversized originally and, in a later production step, reduced to final size. This will optimize the quality and clarity of the artwork. The comprehensive will not be used to generate final artwork for the camera. Its only purpose is to provide clients and the production staff with a full-size representation of the final job.

As the art and photographs are being produced, the manuscript, in its final edited form, will be set in type. **Type** is the term for letters and other symbols used to convey information. Type comes in a variety of classifications and sizes. The process of converting manuscript copy into typeset copy is called **typesetting**. Before the typesetting process can begin, specifications must be added to the manuscript copy and to the rough layout. These specifications describe, in detail, just how the manuscript or layout is to appear after typesetting.

Copy Markup

Before the rough layout and manuscript can be typeset, detailed information must be added by the editors to control the typesetting process. This process is called **copy markup** or just markup. Figure 7–9 shows a page of manuscript copy that has been marked up. Written instructions are added to the manuscript copy specifying the **type size, type style,** type line length, and line spacing to be used. The choice of type specifications contributes greatly to the overall impression the typeset copy will have on the reader, Figure 7–10.

FIGURE 7–10 The type specifications for this movie poster were chosen to appeal to a specific audience. *(Courtesy of Gore Graphics)*

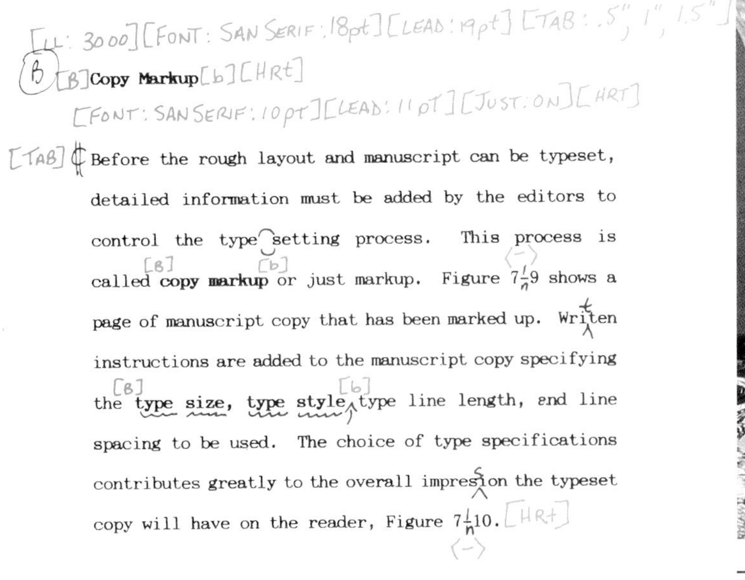

[LL: 3000][FONT: SAN SERIF: 18pt][LEAD: 19pt][TAB: .5", 1", 1.5"]
[B]Copy Markup[b][HRt]
[FONT: SAN SERIF: 10pt][LEAD: 11pt][JUST: ON][HRt]
[TAB] ¢ Before the rough layout and manuscript can be typeset,
detailed information must be added by the editors to
control the type setting process. This process is
called [B]copy markup[b] or just markup. Figure 7-9 shows a
page of manuscript copy that has been marked up. Written
instructions are added to the manuscript copy specifying
the [B]type size, type style,[b] type line length, and line
spacing to be used. The choice of type specifications
contributes greatly to the overall impression the typeset
copy will have on the reader, Figure 7-10. [HRt]

FIGURE 7–9 Manuscript copy that has been marked up for the typesetter.

C.F. GRAPHICS———*Typestyles*———

E6 pt. E8 pt. E10 pt. E12 pt. E18 pt. E24 pt. E 36 pt. E 48 pt. E 60 pt. E 72 pt.

The selection of the proper typeface for a particular piece of work can only enhance the message you are trying to convey. Be selective; don't hesitate to mix styles and families if that is appropriate. And, please, ask for assistance. We want your job to be right the first time.

Antique Olive
Antique Olive Medium
Antique Olive Bold
Antique Ol. Compact
Avant Garde Extra Light
Avant Garde Gothic Book
Avant Garde Gothic Med.
Avant Garde Gothic Bold
Century Schoolbook
Century Schoolbook Italics
Century Schoolbook Bold
CG Times
CG Times Italics
CG Times Bold
CG Times Bold Italics
Cooper Black
Garamond Book
Garamond Book Italics
Garamond Bold
Garamond Bold Italics
Goudy Sans Medium
Goudy Sans Medium Italics
Goudy Sans Bold
Korinna Regular
Korinna Kursiv
Korinna Bold
Korinna Bold Kursiv
Lisbon
Lisbon Italics
Lisbon Bold
Old English
Oracle II
Oracle II Italics

Oracle II Bold
Oracle II Bold Italics
Palacio
Palacio Italics
Palacio Bold
Palacio Bold Italics
Park Avenue
Romic Light
Romic Light Italics
Romic Bold
Serif Gothic Regular
Serif Gothic Bold
Serif Gothic Extra Bold
Signet Roundhand
Souvenir Light
Souvenir Light Italics
Souvenir Demi Bold
Souvenir Demi Italics
Stymie Medium
Stymie Medium Italics
Stymie Bold
Stymie Bold Italics
Triumvirate Light
Triumvirate Light Italics
Triumvirate
Triumvirate Italics
Triumvirate Bold
Triumvirate Bold Italics
Triumvirate Bold Outline
Triumvirate Condensed
Triumvirate Bold Condensed
Triumvirate Bold Cond. Italics
Triumvirate Black
Triumvirate Black Condensed

FIGURE 7–11 Several of the thousands of type styles available to the designer. *(Courtesy of C.F. Graphics)*

Type size is the size of the type to be used. It is determined by the needs of the reader. Small children just learning to read require larger type sizes, as do older adults who may have vision problems.

Type size is specified in **points**. A point is 1/72 of an inch. This system of type measurement is older than the metric system and was developed to conveniently express small sizes. Most textbooks are set in 10- or 12-point type. Ten-point (10 pt.) type is approximately 9/64 of an inch. Not only is this difficult to say, but it is much more cumbersome to specify 9/64 type than 10-pt. type.

Johannes Gutenberg, a German goldsmith credited with the development of the mechanical printing process, developed movable type in the fifteenth century. Since this time, type designers have developed thousands of different, unique styles. Each style has a certain character or "look" that provides an overall impression to the reader. Figure 7–11 shows several type styles.

Matching the appropriate type style to the overall style or "feel" of the manuscript is an important design consideration.

Type line length or measure is the maximum length that a typeset line is allowed by the design. The line length is determined by the overall page size of the printed job as well as the desired amount of white space (margins) around the typeset copy. Line length is specified in **picas**. A pica is equal to 12 points. There are 6 picas to the inch. Picas are used to specify line length, margins, and space between columns.

Line spacing or **leading** is the amount of blank space between typeset lines. Leading is specified in points. This text was set in 10-point type on a 12-point line. The usual amount of leading for textbooks is two points. This provides visual separation of the lines, and makes them easy to read. Notice how readability is reduced when lines are set close together without leading between them. Too much space between the lines may also make reading difficult.

Typography

Typography is the art of using type in such a way that effective communication takes place. A tradesperson skilled in the art of **typography** is a compositor. Typography is more than just the style and shape of individual symbols. It affects the look or impression that the page of type, taken as a whole, imparts to the subconscious mind of the reader.

There are numerous rules of typography, some of which are easily seen in this book. For

FIGURE 7–12 The Compugraphic Corporation 9600 Laser Imagesetter. *(Courtesy of Compugraphic Corporation)*

example, good typography requires that typeset pages be of the same size, that the columns of type do not begin with the end of a paragraph (called a **widow**), and that columns do not end with the first line of a paragraph (called an **orphan**). Typographical rules dictate where words are to be hyphenated. No fewer than two characters may be left on the preceding line and no fewer than three characters may be moved to the next line. There are other rules of typography concerning the type styles that can be used to-

gether on the same page. There are now typesetting systems that apply the rules of typography automatically. However, there will always be the need for human intervention to catch those exceptions to the rules that are found in the English language.

WYSIWYG

Modern typesetting systems include **WYSIWYG** previewing and pagination. WYSIWYG (pronounced wiz-e-wig) is an acronym for the

"Special codes direct the phototypesetting machine to set the manuscript in the specified type size and style, to the specified measure with the specified leading."

"Special codes direct the phototypesetting machine to set the manuscript in the specified type size and style, to the specified measure with the specified leading."

"Special codes direct the phototypesetting machine to set the manuscript in the specified type size and style, to the specified measure with the specified leading."

FIGURE 7–13 The same copy can look very different depending upon how it is set. *(Courtesy of C.F. Graphics)*

words "What You See Is What You Get." This means that the computer monitor shows the type as it will actually appear once it is typeset. This feature helps the compositor to determine how line art and type will fit on a page and if the page follows the rules of good typography.

Pagination

Pagination refers to the flow of copy from page to page in documents such as this text. Any change in type size, line length, or line spacing will affect how the copy flows from page to page. Additions or deletions made by the author will also affect this flow of copy. If the typesetting system has pagination capabilities, any changes made will be adjusted throughout the document.

Image Generation

Today's typesetting machines do more than set type; they can set images. They can produce lines of various thicknesses, and reproduce line art and some halftone art. **Line art** or line copy is artwork that consists of lines of various thicknesses. A pen and india ink drawing is an example of line art. Halftones are pictures made from dots of varying sizes. Look closely at a newspaper photo and the dots will be visible. An optical illusion fools the eye into seeing different shades of gray. A more detailed explanation of halftones will be found in Chapter 9, "Image Preparation."

Because typesetters no longer produce just type, they are now being referred to as "Imagesetters," Figure 7–12. **Image generation** is the term that describes the production of type and illustrations the author has decided to use in transmitting the message to the reader. Image generation consists of three parts: input, editing, and output. Traditionally, this process has been called **phototypesetting**. It involves a computer-controlled workstation similar to a word processor.

Inputting and Editing

The **compositor**, as the operator is called, sits at a computer keyboard and inputs the copy as he or she reads from the typewritten manuscript. Copy refers to any material that has been typeset. Copy also refers to artwork or photographs. As the manuscript copy is input, the compositor edits the manuscript by correcting any errors. The compositor also inserts special codes. These codes direct the phototypesetting machine to set the manuscript in the specified type size and style, to the specified measure, and with specified leading. Figure 7–13 shows the same copy set with 3 variations. Inserting these codes is part of the editing process. The compositor knows what codes to add to the copy by reading the instructions written on the manuscript in the markup operation.

One of the codes affects the justification of the right- and left-hand margins. **Justified** lines are lines of type having equal right-hand and left-hand page margins rather than the ragged right margins common to typewritten pages. Justified lines give a more formal look to the typeset pages.

Gutenberg and the Printing of Books

Until about the year 1450, books were transcribed by craftsmen called scribes. These scribes, often monks of various religious orders, would copy books by hand. If numerous copies were to be made, one scribe or monk would read aloud as others would write. This process made the reproduction of books a very expensive endeavor that resulted in limited distribution of these writings. As a result, only those who were very wealthy, royalty, or members of the Church had access to books. Johannes Gutenberg, a young student in Mainz, Germany, began to think about printing books in quantity by mechanical means. He knew that wooden blocks were used to print playing cards, and that metal stamps were used to imprint coins.

The solution Gutenberg devised was to use separate letters and make them movable. For each letter, Gutenberg made a mold into which he poured molten metal. The letters produced by this casting process were made of an alloy of lead, tin, antimony, and bismuth. He then developed a complete printing system composed of a press, paper, ink, and movable type.

Gutenberg had to solve other problems as well. It was the practice of the time for scribes to adjust the width of their characters and the space between words so that each line ended at the same position. Thus they created lines of the same length, or **justified** lines.

Justification was one of the problems that Johannes Gutenberg solved in his development of the printing process. Since the standard of the time was even right-hand and left-hand margins, Gutenberg had to develop a system of typesetting that would create this appearance. His success helped lead to the acceptance of the printing process and also set the standard for the ''look'' of the printed page that has lasted since the fifteenth century.

Johannes Gutenberg invented the mechanical printing process. *(Courtesy of Inter Nationes)*

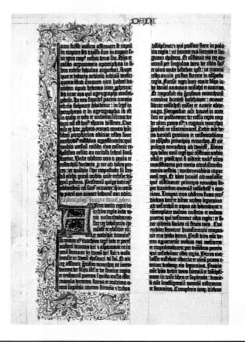

Gutenberg used his ingenuity to invent a system of typesetting that would result in justified lines, as shown by this page from the Gutenberg bible. *(Courtesy of The Pierpont Morgan Library, New York, PML 818 f.131v.)*

Outputting Copy

Output is the actual production of the typeset copy, often in complete pages, on a light-sensitive (photosensitive) paper. When photosensitive paper is exposed to light, it will produce dense black letters on a white background.

Newer phototypesetters use lasers as the exposing unit and can produce the images on film or paper. Since the document is in electronic form prior to output, it can easily be transmitted via modem over telephone lines to regional printers throughout the country.

The output of traditional phototypesetting machines is in galley form. **Galleys** are long strips of phototypesetting paper with the typeset images in proper size, style, and measure. The copy itself (dense black images on a white background) is of a quality suitable for the next stage of printing production. However, it is not yet in page form. The galleys must first be cut and pasted into pages.

Proofreading

Feedback about the quality of the image generation process is provided by a proofreader. The proofreader reads the typeset copy and checks for errors (typos) such as misspellings, transposed letters, and missing words. The proofreader also compares the typeset copy with the manuscript, checking for proper type size and style. Errors are indicated on the typeset copy by a kind of shorthand, called proofmarks. This system of symbols communicates, to the compositor, just what the error was and how to correct it. Figure 7–14 illustrates a few proofmarks that you may find useful for editing draft copies of term papers and reports.

There may be several cycles of proofreading, depending upon the complexity and importance of the job. The first cycle takes place when the copy comes off the typesetter. The copy is compared to the manuscript and checked for spelling.

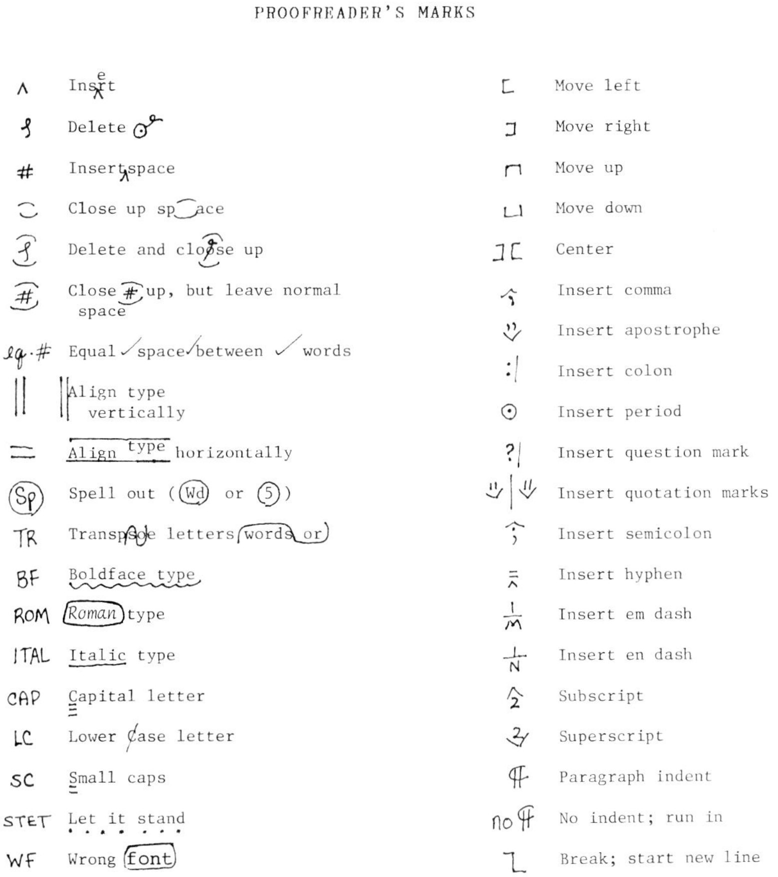
PROOFREADER'S MARKS
Insert
Delete
Insert space
Close up space
Delete and close up
Close up, but leave normal space
eq.# Equal space between words
Align type vertically
Align type horizontally
Sp Spell out (Wd or 5)
TR Transpose letters words or
BF Boldface type
ROM Roman type
ITAL Italic type
CAP Capital letter
LC Lower case letter
SC Small caps
STET Let it stand
WF Wrong font
Move left
Move right
Move up
Move down
Center
Insert comma
Insert apostrophe
Insert colon
Insert period
Insert question mark
Insert quotation marks
Insert semicolon
Insert hyphen
Insert em dash
Insert en dash
Subscript
Superscript
Paragraph indent
no No indent; run in
Break; start new line

FIGURE 7–14 Proofreader's proofmarks.

IMAGE GENERATION SYSTEM

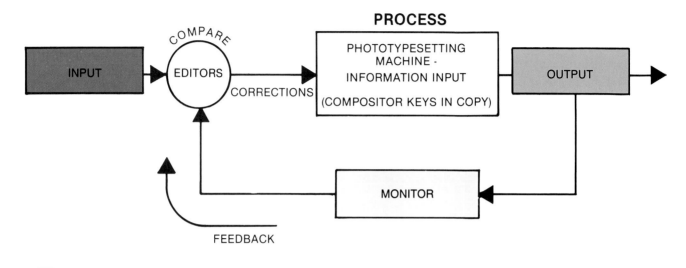

FIGURE 7–15 System diagram of the image generation process.

The second level involves the editors and authors. The spelling of technical words is checked, and the copy is read for accuracy. Any changes in the copy at this point can add to the expense of the job. Significant errors that show up at this stage may require portions of the copy to be reset.

Multiple levels of proofreading above the second level are necessary for printed products such as legal documents, medical formulas and dosages, and financial offerings like stocks and bonds. Printing of this nature requires absolute accuracy. To ensure this degree of accuracy, the job is proofread at each level, every time a correction or change is made. Once the copy has reached an acceptable level of accuracy, it is passed on to the next production stage, Figure 7–15.

Image Assembly

Image assembly, or paste-up as the process is often called, is the process of combining different elements of the job to be printed. The paste-up begins with the output of the image generation process. Galleys of typeset copy and other specified elements, such as line art, are combined to make up the pages to be printed. The output of the paste-up process is **camera-ready copy**, often called a **mechanical**. Camera-ready copy is copy that is ready for the next stage of printing production, photo conversion. A halftone blockout saves a space for the later addition of pictures or illustrations.

The paste-up process requires tools similar to those used by the person doing mechanical drawing or drafting; a T square, triangle, and ruler are standard tools of the process. These tools, along with a special pencil, are used to draw guidelines on a base of heavy paper (illustration board). The guidelines represent the page size and the image size and help the paste-up artist keep each page the same size.

Today's Typesetting Industry

In the past, most typesetting was sent to typesetting trade houses that specialized in phototypesetting. After the type was set in galley form, the agency would complete the mechanicals.

The lower cost of quality phototypesetting equipment and/or desktop publishing (see

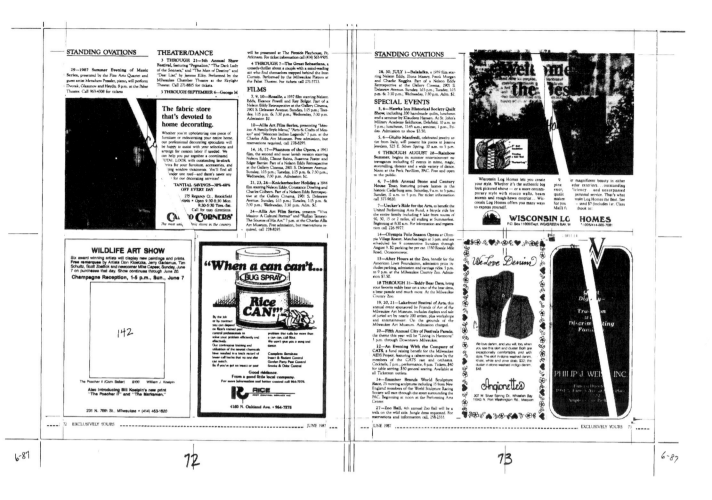

FIGURE 7–16 A mechanical created by a paste-up artist.

Chapter 8) has resulted in more agencies doing typesetting themselves. This results in closer control of the process, faster project completion, and the capacity to make last minute changes. Large phototypesetting jobs are still sent out. This is because trade houses have more equipment and can produce a greater volume, in a shorter time period, than can an agency with only one or two typesetters.

Whether type is set in-house or by a trade house, the most important aspect of producing a quality job is excellent planning. The designer must have fully assessed the needs of the client, matched those requirements to the proper printing processes, and done so in a way that utilizes the superior qualities of the selected process. If

this is not done, the job will not meet the client's expectations.

Adhesives

The elements (text, halftone blockouts, headlines) of the page are held in place by wax or rubber cement. The wax is applied to the back side of the images to be positioned on the paste-up. Wax is a good medium to use to fasten type elements because it holds the elements firmly, yet they can be removed and repositioned if necessary. Figure 7–16 shows a mechanical created by a paste-up artist.

As the paste-up artist positions the job elements, care is taken that the bottoms of all columns come to the same guideline. Also, if there is a two column photo or illustration, the tops of both columns of type should begin at the same level. Take a minute to thumb through several pages of this text to see how this typographic rule was applied.

Feedback to the paste-up process consists of copies of the paste-up pages being sent to the customer. While it is expensive to make changes at this point, the fact that the copy is waxed into position and can be moved if necessary makes those changes feasible. A change at this time is much easier and cheaper than a change later in the production cycle.

Summary

When designing a printed job, the needs of the customer must be carefully considered. The purpose of the printed piece, the intended audience, and the budget will dictate the formula and use of color in the final product. Spot color is used to highlight important parts of the book. It is less expensive to use than process color. Process color involves printing photographs and illustrations in full color. All the colors of the rainbow can be separated into four process colors by a process called color separation. The substrate chosen will also be determined by the end use of the product, and the limits of the budget.

The design process includes making thumbnail sketches, rough layouts, and a comprehensive layout. The actual artwork is produced oversized to optimize quality and clarity, and reduced to final size at a later production stage. The design of a printed job involves choosing and arranging artwork, color, amount of copy, type style, and type size.

Typography is the art of using type in such a way that effective communication takes place. A tradesperson skilled in the art of typography is a compositor. The compositor inputs manuscript copy into a phototypesetting machine. Instruction codes directing the phototypesetting machine are inserted as part of the editing process. The output of the phototypesetting machine is typeset copy on photosensitive paper. The typeset copy is read carefully by a proofreader who looks for errors and marks them with proofmarks. The compositor corrects these errors and produces a new version of the typeset copy.

Once correct copy has been obtained, the text and visuals are assembled into a mechanical (a paste-up) which is camera ready. The elements of the mechanical are positioned with adhesives such as rubber cement or wax, and can be repositioned if necessary.

REVIEW

1. Describe the relationship between the design questions presented in Chapter 1 and the design process presented in this chapter.
2. Explain the differences among thumbnail sketches, a rough layout, and a comprehensive layout.
3. Explain the purpose of copy markup.
4. List several rules of typography.
5. Explain what is meant by a WYSIWYG image.
6. Analyze the relationship between the rough layout and the paste-up.
7. Explain how proofreading provides a feedback mechanism for the process of image generation.

CHAPTER 8

Desktop and Electronic Publishing

OBJECTIVES

After completing this chapter you will know that:

- Electronic publishing links special software to personal computers or computer workstations. It combines input, editing, and output into a single or networked system and produces the finished product on laser printers or phototypesetters.
- Illustrations (graphics) may be included in a document produced on an electronic publishing system by being created by the operator, by inputting existing digitized graphics, or by digitizing original art or photographs on a digitizing scanner.
- The production of a page using electronic publishing methods is similar to using traditional paste-up methods. However, electronic publishing handles the placement of text and graphics electronically while paste-up requires manual placement of text and graphics.
- Desktop publishing is a low-cost form of electronic publishing that produces documents of near typographic quality.
- Corporate publishing is a very flexible and complicated form of electronic publishing that networks specialized workstations together.
- Electronic publishing shows, on the computer screen, how the final product will appear.

KEY TERMS

boldface	electronic publishing	networks
clip art service	electronic stylus	page oriented
corporate publishing	graphics frame	scanner
desktop publishing	italic	stats
document oriented		

Introduction

This chapter will build upon the concepts introduced in the discussion of image assembly in Chapter 7. It will also compare two types of electronic publishing systems that produce camera-ready copy electronically rather than manually: desktop publishing and corporate publishing.

123

FIGURE 8–1 Monitors show what the complete page will look like, including pictures and illustrations. *(Courtesy of Princeton Graphics Systems.)*

Electronic publishing is having a noticeable impact on corporate America. The average Fortune 1000 company spends from 2 to 10 percent of its annual sales dollars on publishing. Documents relating to the corporation's activities can be produced rapidly with the use of in-house, electronic publishing.

The Process of Electronic Publishing

Electronic publishing attempts to do everything that is and has traditionally been done in the image assembly operation up to paste-up. Electronic publishing combines typesetting with the electronic creation and manipulation of line art and halftones in one system. Electronic publishing outputs complete, camera-ready pages, Figure 8–1. This process has been available for the last twenty years. However, electronic publishing technology is presently being adapted to microcomputers such as the Apple MacIntosh® and IBM-PC. This adaptation is changing the way typesetting is being done.

In order to illustrate this change, two categories of electronic publishing systems, desktop publishing and corporate publishing, will be examined. Electronic publishing systems range from very simple to very complex. With each new revision of publishing software, the power and flexibility of desktop publishing increase.

Desktop Publishing

Desktop publishing (DTP) is adding near typographic quality to documents that used to be composed on a typewriter or dot matrix printer. It is providing acceptable quality at a low price for those users who do not need high-quality typography.

DTP links specially designed computer software that runs on a personal computer and combines the three stages of image generation (input, editing, and output) into one system, Figure 8–2. The system contains a monitor, keyboard, mouse, computer, and laser printer. The documents produced on this type of system are mostly text with some charts or line illustrations and are usually for smaller jobs. They are generally written by one person who enters the text into the system, makes format and design decisions, creates any needed graphics, and outputs the finished document.

Input

The desktop operator inputs the copy into the system via input devices such as a digitizer, scanner, mouse, or a computer keyboard. As the format decisions have already been made, the operator only needs to input the manuscript copy. Depending upon the system capabilities, the operator may select from several type style variations. **Boldface** (a heavy typeface) and *italic* (a slanted typeface) are two common variations.

Graphics

The operator can create graphics used to illustrate the contents of the document on the same system used to input copy. Graphics can be

FIGURE 8–2 Desktop publishing systems are used in many offices. They provide a typographic look to material that would otherwise be typewritten. *(Photo of Aldus Pagemaker Software courtesy of Aldus Corporation, Seattle, WA)*

Should the operator wish to do freehand drawing, the grid can be turned off. Freeform drawings can be created by use of the mouse or other digitizing method. The screen will trace the path of the mouse as it is moved. The line may be turned on or off by means of buttons on the mouse. Drawing with a mouse can be awkward, so some systems allow for the use of an **electronic stylus**, Figure 8–4. The stylus works like a mouse but is in the shape of a pen. The pen shape allows for a more familiar means of drawing or tracing existing art.

Some software will allow graphics to be imported from other graphic production software. This ability provides for the use of a digital **clip art service.** The clip art service contains a variety of illustrations that may be copied, from

MARKER	STROKE	LINE	SHAPE
Brush	Point Paint Draw		

SHADE	TEXTURE	RULING	EFFECT
			Opaque Clear Reverse Replace

EDIT		SUPPORT
Scale Stretch Shear Rotate Invert	Size Canvas Add Text Set Symmetry Set Grid Erase All Cursor Off	Erase Fill Zoom Mask
Copy Screen		Reset

FIGURE 8–3 A palette of useful shapes helps the designer produce graphics.

placed within the text of the document or can be created as separate pages. Because the computer treats graphics differently than text, a special area within the text must be created for the graphics. This area is called a **graphics frame**. It is this graphics frame that allows for the creation of graphic designs.

A palette of useful shapes is usually provided, Figure 8–3. These shapes include boxes, circles, ovals, and lines. The operator may select a shape, place it electronically at the desired location, and then modify the shape. The box may be stretched into a rectangle, and the circle or oval expanded to fill a given area. Shapes that are needed but not provided within the program can be created by the operator with the line option. The line option provides for the creation of graphics on a line by line basis.

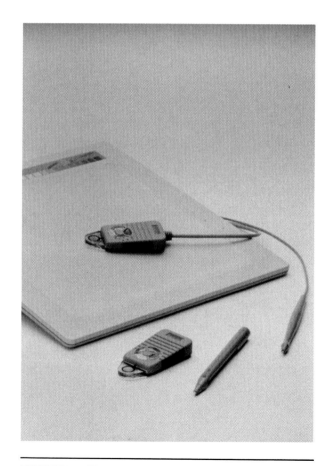

FIGURE 8–4 The graphic designer has many additional tools which help to produce a higher quality graphic. A mouse (puck) or stylus pen allows the operator to do freehand drawings. *(Courtesy of Kurta Corporation)*

the library of artwork stored on a disk, into the document, Figure 8–5. With this service, quality artwork is available to those who have little artistic talent.

Preexisting artwork may be entered into some systems through the use of a **scanner**. A scanner is a device that will read existing art and convert it into a graphic, Figure 8–6. This graphic may then be modified as needed.

Editing

Depending upon the software used, the system will do one of two things. It will either paginate the copy so that any changes will automatically be adjusted for in the rest of the document, or it will flow the text only from column to column on one page. **Page oriented**

FIGURE 8–6 Scanners allow preexisting graphics to be entered into the publishing system, modified, and added to the document being produced. *(Courtesy of Brodock Press Inc./James Scherzi Photography)*

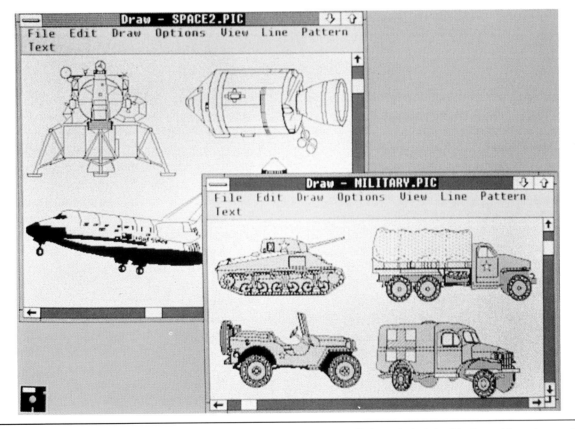

FIGURE 8–5 A disk-based art service is available for those who have little artistic talent but want to enhance their documents. *(Copyright Oct./Nov. 1987 Electronic Publishing and Printing. A MacClean Hunter Publication. All rights reserved.)*

FIGURE 8–7 This phototypesetter can take a disk and create high-quality phototypeset material. *(Courtesy of Compugraphic Corporation)*

software requires the operator to continue the copy onto the next page, as format decisions must be made on a page by page basis. Software that allows text to flow from page to page is referred to as being **document oriented**.

Desktop publishing allows the operator to view the copy in its final form or to preview the copy prior to output. Most DTP systems now provide WYSIWYG (What You See Is What You Get) viewing. WYSIWYG displays the page or portion of the page as it will appear upon output. Various type sizes, type variation, page formats, and graphics will be shown on the monitor just as they will appear on the finished page.

The operator can readily make any corrections required of mistakes found when proofreading. He or she can also make changes due to revisions or can use a different format.

Output

The output of a desktop publishing system utilizes a laser printer. The laser printer produces low cost, camera-ready originals on plain paper. The output cost of a plain paper laser printer is less than that of traditional phototypesetting. In addition, the cost of DTP systems can be 80 to 90 percent less than even inexpensive phototypesetting systems.

Laser output from a desktop publishing system is a big improvement over output from a dot matrix printer, but it does not compare favorably to phototypeset quality. Newer laser printers are increasing their resolution to 600 or 800 dots per inch. This is an improvement, but it does not reach the typographic quality of 2,000 to 5,000 dots per inch.

Desktop publishing systems provide much more flexibility in the "look" of a job that once was produced on a typewriter or word processor. To a certain extent, these systems are also taking work away from traditional typesetting systems. However, some typesetting machines are capable of accepting input from desktop publishing systems and outputting typeset quality onto photographic paper or film.

Some "quick print" operations are catering to the need for quality output from desktop publishing systems. They provide the photo-typeset output from a floppy disk brought in by the customer. The customer creates the document, proofs it on his or her laser printer, and then brings the disk to the local quick-print shop for quality, phototypeset output, Figure 8–7.

The cost factor has contributed to the growing popularity of this type of system.

A strength of desktop publishing is the menu-driven choices which make learning to use the system easy. Menu-driven choices can become a drawback once the system is learned, however, since they are not very efficient when fast production is required. An experienced operator can enter the codes faster than he or she can select a menu and then select an operation within that menu. Some desktop systems now allow the user the option of selecting functions via the mouse and menu, or through the use of special function keys. This flexibility makes the system easier to learn, and more productive once it is learned.

Another problem with DTP is that of limited choices. There may seem at first to be many format choices available. However, the requirements of printed jobs are so varied that the system's capabilities will soon limit what can be done.

Whenever software is written to do a specific task, it usually does not do associated tasks very well. Desktop publishing software combines text and graphics into finished pages rather well. DTP software can do word processing, but it often does not perform that function as well as many of the word processing programs. The same is true of creating graphics. DTP can create graphics, but not as well as many special purpose graphics programs.

Corporate Publishing

As the creation of the document moves from being the responsibility of a single person (in desktop publishing) to being the responsibility of groups of specialists, electronic publishing becomes known as **corporate publishing**.

Corporate publishing systems (CPS) answer the need for flexibility as well as the ability to impose a corporate identity on all documents produced within a corporation.

Document Design

The purpose of the document must be thoroughly evaluated before production can begin. Large corporations are likely to employ trained designers whose function is to design the documents published by the corporation. The designers will provide a "corporate look" for all the documents produced.

Several people are likely to be involved in the creation of a document within a CPS. It is important for the creation of a document to begin with a design concept of what the finished document will look like. A rough layout of the document will serve as the unifying link for the different people who work on the production of the document.

Regardless of the type of document, corporate publishing automates the assembly of text, line art, and photographs. High-quality, finished output is produced within a short time period. The needs of the corporate publisher dictate that electronic publishing systems provide more control over the production process and more versatility in document formats than can be provided by desktop publishing. Shorter production schedules, the ability to make last minute changes, and reduced costs are additional advantages of corporate publishing systems.

Input

Corporate publishing systems are networks of different types of workstations that provide for specialists contributing to the final document. Writers enter text at a keyboard. The writer's workstation is more of a word processing station than an illustration or editing station. The equipment the writers work on may be terminals connected to a mainframe or minicomputer, or microcomputers connected with the CPS.

Illustrators can generate original art at a workstation, call up illustrations and drawings created

FIGURE 8–8 Graphic work stations are designed to make the illustrator's job as easy as possible. The keyboard is mounted on an arm for easy movement out of the way of the digitizing tablet. *(Courtesy of Brodock Press Inc./James Scherzi Photography)*

in a CADD system, or scan drawings or photographs from a library of existing art, Figure 8–8. The illustrators have the ability to edit, crop, and size the drawings to suit the needs of the document. Their workstations are equipped with an electronic stylus, light pens, digitizing tablets, and mouse. These tools allow great flexibility in the work that can be produced on these stations.

Editing

The advantage of a corporate publishing system is the ability to connect all the separate creative workstations within the system. These connections are called **networks**, and allow editors and document specialists to call up the document as it is being created. Illustrators can call up the text and see what the writers need in the way of illustrations.

Summary

Electronic publishing links special computer software to personal computers or computer workstations. It combines input, editing, and output into a single system, and produces high-quality output using laser printers or phototypesetters. Electronic publishing systems combine text and graphics, and output complete camera-ready pages.

Desktop publishing provides acceptable quality at a low price. The system contains a monitor, keyboard, mouse, computer, and laser printer. Graphics can be created by the operator, or imported clip art from other graphic software can be used. Preexisting artwork may be entered into the system through the use of a scanner. The output of a desktop publishing system generally utilizes a laser printer. Although laser printers provide a higher-quality output than do dot matrix printers, they do not yet compare favorably to phototypesetters.

Corporate publishing offers greater flexibility and productivity, compared to DTP, but it is challenging to learn. Since these systems are used by specialists, this handicap is soon overcome. The system's ability to proof on plain paper or output on typographic paper provides the flexibility desired by corporate users.

Corporate publishing systems network a variety of workstations that permit a number of individuals to contribute to the final document. High-quality documents of many pages can be produced by these systems.

REVIEW

1. Define the term electronic publishing.
2. Compare the processes of desktop and corporate publishing.
3. List the general characteristics of desktop publishing.
4. Describe two methods of including (generating) illustrations (graphics) in a document produced by one of the electronic publishing processes.
5. Compare the production of a page using traditional paste-up methods and desktop publishing.
6. Compare the duties of a person responsible for document preparation on a desktop publishing system with the duties of the people responsible for document preparation on a corporate publishing system.
7. Desktop publishing systems are becoming more flexible and powerful and corporate publishing systems are becoming more user friendly. Do you foresee much distinction between the two types of electronic publishing systems in the future? Explain your answer.

CHAPTER 9

Image Preparation

OBJECTIVES

After completing this chapter you will know that:

- The four prepress processes are photo conversion, film assembly, proofing, and image carrier production.
- The three types of copy that can be reproduced on a process camera are line, continuous tone, and color.
- Continuous tone photographs must be converted into a pattern of dots called halftones in order to be printed by the printing process.
- The four process colors of magenta, yellow, cyan, and black can reproduce all of the visible colors.
- Page imposition is the process of determining which pages will be carried on a printing plate so that proper page order will be maintained when the printed pages are combined with other pages, folded, gathered, and bound together.
- Contacting is the process of placing unexposed film in direct contact with a flat or processed film, and exposing the combination to light.
- Electronic page makeup is an electronic method of performing those tasks done manually in the stripping operation.
- Electronic pagination electronically creates entire documents. Electronic page makeup electronically creates single pages consisting of color pictures originating from different sources.
- Four types of proofs are called scatter, page, plate, and press, and each has a different purpose.

KEY TERMS

color separation photography
composite film
contacting
continuous-tone photography
electronic color page makeup
electronic plate exposure
filter
fit
flat
halftone photography

image preparation
line copy
line negative
line photography
nonimage area
page imposition
plate proof
page proof
platemaking
preparation

press proof
process camera
process colors
process photography
proofing
reproduction camera
scatter proof
signature
spot color
stripping

Introduction to the Preparatory Processes

The pages of a document produced by an electronic publishing system or by traditional paste-up procedures are not in a form suitable for mass production. Paste-ups and the output from electronic publishing systems must be converted into a form suitable for use in the production of a printing plate. The plate can then be mounted on a printing press and the inked image carried to paper.

Each printing process—offset lithography, letterpress, gravure, flexography, and screen process—uses photography to produce the kind of printing plates required by that process.

If the document contains only one color, no photographs, and is to go to only a few people, multiple copies can be made by the electronic publishing system or a photocopier. Current laser printer output devices are neither fast enough nor economical enough to produce the multiple copies necessary for large-scale distribution. Also, if color and photographs are to be added and a large number of copies is to be produced, then reproducing the document one page at a time is not economical. The document must go through additional preparation in a process called **preparation** or prep.

The Preparation Processes

The preparation processes are those processes which convert typeset pages, line art, photographs, and color art into a form suitable for use on a printing press. They include the processes of photo conversion (camera and color scanner), film assembly (stripping), proofing, and producing the image carrier (platemaking).

Until the actual image carrier (printing plate) is made, all of the prep operations for each printing process are similar.

Photo Conversion (Process Photography)

The traditional method of preparing the document for printing in proper page order is first, photographing the text pages, artwork, and photographs on a **process camera**, Figure 9–1,

FIGURE 9–1 The purpose of a process camera is to reproduce exactly the image placed on the copy board. *(Courtesy of DS America)*

using a process called process photography. **Process photography**, also known as reproduction photography, differs from the type of photography explained in Section 4. Process photography is a very high-contrast form of photography which produces a negative without continuous tones of white, gray, and black. In process photography the negative areas are either completely black or completely clear. This negative is called a line or halftone negative. Process photography is the first of the four preparation processes.

Line Photography

All textual copy to be printed must first be converted to a **line negative**. A line negative is clear in areas that are to be printed (called **image areas**) and will be dense black in those areas that are not to be printed (called **nonimage areas**). A separate negative is used for each color that is printed. Printing processes can print only one shade of ink at a time. A characteristic of the high-contrast, photographic film used in this process is the film's inability to ''see'' a difference between the light blue guide lines used in the paste-up process and the white of the paste-up paper.

Producing and printing a high-contrast image of text (also known as **line copy**), which is a black image on white paper, presents no major problems. This is not true for black-and-white photographs where areas of gray are involved. When converting continuous-tone photographs to high-contrast negatives, a special process must be used.

Halftone Photography

Typical photographs include a range of tones from white to gray to black. This is called **continuous-tone photography**. Because a printing press can only transfer a solid color of ink to a sheet of paper, it cannot produce shades of gray or of colors. The press can produce a shade of gray only when that shade of gray ink is put in the printing press. For any other colors or shades of gray, the press has to be cleaned and a new color or shade of gray placed in the press.

The method for printing a continuous-tone photograph is to convert the shades of gray into areas made up of dots.

The process of converting a continuous-tone photograph into a pattern of dots is called **halftone photography**, Figure 9–2. The dots vary in size with the amount of gray in the photograph. The darker the shade of gray, the larger the printing dot and the more of them. The lighter the shade of gray, the smaller the printing dot and the fewer of them. This pattern of small dots, when printed and viewed, fools the eye into seeing the same shades of gray that were in the original, continuous-tone photograph.

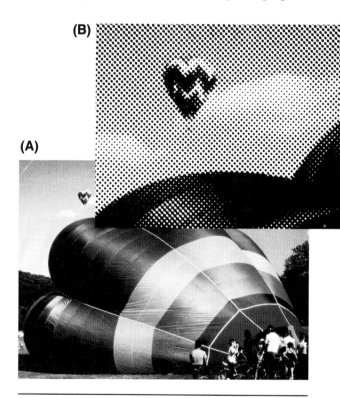

FIGURE 9–2 When a picture (A) is enlarged, the halftone dots are easily seen (B). *(Photo by Ruby Gold)*

Spot Color

Color for subheads and for illustrations is added to make the document more interesting and to help highlight important information. This use of color is called **spot color**. If spot color is to be added to the document, a separate line negative must be made of the work to be printed in each color. The paste-up artist will place the

FIGURE 9–3 This paste-up includes overlays for spot color. *(Reprinted from PRINTING TECHNOLOGY, 3rd Edition by Adams, Faux, and Rieber, © 1988 Delmar Publishers Inc.)*

images to be printed in color on a separate portion of the paste-up, called an overlay. An overlay is a clear plastic sheet, placed over the paste-up, which carries the images to be printed in color. There should be one overlay for each color contained on the page. The clear plastic allows the paste-up artist to view the color images in relation to the main black-and-white images, Figure 9–3. The camera operator will produce a line negative of each overlay.

Process Color Photography

To reproduce all of the colors that appear in a color photograph, a variation of the halftone process is used. The principle of reproducing the picture as a series of various-sized dots is the same; however, four negatives of the color picture must be made. Each one of the four halftone negatives will be exposed through a color filter. Each will be a record of specific colors found in the original photograph. A **filter** is a glass or plastic sheet that has been coated with a dye. The dye absorbs some colors of light and passes or transmits other colors of light. A filter is named by the color of light that it transmits. A blue filter appears blue because it absorbs all the colors of light and transmits only blue light, Figure 9–4.

All the colors of the rainbow can be recreated by printing various-sized dots of only three ink

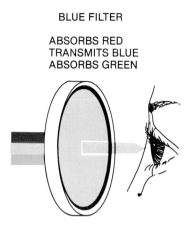

BLUE FILTER

ABSORBS RED
TRANSMITS BLUE
ABSORBS GREEN

FIGURE 9–4 White light contains each color of light. A blue filter absorbs other colors and allows only blue light to pass through. *(Reprinted from PRINTING TECHNOLOGY, 3rd Edition by Adams, Faux, and Rieber, © 1988 Delmar Publishers Inc.)*

colors. These colors are called **process colors** and are cyan, magenta, and yellow, Figure 9–5. To produce a cyan printing plate, the film must be exposed through a red filter. The magenta print-

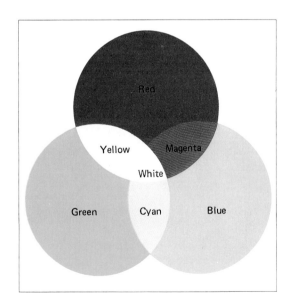

FIGURE 9–5 Cyan is a combination of green and blue light. Magenta is a combination of red and blue light. Yellow is a combination of red and green light. *(Reprinted from PRINTING TECHNOLOGY, 3rd Edition by Adams, Faux, and Rieber, © 1988 Delmar Publishers Inc.)*

ing plate is exposed through a green filter, and the yellow is exposed through a blue filter.

If the inks that printers used were perfect in the way they absorb and reflect the various colors of light, then printers could reproduce all the colors of the rainbow with the three process colors. Unfortunately, this is not the case. Ink manufacturers constantly try to improve printing inks, but they have not yet been able to produce process inks that react as theory predicts they should. In theory, printing a solid cyan over a solid magenta over a solid yellow should produce black. Because the inks are not pure color, a dark brown is produced instead of black. To compensate for the ink impurities, a fourth separation film is necessary. This is called a black printer (printing plate) and is made by separate

exposures through each of the color separation filters—red, green, and blue. The black printer compensates for the dark brown produced by the three process colors.

The process of creating four film negatives of the same color picture, each a record of different colors within the picture, is called **color separation photography**. While the process can be performed on a process camera, 95 percent of all color separations are produced on an electronic color scanner, Figure 9–6. Color separations produced on a process camera use the light that comes from the color copy to expose the separation film. Long exposures may result, requiring two hours of production time to produce a set of separations. The electronic color scanner converts the light from the color copy into an electrical current, Figure 9–7. The electrical current can be modified by a computer to correct for ink impurities. The corrected signal is used to control a laser beam which exposes the separation films. The electronic scanner is capable of producing a complete set of separation films in less than thirty minutes and is therefore 400 percent more productive than a process camera.

FIGURE 9–6 The color scanner can produce color separation films of color prints as well as of 35mm slides. *(Courtesy of Quad/Graphics, Inc.)*

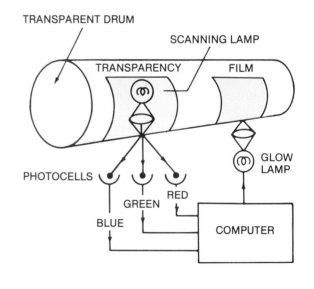

FIGURE 9–7 The scanner converts light into electrical energy which can be modified by the color computer. It can then control a laser as the laser exposes the separation film. *(Courtesy of Eastman Kodak Co.)*

Laser, Flatbed Scanners

High-quality, photographic images are consistently produced by color separation scanners which utilize electronics and a laser to produce film images.

A scanner is used because it converts light, reflected by the copy, into an electrical signal that can be manipulated by the scanner's computer.

In comparison, a camera is an optical device that uses light reflected from the copy (picture) to expose photographic film.

The flatbed scanner is a versatile electronic camera that can electronically manipulate picture information to produce the required results.

The flatbed scanner can stretch, squeeze, enlarge, or reduce both line and continuous-tone copy to fit into a required spot on the job layout. The flatbed scanner also has several special effect halftone screens as well as conventional halftone screens. These range from 65-line screens used for newspaper halftones to 175-line screens used for high-quality magazine halftones.

The usual output of the flatbed scanner is to photographic film or paper. However, it can also output directly to an electronic publishing system.

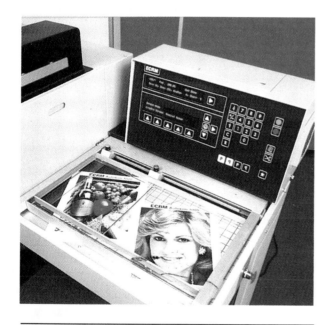

The Autokon 1000 Laser Black-and-White-Scanner uses a pressure-sensitive control panel for entering job specifications, such as size. *(Courtesy of ECRM, Inc.)*

There's a better way to reproduce

There's a better way to reproduce

G

There's a better way to reproduce

Get the Picture

Copy can be stretched, squeezed, enlarged, or reduced to fit layout specifications. *(Courtesy of ECRM, Inc.)*

Film Assembly

Stripping

After all the copy has been converted to film, these films must be combined into flats. A **flat** is a paper or vinyl carrier sheet which has been cut to frame the negatives and place them into the proper positions for printing, Figure 9–8. The process of assembling the films into proper page order on the carrier sheets is called **stripping**. Separate flats are usually required for the films that contain each process color, additional artwork, or halftones and line copy. Special care must be taken in the stripping process so that all the artwork, process color, spot color, halftones, and text will be printed in the correct places on the paper. A great deal of skill is involved in the proper placement of the four films of a set of color separations. The term for this placement is **fit**. A poor fit results in a blurry product, Figure 9–9.

Rather than printing a page at a time, it is much more economical to print pages in 8-, 16-, or 32-page forms. These forms are called signatures, and the pages are put together on the flat in a process called **page imposition**. Page imposition is important because pages cannot be printed in reader spread form. They must be printed as a signature. A **signature** is a set of pages which, after being printed on both sides and folded, will be in proper page order, Figure 9–10. That is, Page 1 is backed by Page 2 which is across from Page 3 and so on. Page imposition is a decision-making process concerning the physical positioning of pages on the flat.

FIGURE 9–8 A flat represents a printing plate and contains all films that will be used to expose that plate. The films must be in **exact position for proper printing.** *(Courtesy of Brodock Press Inc./James Scherzi Photography)*

A.

B.

FIGURE 9–9 Photo A was stripped so the magenta did not fit well with the cyan, yellow, and black. Notice how blurry this looks compared with the properly fitted **Photo B.** *(Photo by Ruby Gold)*

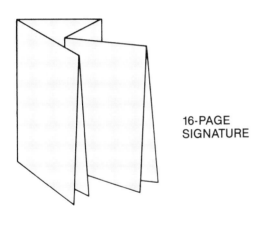

16-PAGE
SIGNATURE

FIGURE 9–10 This is called a **16-page signature. It is printed with 8 pages on one side of the sheet and 8 pages on the other side. When folded and bound together with other signatures, it will be part of a text.** *(Reprinted from PRINTING TECHNOLOGY, 3rd Edition by Adams, Faux, and Rieber,* © *1988 Delmar Publishers Inc.)*

Contacting

The production of a single page of advertising may require stripping a large number of flats. The advertisement shown in Figure 9–11 required ten flats. It is not practical to use that many flats to produce the four printing plates necessary to print the ad. A better method is to produce a

FIGURE 9–11 This advertisement is a combination of several different film separations, black-and-white text, and tints in different colors. *(Courtesy of The Pattern Company, Inc.)*

single piece of **composite film** for each color. Composite films are produced in an operation called **contacting**.

Contacting is the process of placing unexposed film in direct contact with a flat and exposing the raw film to light. For example, a complicated job can require ten flats, each containing information that is to be printed with magenta ink. A composite magenta film would be made by exposing each of the magenta flats, one after the other, onto the raw film. The developed film would then contain all the information contained separately on the ten magenta flats.

Electronic Page Makeup

The preparation processes of printing are very labor intensive and as such are prime targets for computer automation. These areas have been difficult to computerize because no two printing jobs are exactly alike. The job differences show up most dramatically in the prep stages of production.

While total computerization is available in some printing companies, computerization of parts of the prep area has been a reality for some years.

Electronic color separation scanners are the dominant tool for the production of color separation films. **Electronic color page makeup** was introduced to the printing industry in 1979. Electronic color page makeup systems have the ability to produce color pages entirely in electronic form. Type can be sent to the system electronically through a text processor or it can be scanned into the system. Color photographs are scanned into the system by an electronic color scanner. Line art is sent to the system by means of a line art scanner.

Once all elements of the page are digitized, the system operator begins to compose the page according to the designer's layout, Figure 9–12. Pictures may be rotated, flopped (changed the direction from left to right, or right to left), and retouched from the system keyboard. Elements of separate pictures can be combined into an entirely new picture without visible merge lines between the elements, Figure 9–13. Imperfections in the photographs can be retouched.

Colors can be changed; backgrounds can be replaced with other backgrounds. The system's capabilities seem to be limited only by the imaginations of the users.

Combinations that once took strippers hours to produce can now be done in minutes. The capabilities of these systems are great, however, they are not replacing the stripping function. Instead they are making possible work that was considered impossible only a few years ago. This results in more work being brought into the companies that use these systems.

For example, those companies that use electronic page makeup systems still use conventional stripping techniques for some work. It is currently more economical to first produce complex color pages on an electronic page makeup system, and then have the stripping department merge the color pages with pages containing type.

For work like national newsmagazines, such as *Time, Newsweek,* and *U.S. News and World Report*, it is possible to make up the entire magazine in digital form, Figure 9–14. The magazine is sent electronically, through satellite transmission, to regional printing companies throughout the United States. (See boxed article for more information on satellite transmission.) National news and advertisements are combined with regional advertising in regional editions of these magazines. This has kept these magazines viable in times that have seen the demise of general interest national magazines such as *Saturday Evening Post* and *Life*.

Proofing

Proofing is the feedback mechanism that controls the quality of the preparation processes. During the proofing process, a copy of the image in its current form is produced. There are several levels of proofing, each designed to provide feedback concerning the quality of the process just preceding the proof.

Scatter Proofs

The first level of proof is the **scatter proof**. Scatter proofs are four-color proofs of the color

FIGURE 9–12 The operator is assembling a full-color page on an electronic color page makeup system (*Photograph courtesy of Scitex America*)

separation process. As soon as the separation films are processed, a scatter proof is made to show the quality of the separation. The scatter proof will show if the scanner operator achieved the proper amount of color correction.

Page Proofs

Page proofs are used for those jobs that are stripped into a single page format. These color proofs show final color and final position of the elements of the page. They are a reliable representation of the way the job will look in its printed form.

FIGURE 9–13 This picture was made up of the three different photographs — A, B, and C. The line drawing D shows

A

C

B

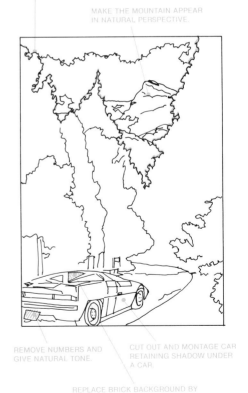

COMBINE TREES, WITH SKY FROM SECOND
PICTURE, USING DENSITY DIFFERENCE MASK

MAKE THE MOUNTAIN APPEAR
IN NATURAL PERSPECTIVE.

REMOVE NUMBERS AND
GIVE NATURAL TONE.

CUT OUT AND MONTAGE CAR
RETAINING SHADOW UNDER
A CAR.

REPLACE BRICK BACKGROUND BY
TREE ONE THROUGH CAR WINDOW.

D

how the final composite was made. *(Courtesy of DS America)*

Satellite Transmission

How can weekly news magazines such as *Newsweek* include up-to-the-minute news and yet print and distribute the magazine, from coast to coast, in a timely manner? The answer is satellite transmission.

Satellite transmission of current news stories from the east coast to five regional printers throughout the country is the way *Newsweek* can provide in-depth coverage of national and international events.

Because editorial offices are in New York City, getting late-breaking stories to press for national distribution on time was not easy prior to the development of satellite transmission. Editors, now using video monitors and keyboards, review and edit stories before they are released to press.

Once the story is ready for publication, it is sent electronically to *Newsweek's* image preparation department in Carlstadt, New Jersey. There the story text files are combined with electronic, four-color process pictures. The contents of each page are created electronically. When the page is ready, it is sent, via satellite transmission, to each of the five regional printing centers.

At the regional printing center, a satellite dish receives the electronic page and records the information on magnetic tape. The magnetic tape is placed in an output device which converts the electronic signal to light and exposes a sheet of film. After processing, the film is stripped, and the page is made ready for the press.

This technology allows *Newsweek* to send its late-breaking stories to the regional printers by 9:00 P.M. Saturday and have complete magazines being loaded onto delivery trucks by 5:00 A.M. Sunday morning.

The regional printers begin receiving advance pages by mail or delivery service on Tuesday. Advance pages are pages containing advertising and feature articles that do not contain last-minute information. These pages can be written in advance and are prepared by conventional means. The printed signatures are stored until they can be combined with the last signatures of the magazine early Sunday morning.

An interesting feature of this system is that the regional printers are the first to see the complete magazines. Prior to the printers outputting the information onto film, all the stories and pictures were in electronic form and were only viewed on the computer screen.

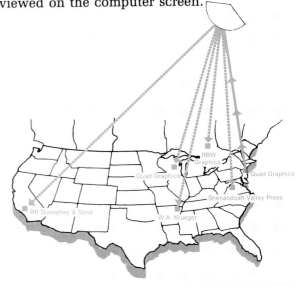

Newsweek's **electronic pages are sent from Carlstadt, New Jersey to each of five regional printing plants across the country.**

The bindery will begin delivering finished magazines shortly after the last signature arrives at the end of the press. (*Photograph courtesy of Scitex America*)

FIGURE 9–14 *Time, Newsweek, U.S. News & World Report*, **and other magazines are transmitted via satellite from New York to regional printing centers.** *(Courtesy of U.S. News & World Report)*

Plate Proofs

After the job is stripped into final form but prior to the production of the printing plates, a **plate proof** must be made. The purpose of this proof is to check for proper page imposition and the proper fit of all the elements of the job. If color is used, a color proof must be made as well. The proofs are checked by the strippers. They compare the proof with the instructions written on the paste-up.

When everything checks out, proofs are sent to the customer for approval. The customer may find errors or other technical problems in the proofs that the printer missed.

Scatter, page, and plate proofs are known as off-press proofing, and all use some photographic means of producing a representation of the final product. Some customers demand an additional proof known as a press proof.

FIGURE 9–15 Press proofs show final image positions. *(Courtesy of Heidelberg Eastern, Inc.)*

Press Proof

A **press proof** is a proof produced on a printing press, using the actual inks and paper that will be used in the final job, Figure 9–15. Using the actual materials produces a more realistic proof of the job.

Press proofs for process color were once the standard of the industry, but this has changed in recent years. Because press proofs require additional time and expense, only about 25 percent of proofing is currently being done on-press. The reduction in the use of on-press proofing is due to the high cost of these proofs. Off-press proofing methods have been developed to reduce the cost and time involved in proofing.

Platemaking

After the customer approves the proofs, the job is ready for **platemaking**. In the platemaking process, a photosensitive plate (image carrier) is exposed to high-intensity light through the film flats produced in the stripping process. Different plates react in different ways depending upon the printing process to be used.

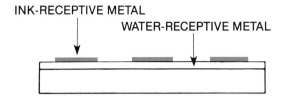

**FIGURE 9–16 The offset plate is ink receptive in im-
age areas and water receptive in nonimage areas.**
*(Reprinted from PRINTING TECHNOLOGY, 3rd Edition by
Adams, Faux, and Rieber,* © *1988 Delmar Publishers Inc.)*

Offset Lithographic Plates

The light exposing a plate being made for the
offset lithographic process causes the photosensitive coating to become water repelling. Those
areas of the plate which are not exposed to light
would, upon processing, be water loving. Offset lithography is based upon the principle that
an oil-based ink will not mix with water. Proper
plate exposure and development will produce a
printing plate that has oil-loving image areas that
will hold ink, and water-loving nonimage areas
that will not, Figure 9–16.

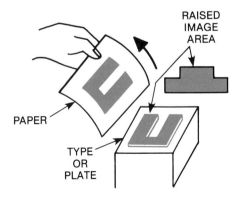

**FIGURE 9–17 When processed, relief plates have the
image areas raised above the nonimage areas.**

Relief Plates

The relief printing process depends upon a
mechanical separation of the image and
nonimage areas of the plate. Light exposing the
image area will harden that area. Subsequent
development will wash away all the nonimage
areas leaving only the raised image area adhered
to a metal base, Figure 9–17.

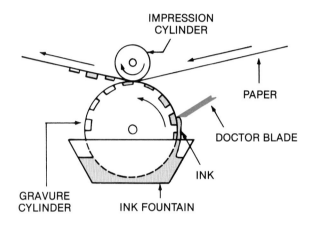

**FIGURE 9–18 The gravure cylinder has image areas
below the surface of the cylinder. A doctor blade wipes
ink from the surface, nonimage areas.** *(Reprinted from
PRINTING TECHNOLOGY, 3rd Edition by Adams, Faux,
and Rieber,* © *1988 Delmar Publishers Inc.)*

Gravure Plates

Gravure printing is the first printing process
to use electronic data to directly prepare the
image carrier. Digital electrical signals from an
electronic page makeup system are used to control a machine that engraves the image areas of
the gravure cylinder with a diamond point. The
surface area that remains is the nonimage area,
Figure 9–18.

Screen Process

The image carrier for screen process printing
also uses light to expose the nonimage areas. In
this process the exposed areas are hardened and
the unexposed, image areas are developed out.

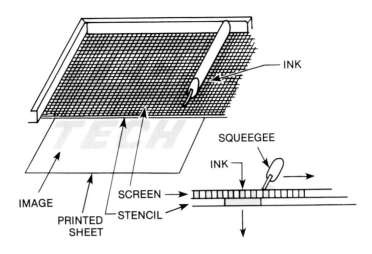

FIGURE 9–19 Open image areas allow ink to pass through the screen onto the paper below.

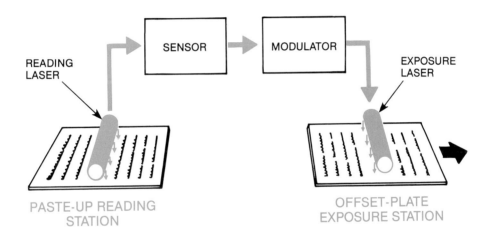

FIGURE 9-20 The computer controls a laser as it exposes the plate a line at a time.

This open image area allows ink to flow through the image carrier and onto the paper, Figure 9–19. Once the image carrier is produced, the job is ready for printing.

Electronic Plate Exposure

Electronic plate exposure is the use of digital data to directly expose a printing plate. Most systems under development use a computer-controlled laser to expose the printing plate. Some newspapers are experimenting with direct computer-to-plate technology.

Laser exposure units use laser output matched to the light sensitivity of the plate. Even though the exposure times are very fast, the laser must expose a plate a line at a time, much like a line printer prints out one line at a time, Figure 9–20. This results in longer total-plate exposure times when compared to conventional methods which expose the entire plate at the same time. This becomes a disadvantage when a plate goes bad on the press and a new plate must be made.

Summary

A paste-up, or copy from an electronic publishing system is a single original copy. If hundreds or thousands of printed copies are to be made, the original must be put into a form suitable for use on a printing press. The techniques used involve photoconversion processes which convert copy into photographic film. Line drawings are converted into line negatives. Continuous-tone photographs are converted into halftones.

Spot color is added to subheads and for highlighting. Process color is used when four-color photographs are used.

Once the copy has been converted to film, the individual pieces of film are stripped into flats with the proper page order.

The pages are assembled in such a way as to print in signatures. When the signatures are printed and folded, the pages will be in the proper sequence.

Electronic page make-up eliminates some of the drudgery associated with mechanical paste-up. Line art, photographs, and text can be combined and arranged electronically.

Once the image is produced, it must be proofed. Following proofing, a photographic plate is made by exposing it to a high-intensity light through the stripped film flats. The light hardens the image areas of the plate and makes them receptive to oil-based ink. The nonimage areas are receptive to water. The offset printing press is based upon the principle that oil and water do not mix.

In the near future, it is conceivable that documents will be prepared on an electronic publishing system, sent electronically to the customer for approval, and then transferred directly to printing plates or to a digitally imaged printing press.

Until that time arrives, it will be the function of the prepress areas to convert various types of copy into a form that can be used to produce a printing plate. The process will continue to utilize a camera to convert the copy into a high-contrast film negative or positive. The films will then be stripped into proper position so that when they are printed, folded, and bound together, the pages will be in the correct order.

REVIEW

1. Write a brief description of each of the four prepress processes known collectively as prep.
2. Describe the three types of copy that can be reproduced on a process camera. List one example of each type of copy.
3. Explain why a continuous-tone picture must be converted into a halftone.
4. Explain briefly how it is possible to print all of the colors of the rainbow with only the four colors of cyan, magenta, yellow and black.
5. Explain why the film stripper cannot just attach films to a flat sheet in any position (page imposition).
6. Discuss the purpose of film contacting.
7. Compare film stripping with electronic page makeup.
8. Describe the process of electronic page makeup.
9. What is the purpose of proofing a printed job?

CHAPTER 10

Image Transfer

OBJECTIVES

After completing this chapter you will know that:
- A press is a collection of subsystems, each with its own function.
- Offset lithography prints from a flat surface.
- The relief process prints from a raised surface.
- The gravure process prints from a sunken surface.
- The screen process uses a stencil and prints through a screen.
- The selection of a printing process should be matched to a given printed job.

KEY TERMS

dampening system	image transfer	registration
delivery system	imprinting	relief printing
dryer system	inking system	screen printing
electrostatic printing	ink jet printing	sheet-fed presses
embossing	letterpress	substrate
fleet graphics	nonimpact printing	toner
flexography	offset lithography	transfer system
gravure printing	press	waste sheets
hot foil stamping	register system	web-fed presses

Introduction

There are many different printing processes used today. The process used depends on the type of job. As printing technology changes, the advantages and disadvantages of these processes for certain types of printed work change. For example, prior to the 1960s, the relief printing process was the process of choice for publication printing. Changes in offset lithography and typesetting technology combined to make offset more economical, faster, and of a higher quality than the relief process. As a result it has been forecast that relief printing will account for about

151

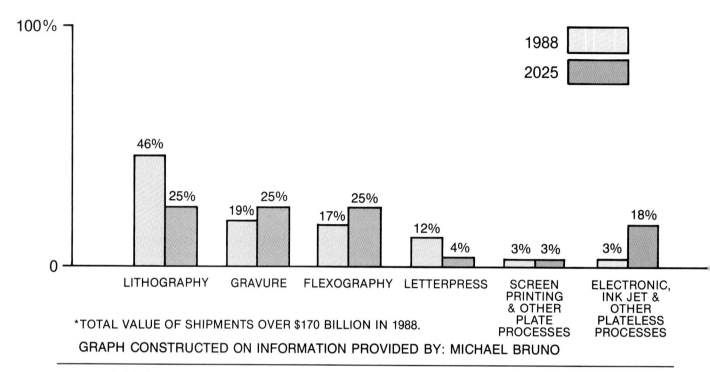

FIGURE 10–1 Shipments by printing processes. *(Reprinted from Status of Printing in the U.S.A., © 1988. Part of WHAT'S NEW(S) IN GRAPHIC COMMUNICATION newsletter.)*

7 percent of printing production by the year 1995, and 4 percent by the year 2025, Figure 10–1.

Image transfer refers to the process of transferring ink from one surface to another. This transfer has traditionally involved the use of pressure. Therefore, image transfer devices are known as **presses**. The surface or object that the image is being printed on is often called a **substrate**. Regardless of the printing process used to transfer the image, all image-transfer systems utilize subsystems in the production of the document, Figure 10–2.

The Press as a System

Before examining how printing processes differ, it is worth looking at their similarities.

Subsystem—Feeder

One method of classifying presses is the means by which paper is fed into the press. **Sheet-fed presses** feed single sheets of cut paper into the press, one at a time, Figure 10–3. These presses are named by the maximum length of paper that the press can run. The largest sheet-fed press will print paper up to 52 × 77 inches long and is called a 77-inch press.

Web-fed presses feed paper into the press from a roll of paper, Figure 10–4. A 38-inch press is a typical commercial web press size. The roll is 38 inches wide and the web (paper) will be cut, at the delivery end of the press, to 22.75 inches.

Feeding from a roll presents several economic advantages. First, paper is less expensive in roll form. Paper companies do not have to cut the

PRESS SUBSYSTEMS

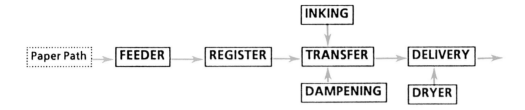

FIGURE 10–2 Press systems are composed of many different subsystems.

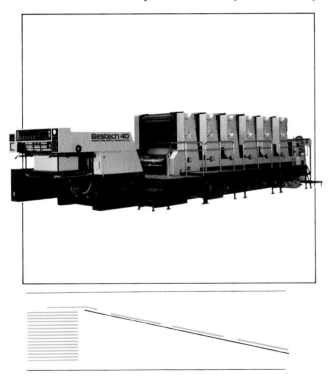

FIGURE 10–3 Sheet-fed presses feed paper, a single sheet at a time, into the printing press. *(Photo of the Akiyama Bestech 40 courtesy of MARUKA-Akiyama, Pine Brook, New Jersey. Line drawing reprinted from PRINTING TECHNOLOGY, 3rd Edition by Adams, Faux, and Rieber, © 1988 Delmar Publishers Inc.)*

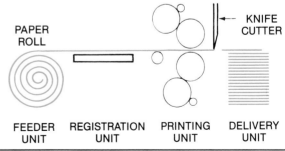

FIGURE 10–4 Web-fed presses feed paper from a roll. The press will cut the paper into sheets after it has been printed. *(Photo courtesy of Heidelberg Eastern, Inc. Line art reprinted from PRINTING TECHNOLOGY, 3rd Edition by Adams, Faux, and Rieber, © 1988 Delmar Publishers Inc.)*

paper, which comes from the paper-making machine in large rolls, into sheets that must then be wrapped and boxed. Another advantage of web feed is higher press speeds. Sheet-fed presses are limited by the speed at which the feeder unit of the press can consistently separate the top sheet of paper from the pile and feed it into the press for printing. Web-fed presses do not have this limitation and can therefore run at faster speeds. Web speeds are instead limited by the capability of the folder at the delivery end of the press to fold and deliver printed signatures.

One aspect of web offset printing is both an advantage and a disadvantage. This is the fact that all the printing that is going to be transferred to the paper must be done in one pass through the printing press. A press large enough to print four process colors, on both sides, plus a special color or two can cost 2.5 million dollars. This becomes a disadvantage when trying to raise enough money to supply a printing plant with this type of equipment. The advantage is that the paper is handled only once. There is no need for intermediate storage as press sheets dry before they are printed with other colors or on the reverse side. In one pass through the press, web presses can print all four process colors and sometimes a spot color, on both sides of the sheet, Figure 10–5.

Whether web or sheet feeding is used, the purpose is the same: to consistently feed the substrate into the printing system.

Subsystem—Register

The purpose of the **register system** is to consistently position the substrate so that each piece of paper is in an exact position, time after time, Figure 10–6. Without this consistency, it would not be possible to print multicolored images so that they fit together. **Registration** is the process of obtaining this exact positioning between the paper and the image. It is controlled manually in smaller presses, and automatically via electric eyes and computers on high-speed publication presses.

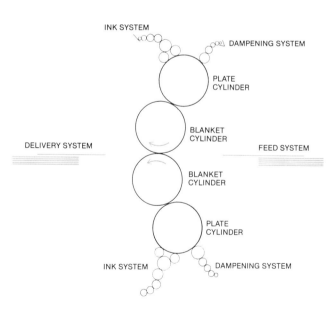

FIGURE 10–5 This printing unit, called a blanket-to-blanket perfector, prints both sides of the web in one pass through an offset press. *(Reprinted from GRAPHIC ARTS TECHNOLOGY by John R. Karsnitz, © 1984 Delmar Publishers Inc.)*

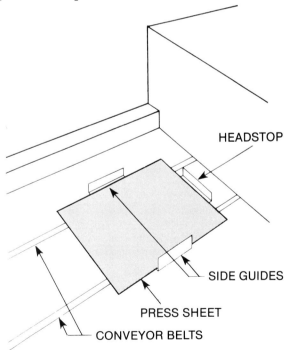

FIGURE 10–6 The paper is guided into exact position, time after time, by the register system. *(Reprinted from PRINTING TECHNOLOGY, 3rd Edition by Adams, Faux, and Rieber, © 1988 Delmar Publishers Inc.)*

Computer Control of Publication Presses

Technology has increased the speed of printing presses used to print magazines, books, and other high-quality material. Modern presses can print at 2,000 feet per minute (22.73 miles per hour). Controlling the exact positioning of each process color image (cyan, magenta, yellow, and black) at this speed presents a problem. If one or more of the colors begins to print out of register, a great deal of paper can be printed incorrectly before the problem can be detected and corrected by the press operators.

Microprocessor-based register guidance systems automatically control left and right, and up and down register to within one thousandth of an inch. They do this by using sophisticated computer software that is capable of recognizing a repeated pattern of register marks in relation to the rest of the images being printed on the paper web.

The system uses an optical scanner for each web surface to be controlled. At the beginning of the press run, the scanner moves across the paper web to locate the register marks. It then locks onto the marks for the remainder of the run.

Data collected by the scanners is analyzed by the main computer and register error is computed. Corrections are sent to the computers controlling the position of the web or plate cylinders and register is restored automatically.

The press operators, using a keyboard, can override the computer to adjust for any errors introduced when the printing plate was made. The correction values will be used to monitor and maintain color register throughout the press run.

Today's technology requires that press operators not only be expert in four-color process printing, but also that they be experienced in the use of computers. In addition to press register control, press technology uses computers to control paper web positioning and ink and water levels.

Once locked onto the register marks, the optical scanner collects register data which is transmitted to the master computer for analysis. *(Courtesy of Quad/Tech, Inc., Division of Quad/Graphics, Inc.)*

DIAMOND SIZE

0.041" 1.054mm   0.041" 1.054mm

Small (less than 0.06") diamond-shaped register marks can fit just about anywhere on a printed job. *(Courtesy of Quad/Tech, Inc., Division of Quad/Graphics Inc.)*

The operator can correct errors produced in platemaking or stripping by entering correction values on a computer keyboard. *(Courtesy of Quad/Tech, Inc., Division of Quad/Graphics, Inc.)*

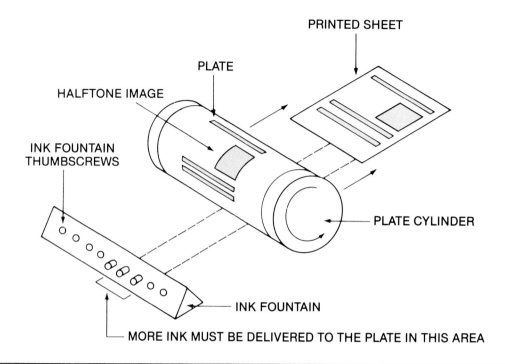

FIGURE 10-7 Ink fountain keys provide ink thickness control across the printed sheet. *(Reprinted from PRINTING TECHNOLOGY, 3rd Edition by Adams, Faux, and Rieber, © 1988 Delmar Publishers Inc.)*

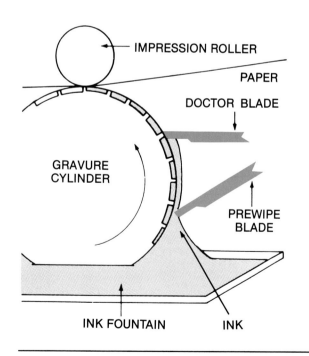

FIGURE 10-8 Gravure and flexography have no ink fountain keys to control ink across the printed sheet. *(Reprinted from PRINTING TECHNOLOGY, 3rd Edition by Adams, Faux, and Rieber, © 1988 Delmar Publishers Inc.)*

Subsystem—Inking

The **inking system** places a uniform layer of ink on the image carrier. This is a continuous process as ink is transferred from the image carrier to the substrate. The inking system of offset and letterpress presses is able to increase or decrease overall ink thickness to the substrate as well as adding ink to one side while decreasing it to the other side of the press sheet, Figure 10–7. Flexography, gravure, and screen process presses only have the ability to control the ink thickness overall, Figure 10–8.

Subsystem—Transfer

The **transfer system** provides the means of transferring the image from the image carrier to the substrate. On printing presses, the transfer system provides a uniform amount of pressure to transfer the image. Nonimpact printing systems use some form of electrostatic device to cause the image to move from an imaging surface to the paper.

Subsystem—Delivery

The **delivery system** receives the printed substrate and stacks it in a uniform pile. Seldom does the printed job go directly from the printing press to the customer. At the very least the job is boxed or wrapped. Other jobs may consist of a combination of many signatures that must be folded, gathered, and bound together in some manner. Any postpress operation is much easier when working from a uniform pile of printed paper.

Additional Subsystems

The offset lithographic press uses water to discriminate between image and nonimage areas. Therefore, this type of press requires an additional subsystem that places a uniform layer of water on the printing plate. This subsystem is called the **dampening system**, Figure 10–9.

High-speed presses, running at 1500–1700 feet per minute, require the addition of a **dryer system** to dry the printed stock. It is the function of the dryer system to provide enough heat to set the ink.

Offset Lithography

Printing on an offset lithographic press is the major method of printing today. **Offset lithography** is a very flexible process. It works well even on inexpensive equipment operated by workers with minimal training.

The Principle of Offset

Prior to 1907, lithography was a direct printing process. That is, the image was transferred directly from the printing plate to the paper. Because paper then was not as smooth as today's paper, the process produced a lower quality product than the other printing methods. In 1907, Ira Rubel discovered that transferring the ink image first to a rubber blanket and then to the paper would result in a superior image. The rubber blanket could conform to the slight irregularities of the paper. The hard zinc printing plate used before Rubel's process could not conform to the paper surface.

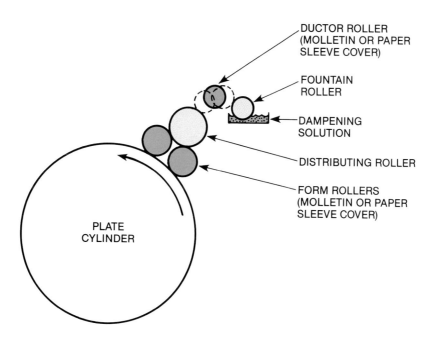

FIGURE 10–9 Offset lithography is the only printing method that requires water to separate image from nonimage areas. *(Reprinted from GRAPHIC ARTS TECHNOLOGY by John R. Karsnitz, © 1984 Delmar Publishers Inc.)*

The Principle of Lithography

Offset lithography involves printing from a flat surface. There is no mechanical (physical) separation of the image and nonimage areas of the plate.

The plate's ability to discriminate between image and nonimage areas is due to a chemical separation of the two areas. This chemical separation is based upon the principle that ink and water do not mix. A properly exposed and processed offset plate can print tiny dots of ink surrounded by paper, and leave small open areas surrounded by solid areas of ink.

In order for the separation of image and nonimage areas to work, the offset press must first coat the plate with a thin layer of water. The nonimage areas will retain this film of water, but the water will be repelled by image areas. The plate is then coated with ink, Figure 10–10. The nonimage areas of the plate that have a coating of water will repel the oil-based printing ink. The image areas, having repelled the water, will accept ink. It is the job of the person operating the press to maintain a delicate balance of ink and water throughout the press run. Too much water will make the ink look washed out. Too little water will make the nonimage areas of the plate begin to accept and transfer ink.

Material Requirements for Offset Lithography

Using the principle that ink and water do not mix presents several advantages and disadvantages. The offset process lays down the thinnest film of ink of all the processes, Figure 10–11. In the early days of the process, this was a disadvantage; black inks appeared gray. The newer ink formulas result in very acceptable ink colors.

Using water to separate image and nonimage areas means that the paper used must have a high degree of water resistance. This minimizes the amount of moisture the paper absorbs, which in turn reduces the change in paper size.

FIGURE 10–11 The relative ink film thicknesses of the four printing processes.

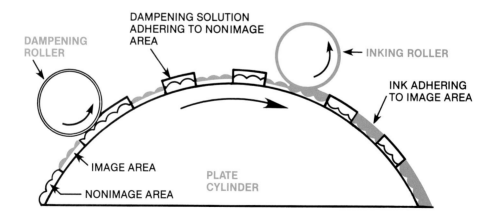

FIGURE 10–10 The plate is first coated with a thin layer of water. Image areas repel the water; nonimage areas attract the water. As the plate passes into the inking system, the nonimage areas with water repel the ink and the image areas attract the ink.

FIGURE 10–12 The customer and press team are evaluating a job as it is being made ready to run on an offset press. *(Courtesy of Quad/Graphics, Inc.)*

Advantages and Disadvantages of Offset Lithography

A disadvantage associated with the use of offset printing is waste. This unavoidable waste of paper is due to the sheets run as the press crew brings the ink and water into balance at the beginning of the printing process. Some of the waste can be avoided with the use of waste sheets in sheet-fed offset. **Waste sheets** are sheets of unacceptable quality left over from a previous job. These sheets are used to set up the next job and are run as the operator brings the ink and water into balance. This practice is not possible on web fed presses.

Large modern presses are equipped with computer aids that help the press crew to preset the press for new jobs. Such controls are about 90 percent accurate, leaving the final setup to the judgment of the press operators, Figure 10–12.

Advantages of the offset lithography process result from the relative ease of creating and handling the image carrier. Both the relief and gravure printing processes require the removal of image (gravure) or nonimage (relief) areas of the plate. This results in more expensive plate-processing equipment as well as more expensive plate material. The offset plate is usually made of relatively thin aluminum. A plate made for a 77-inch offset press can be mounted on the press by two people. An equivalent size gravure cylinder requires overhead hoists or wheeled carts and other support equipment to mount the cylinder on the press.

Applications of Offset Lithography

Offset lithography is the process of choice for printing on paper with runs up to about one million copies. It is used, for example, to print newspapers, magazines, books, greeting cards, catalogs, and labels. Nonpaper substrates include metal for printing on cans and metal tags.

Relief Printing

Relief printing (letterpress) is the printing process that prints from a raised surface. In 1987, 31 percent of American newspapers were printed by the letterpress process. Twenty years ago, 100 percent used that process. While the percentage of jobs printed by letterpress is declining, it is the only one of the printing processes suitable for some types of specialty printing. Wedding invitations, jobs requiring numbering and perforating (such as raffle tickets), hot foil stamping, embossing, and some die cutting are examples of the uses of the letterpress process.

Imprinting

Letterpress is often used to imprint preprinted material. **Imprinting** is the process of printing variable information onto material that has already been printed. For example, a large insurance company may have 100,000 calendars printed which show the company name. The company sends 1,000 of these to a local agent who contacts a printer to imprint the agent's name and address on each one. This usually involves a limited number of copies and the printing involves straight text matter. The relief printing process is the most cost-effective method for producing the job. The relief printing process is cost-effective for this job because a text-only printing plate can be quickly and inexpensively produced. Additionally, a run of a thousand calendars, bound in their final form, can be accommodated in a manually fed, relief printing press.

Hot Foil stamping

Other operations, such as hot foil stamping, can only be done by the relief printing process.

Hot foil stamping is the process that imprints the decorations and text on books such as the *Encyclopedia Britannica*, business cards, greeting cards, and other products. In this process, a metal relief plate is heated and pressed against a plastic foil carrier ribbon. This ribbon, which comes in a variety of gold, silver, brass and other colors, is in turn pressed against the material that is to receive the image. The proper combination of heat, pressure, and contact time will transfer the image from the plastic carrier ribbon onto the receiver material. The letterpress process is ideal for this type of printing, because the relief printing plate can easily be heated and the press speed can be controlled, Figure 10–13. There must be sufficient time, when the relief plate, the foil, and the substrate are in contact, for the foil to be transferred to the substrate.

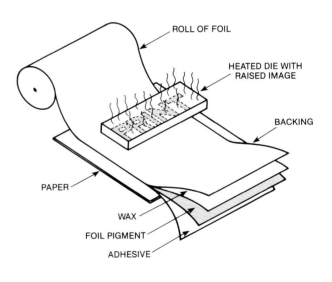

FIGURE 10–13 The hot foil stamping process involves the use of a press, relief image carrier or die, heat, and metallic or colored foil.

Embossing

Embossing is the process of compressing the fibers of a sheet of paper between a negative and positive die. The compression causes the paper to conform to the die, resulting in a relief image in the paper. If there is no other printing or foil stamping in this area, the process is known as dry embossing.

Raw Paper to Finished Book in Continuous Operation

The Cameron Belt Press is more than a printing press. It is a complete book manufacturing system. The system begins with rolls of paper and ends with completely bound and packaged books.

The Cameron Belt Press prints by web-fed, rotary letterpress, using flexible plates mounted on two continuous printing belts. Each plate is the size of a book page. Plates on one belt print the pages of half of the book. The other belt contains the plates for printing the backside of the page. All pages of the book are printed, both sides, in one pass through the press.

After printing, the paper is folded and cut into two- or four-page signatures that are continuously sorted into books. When a book has all its signatures, it is delivered to an adhesive binder. The signatures are then bound with a soft cover, trimmed, delivered to a counter-stacker, and packed in shipping cartons. A complete book can be produced in less than two minutes from the time the paper enters the press.

One of the unique features of the press is its ability to print books of different page lengths and page sizes. If a book requires more pages, the extra pages are placed on longer printing belts. By using different belt lengths, page counts can vary from 32 to 1152.

The plate belt can vary in size to accommodate books of different page lengths. *(Courtesy of Somerset Technologies, Inc.)*

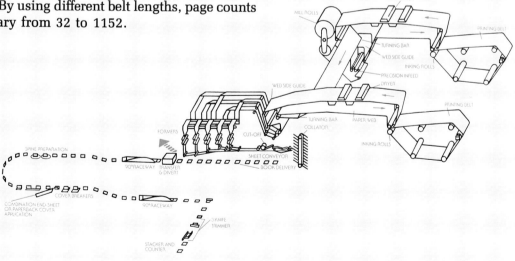

A web of paper enters the Cameron Belt Press and leaves as complete books. The press is so efficient that it can produce 2,000,000 160-page paperback books per week. *(Courtesy of Somerset Technologies, Inc.)*

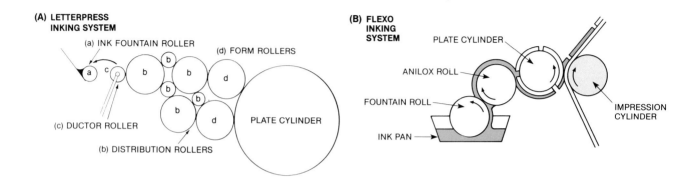

FIGURE 10–14 Though letterpress and flexography are classified as relief printing process, a look at their inking systems shows a major difference in the two forms of the relief process. *(Courtesy of Graphics Arts Technical Foundation)*

Flexography

A form of relief printing that is considered by some to be a separate form of printing is called flexography. **Flexography** is best known in its simplest form: a rubber stamp. The process got its start in the early 1900s as a means of printing on paper bags. As various plastics were developed as packaging material, the advantages of flexography became obvious. Today, flexography accounts for more than 85 percent of all packaging printing done in the United States, and by the year 2025 will account for about 25 percent of total printing volume.

In flexography, molded rubber has been replaced with light-sensitive plastic materials called photopolymers. The photopolymer plates, when exposed to light, harden in image areas. When developed, the unexposed nonimage areas wash away, leaving a raised surface ready to accept ink.

A normal letterpress cannot be used to print on plastic films because the pressure that is necessary to transfer ink from one surface to another must be absorbed by the substrate. When printing on paper, this is no problem. However, when printing on a hard surface such as plastic, the pressure distorts the material or punctures it. The flexographic image carrier is a rubber duplicate made from a mold of a letterpress image. When a plastic material is printed from this rubber plate, the plate itself absorbs the printing pressure.

An additional advantage of the flexographic process is the type of ink used by the process. It requires only a very simple inking system that has few parts, Figure 10–14. Also, the ink is very fast drying. This promotes fast setting on materials that will not permit absorption of the ink as a drying method. Ink cannot dry by absorption into the substrate when printing on a plastic film. An ink that dries by evaporation must be used. The flexographic inking system is capable of applying this very fluid type of ink.

In addition to plastic packaging materials and corrugated cardboard boxes, flexography is also used to print labels for consumer goods. The ability of the process to print a heavier film of ink than offset results in brighter product labels. These labels seem to draw more consumer attention than those printed by the offset method. Since more and more products are competing from market shelves for purchase by consumers, this is an important feature.

Flexography is being used more and more to print newspapers, newspaper inserts, directories, and the yellow pages. Several of the newspapers trying the flexographic process have converted their old letterpress presses to flexography. One of the advantages of using flexography for newspaper printing is the ability of the process to use water-based inks. Water-based inks can be cleaned up with water. This makes for a faster and safer press washup.

Material Requirements of Relief Printing

Letterpress requires a coated paper in order to print high-resolution images. When the reproduction of fine detail is not important, very inexpensive paper, such as newsprint, can be used.

Flexography can print on a wide variety of papers and plastics. It is ideal for printing on corrugated boxes and nondimensionally stable plastic films such as bread wrappers.

Gravure

Gravure printing is the process that prints from a sunken surface, Figure 10–15. Gravure accounts for about 20 percent of all printing. Unless there are some remarkable technological advances, it will continue with that share through the year 1995.

FIGURE 10–15 These gravure cylinders will be chrome plated for longer press life. *(Courtesy of Quad/Graphics, Inc.)*

The Gravure Process

In gravure printing, a cylinder is etched with the image to be printed. Ink is applied to the cylinder and then wiped off. Some ink remains in the etched portions of the cylinder. When the cylinder is rolled across the paper under pressure, the ink is deposited on the paper in the image pattern.

Like flexography, gravure uses a fluid ink and a very simple inking system. Unlike flexography, the image carrier is quite expensive, often costing $1,000 per cylinder. This cost is typically spread over 96 pages per cylinder, bringing the per page cost to a price more competitive with other printing processes. The production of a gravure cylinder is quite time-consuming. (See Chapter 9 for more details.)

Because the inking system is so simple, the printing cylinder must be perfect. Offset and letterpress presses allow for the adjustment of ink density both overall and across the printed sheet. In gravure printing, ink density is controlled by the depth of the wells in the gravure cylinder. This requires that the gravure cylinder be proofed prior to being placed on the printing press. If errors are found, the cylinders must be manually corrected. This is done by skilled engravers who etch the copper cylinder by hand for small corrections, or who paint the correct portions of the cylinder with a material that resists acid and then etch, with acid, the portions needing corrections. This is a delicate operation. Sometimes a mistake can mean that the entire cylinder must be remade.

The length of time required to produce a gravure cylinder is the major drawback of this printing process. This process is generally not considered cost-effective for runs of less than 60,000 to 70,000.

For long runs, the gravure process is ideal. The simple inking system, with control of the ink density built into the gravure cylinder, means that the job can be made ready and run with very little paper waste. The gravure process has less than half the spoilage rate of the offset process. Low waste, high-press speeds of 45,000 impressions an hour, a cylinder image life of several million impressions, and high-quality printing on low-quality printing papers make the gravure ideal for long-run jobs. Some of these jobs include *TV Guide, National Geographic,* Sears and J.C. Penney's catalogs, and a wide variety of packaging materials.

Material Requirements of Gravure Printing

Gravure presses can print on various substrates, ranging from smooth-surfaced papers to dimensionally stable plastic films and laminated paper/foil combinations. The only requirement is that the surface be very smooth.

Gravure prints a thicker ink film than either offset or relief. This additional ink thickness means that an opaque, white ink can be printed. Because of the thinner ink-film thicknesses transferred by offset and relief, those processes cannot print a white ink.

Screen Process

Screen printing process is an extremely flexible process that can print on any material in almost any shape. It can print designs on fabric before it is manufactured into shirts or after it has been sewn into caps or jackets. Despite this advantage, screen process accounts for less than 6 percent of total printing production. This small market share is because the process is slower than the other processes.

The Screen Printing Process

Screen process gets its name from the screen —either silk, polyester, or nylon—that holds a stencil in place. The stencil is cut in the shape of the image to be printed. The stencil, either hand-cut or photographically produced, is adhered to the screen. Ink is pressed through the screen onto the substrate below only in those areas allowed by the stencil, Figure 10–16.

In all printing processes, the designer must first consider the nature of the work to be printed, the material to be printed on, and the end use of the printed piece. This is especially true with screen process printing. Special care must be given to the correct selection of ink, stencil material, and screen type for the material that is to be printed. Paper, wood, glass, plastic, metal,

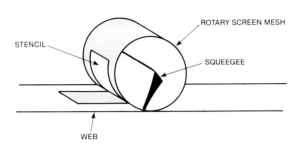

FIGURE 10–16 Screen process is able to print continuous images on a web of paper or other material by using a rotary screen printing press.

leather, and cloth all require different ink formulations. The end use of the printed product will also determine the type of ink to use. Inks must be compatible with the stencil medium used. Water-based inks will dissolve a water-based stencil, so the two are not compatible. Screen process inks are often specially formulated for specific jobs.

The applications of screen process printing seem to be without limit. It is used for products ranging from T-shirts to printed circuit boards, Figure 10–17. A typical characteristic of the process is the thickness of the ink film that can be printed. While new ink formulations allow a thinner ink film, the ink on a typical printed piece can be felt as a definite thickness sitting on the paper or other substrate.

Fleet Graphics

A new product for screen process printing is **fleet graphics**. Fleet graphics refers to printing logos and other company-related advertising on the sides of large trucks, Figure 10–18. A look at fleet graphics will highlight the strengths of screen process printing.

Fleet graphics jobs are often four feet by twelve feet in size and contain spot as well as process color. Fleet graphics are printed on four-by-twelve-foot panels of pressure-sensitive material. (Bumper stickers are often printed on pressure-sensitive paper.) Ten of these panels are required to cover the entire side of a truck.

FIGURE 10–17 Screen process can print on a wide variety of materials. *(Courtesy of Primo Enterprises Inc.)*

FIGURE 10–18 Fleet graphics makes possible the printing of logos and other material on the sides of large trucks. *(Courtesy of Mandel Screentech Division)*

Another advantage of the screen process method of printing that makes it suitable for this very specialized market is the availability of inks that can withstand the effects of sun, rain, salt, and freezing temperatures. Ink formulations that resist the elements produce a thick ink film which can be best transferred using the screen printing process.

Nonimpact Printing

Nonimpact printing uses a form of electrostatic attraction/repulsion to control the placement of tiny droplets of ink or the placement of a toner. Digital information from a computer controls the placement of the ink droplets or toner and creates a new image for each sheet.

The system of nonimpact printing that uses deflected droplets of ink is called **ink jet printing**. In this system, microscopic droplets of ink are created and sprayed toward the paper. Each droplet is given an electrical charge. A deflector plate, using an electrical charge controlled by computer information, deflects the droplets just before they contact the paper. The plate then guides them to the proper position on the sheet to form an image, Figure 10–19.

The system of nonimpact printing using a toner is called **electrostatic printing**. An electrostatic printer uses a laser, controlled by a computer, to selectively charge a rotating drum. A **toner** adheres to the charged areas of the drum. Paper receives a charge opposite to that of the toner and drum. The opposite charge attracts the

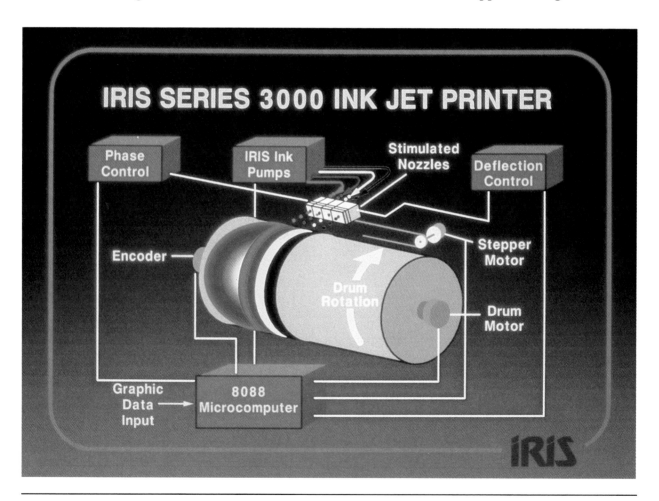

FIGURE 10–19 Ink jet printing can be used as a means to proof digital information. *(Copyright Electronic Publishing and Printing, 1988. Maclean Hunter Publishing Co. All rights reserved.)*

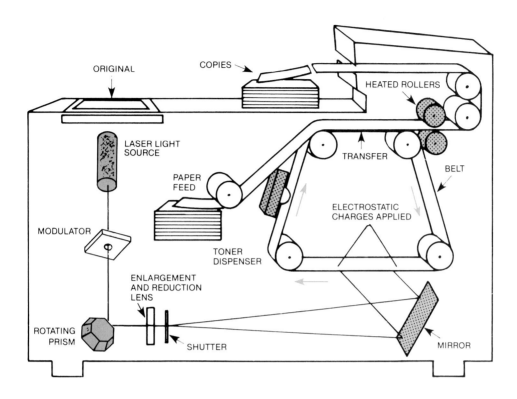

FIGURE 10–20 A computer controls the positioning of a rotating mirror which deflects the laser. The laser light produces an electrical charge on the drum. *(Modified from PRINTING TECHNOLOGY, 3rd Edition by Adams, Faux, and Rieber, © 1988 Delmar Publishers Inc.)*

toner from the drum to the paper. The toner is then fused to the paper surface with heat, Figure 10–20. Office photocopiers function by means of electrostatic printing.

Generally, nonimpact printing cannot yet compete with impact printing processes because it is slower. The process is limited to the speed at which the printer can reimage the next sheet. Current technology can produce acceptable quality for addresses and short customized messages, but is incapable of imaging an entire press size sheet at current press speeds.

However, nonimpact printing can compete with small offset lithographic presses for short run, general quality work. Nonimpact printing is a computer-to-paper system without the need

for photoconversion, stripping, and platemaking. Therefore, runs of up to 1,000 copies are economical on this equipment. Longer runs can be produced more economically by other printing processes, especially if multicolor printing is desired.

The Digital Press

Many advances in electronic technology have already been made in typesetting and prep. With these advances, it is natural to wonder if the computer will someday be able to transfer a full-color image directly to paper without any intermediate handling.

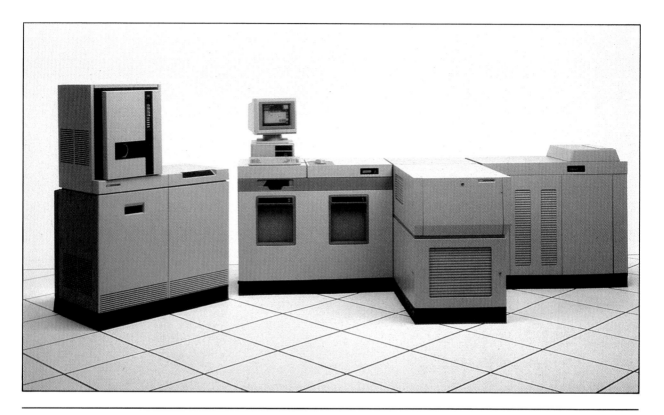

FIGURE 10–21 The Xerox 9700 family of printers may be the forerunners of computer-to-press technology. *(Courtesy of Xerox Corporation)*

A direct computer-to-press linkage would eliminate many expensive labor costs, and produce savings on typesetting paper, film, and plates. Convenient last minute image changes could be made. Books could be printed on demand, thus eliminating inventories and warehousing costs.

The technology exists today, in a very basic form, Figure 10–21. The system is able to accept computer input and produce a finished product. One such application is the combining of variable information, such as a customer's personal data on an insurance form. The basic insurance form is stored in the print system. The program calls up the insurance form and adds client information as each piece is produced. Each copy is an original.

The problems inherent in this process begin as output size changes from the basic 8½ × 11 format to press sizes. Each impression requires that the image carrier be reimaged. Computers have little problem refreshing an image area of 8½ × 11 inches (.6 square feet of image area) at a rate of 30 sheets per hour. However, it would take a very large computer to process the information for a 36-inch web press (4.6 square feet of image area) running at 300 impressions per hour. The problem is compounded when dealing with color.

At this point in time, there are some problems that do not seem to have a solution. Computer-to-press printing would have to compete with modern printing equipment. Current imaging toner technology cannot process paper at anywhere near the speed of an offset or gravure press. Ink jet printing at high speeds results in splatter that makes the image unreadable.

Another consideration is that of quality. Current electronic reproduction has loose quality requirements. Four-color electronic color reproduction is now available. However, applying this technology to a large press size with the speed requirements is not currently possible.

Summary

A wide variety of printed products is available today. While there is an overlap in applications for some printing processes, each process excels in specific applications.

Image transfer is the process of transferring ink from one surface to another. Because it is done under pressure, image-transfer devices are often called presses.

All presses are made up of subsystems. These subsystems include feeder systems, registration systems, inking systems, image transfer systems, and delivery systems.

Offset lithography is the major method of printing in use today. Offset refers to transferring the image first to a rubber blanket, and then to the paper. Lithography refers to the separation of image and nonimage areas by chemical means. In lithography, nonimage areas are coated with water, and the image areas are coated with an oil-based ink. Your school newspaper and your math textbook are both probably produced using the offset method.

Relief printing (letterpress) is now best-suited for specialty operations such as imprinting, numbering, perforating, embossing, and hot foil stamping. There are still some letterpress newspaper presses, but the number of these is declining.

Flexography, while technically a form of relief printing, is different enough to be considered by some to be a separate process. Flexography specializes in packaging printing. It is also replacing older letterpress newspaper presses.

In gravure printing, a cylinder is etched with the image to be printed. Ink is deposited on the paper in the etched image pattern. Gravure is offset's largest competitor. The two processes are both used to print those publications that require a million or more copies. For an extremely long run requiring high-quality work, no other process can produce the product as cost-effectively as gravure.

In screen printing, ink is pressed through a screen onto the substrate below. The image is created by a cut stencil that allows the ink to go through the screen only in certain areas. Screen process is ideal for short runs. No other process can be as cost-effective for quality multicolor work on a very short run. Screen process is also the only process that can print on a wide variety of already manufactured goods.

Nonimpact printing uses electrostatic attraction/repulsion to control the placement of ink or toner. Nonimpact printing using deflected droplets of ink is called ink jet printing, while nonimpact printing using toner is called electrostatic printing. These systems are often directly controlled by computers.

REVIEW

1. Name four press subsystems and briefly describe the purpose of each.
2. Name four printing processes.
3. List the kinds of substrates that each of the four processes named in Question 2 are best suited for.
4. Describe how offset printing works. What is its advantage?
5. List two printed products unique to relief printing.
6. List two printed products unique to gravure printing.
7. List two printed products unique to screen process printing.
8. List the printing process that would be best suited to print 100,000 copies of a 16-page, full-color brochure that contains text and line illustrations.
9. List the printing process that would be best suited to print 500 bumper stickers in two colors with no photographs.
10. List the printing process that would be best suited to print a multicolored, flexible plastic wrapper for loaves of bread. The job must be printed from a roll and delivered in roll form.

Finishing

KEY TERMS

adhesive binding	guillotine cutter	punching
banding	hardcover	right-angle folds
binding	in-line finishing	saddle stitching
buckle folder	inserting	scoring
casebound	knife folder	soft-cover
collating	loose-leaf binding	shrink-wrap
die cutting	numbering	side stitching
drilling	parallel folds	spiral binding
finishing	perforating	three-knife trimmer
folding	plastic comb binding	trimming
gathering		

Introduction

Seldom does a printed job go directly from the press and into the hands of the customer. There are a number of operations that a particular job may have to go through before it is ready for delivery to the customer.

The **finishing** operations that jobs such as books, booklets, and magazines require are folding, gathering, binding, trimming, and packaging. Other jobs such as raffle tickets, advertising brochures, printed cartons, and catalog sheets may require drilling, numbering, perforating, scoring, and die cutting.

Finishing operations are becoming increasingly automated. This phase of the printing operation, which received little attention in the past, is now recognized as another area in which the use of modern technology lowers the overall cost of the product.

Folding

Folding is the process of bending a large, press-size sheet into smaller, page-size units. In Chapter 9, image preparation was introduced. The process of printing more than a single page at a time increases production. In addition, printing 8-, 16-, or 32-page signatures increases the efficiency of combining and fastening pages together. Folding is done by a machine called a folder. The folder's function is to crease large sheets of paper at precise locations.

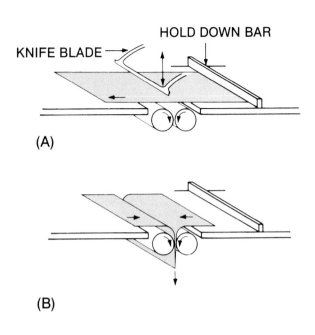

(A)

(B)

FIGURE 11–2 Knife folders can be used to fold more than one sheet at a time. *(Reprinted from PRINTING TECHNOLOGY, 3rd Edition by Adams, Faux, and Rieber, © 1988 Delmar Publishers Inc.)*

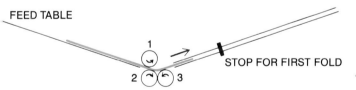

(A) PAPER IS MOVING TOWARD PRESET FOLD GUIDE.

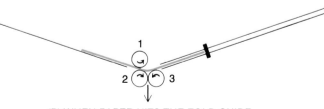

(B) WHEN PAPER HITS THE FOLD GUIDE, A BUCKLE IN THE PAPER IS CREATED ABOVE THE TWO EJECTION ROLLERS.

FIGURE 11–1 Continuous driving of rollers 1 and 2 causes the sheet to buckle between rollers 2 and 3. *(Reprinted from GRAPHICS ARTS TECHNOLOGY by John R. Karsnitz, © 1984 Delmar Publishers Inc.)*

Types of Folders

There are two types of folding machines in use, the buckle folder and the knife folder. **Buckle folders** drive the sheet between two plates until the sheet reaches an adjustable stop, Figure 11–1. Though the stop prevents the sheet from further forward movement, the drive rollers continue to push the sheet. The continued pressure causes that sheet to "buckle" into two adjoining rollers. These rollers grab the sheet at the fold, compress the fold, and move the sheet into the next fold section or into the delivery section.

Knife folders use a thin knife blade to force the sheet, which has been stopped by an adjustable stop, between two rollers. The rollers compress the fold and move the sheet into the next fold section or delivery, Figure 11–2.

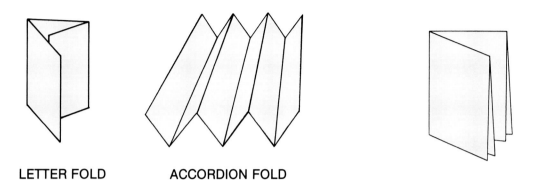

LETTER FOLD   **ACCORDION FOLD**

FIGURE 11–3 Two types of parallel folds: accordion and letter. *(Reprinted from PRINTING TECHNOLOGY, 3rd Edition by Adams, Faux, and Rieber, © 1988 Delmar Publishers Inc.)*

FIGURE 11–4 An 8-page signature. *(Reprinted from PRINTING TECHNOLOGY, 3rd Edition by Adams, Faux, and Rieber, © 1988 Delmar Publishers Inc.)*

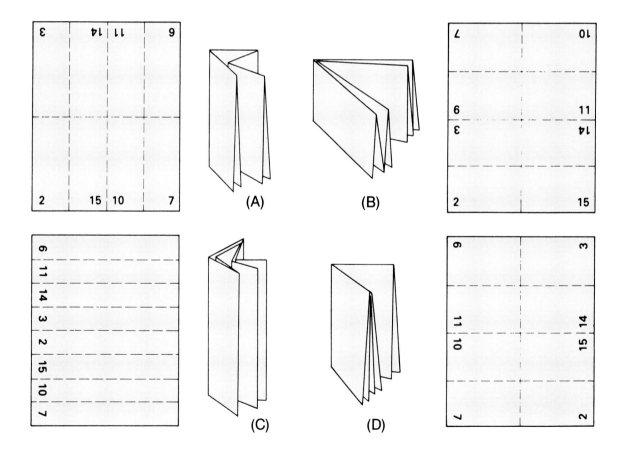

FIGURE 11–5 Four different folds resulting in a 16-page signature. *(Reprinted from PRINTING TECHNOLOGY, 3rd Edition by Adams, Faux, and Rieber, © 1988 Delmar Publishers Inc.)*

Types of Folds

Folds that are made parallel to the direction of sheet travel in the folder are known as **parallel folds.** Accordion folds and letter folds are examples of parallel folds, Figure 11–3.

A fold made at a right angle to a previous fold is known as a **right-angle fold.** One or more right-angle folds are required to produce multiple-page signatures. An 8-page signature may use one parallel fold and one right-angle fold, Figure 11–4. A 16-page signature may use one parallel fold and two right-angle folds as in Figure 11–5D. Figure 11–5A, B and C show three more ways to fold a 16-page signature. There are many other options. Each fold combination is suitable to a specific application.

The Folder and Page Imposition

At the beginning of the job, the designer must decide how the job will be folded. This decision must be included in the job specifications so the stripper can plan the page imposition (see Chapter 9) with the requirements of the folder in mind. Even though there are hundreds of folds that can be produced on a folder, the folder has definite limitations. These limitations must be considered if the printed job is to have the proper page sequence after folding. Deciding how to fold the job after it has been printed could be a costly decision.

Combining Printed Forms

Many jobs, such as books and booklets, need more than one signature to make up the required number of pages contained in the job. Other jobs, such as parts catalogs, may be put together using printed material from a variety of sources. This material may be in the form of individual sheets rather than folded signatures. Still other jobs, such as magazines, may combine folded signatures containing editorial material with individual printed sheets containing product advertising. The processes of combining printed material into final page sequence are called collating, gathering, and inserting.

Collating is the process of combining indi-

FIGURE 11–6 Signatures are placed side by side in an operation called gathering. *(Reprinted from PRINTING TECHNOLOGY, 3rd Edition by Adams, Faux, and Rieber, © 1988 Delmar Publishers Inc.)*

vidual printed sheets into proper page order. Loose catalog sheets that are drilled with three holes and placed in a three-ring notebook binder are an example of collated pages.

Gathering refers to the process of combining folded signatures into proper page order, side by side, Figure 11–6. For example, four 32-page signatures are gathered to form a 128-page booklet.

Inserting is the process of placing one folded signature inside another, Figure 11–7. Many

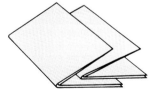

FIGURE 11–7 Inserting places one signature inside the next. *(Reprinted from PRINTING TECHNOLOGY, 3rd Edition by Adams, Faux, and Rieber, © 1988 Delmar Publishers Inc.)*

FIGURE 11–8 Inserting is commonly used to put together magazines. *(Courtesy of Quad/Graphics, Inc.)*

magazines use this process to combine signatures into the proper page order, Figure 11–8.

Binding

Once the individual sheets or folded signatures are combined, they are joined together using a process called **binding**. The designer considers several factors in the selection of binding

method. Some questions that should be asked are:

1. Will the job be used as a reference book? If so, the binding method must allow the work to lie flat when opened.

2. Will it be necessary to replace pages in the document so the useful life of the material will be prolonged? A book of paper prices is an example of this need. As some prices change, the new price sheets can replace the old.

3. Is the printed material considered to be permanent, or will it be used for a limited time and then discarded?

4. How thick (how many pages) will the finished job be?

5. What are the budget considerations?

The answers to these questions will determine the best method of binding to be used.

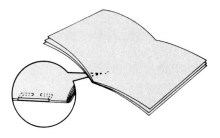

FIGURE 11–9 Inserted signatures are held together by the saddle-stitch method. *(Reprinted from UNDERSTANDING GRAPHIC ARTS by Kenneth F. Hird, © 1982 South-Western Publishing Co.)*

Signature Binding

An inexpensive method of binding folded signatures is saddle stitching, Figure 11–9. **Saddle stitching** uses wire, formed into a staple, to fasten together one or more signatures. Two or more wire staples are placed so the staple runs through the fold. If more than one signature is

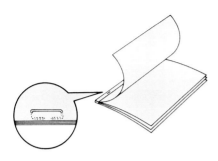

FIGURE 11–10 Side stitching can be used to fasten signatures or individual sheets, called inserts. *(Reprinted from UNDERSTANDING GRAPHIC ARTS by Kenneth F. Hird,* © *1982 South-Western Publishing Co.)*

to be saddle stitched, the signatures must be inserted within each other. Only folded signatures can be bound by this method.

Saddle stitching works well for reference materials that are no thicker than one-half inch and must lie flat when opened.

Generally, saddle-stitched jobs are designed to be read and then discarded. Weekly newsmagazines use the saddle-stitch method.

Side stitching uses a wire that is cut and formed into a staple to fasten folded signatures and/or individual sheets. Signatures to be side stitched must be gathered, side by side, Figure 11–10.

An advantage of the side-stitch method is that individual sheets of different paper types may be included with the signatures and fastened together with the wire staple.

Side-stitched material will not lie flat when opened. Therefore, the process is not recommended for jobs that will be used as reference material.

Neither saddle stitching nor side stitching is an appropriate binding method for products that must allow new information to be added because the wire staples hold the sheets together. There is no easy method to remove the staples, add or replace pages, and restaple the product.

Individual Sheet Binding

A method of fastening together individual sheets while allowing new sheets to replace old

is **loose-leaf binding**. The most familiar type is the three-ring binder. Ring binders are available in several sizes having from two to twenty-two rings.

The ring binders hold material up to four inches thick, allow the material to lie flat, and provide an easy method of adding new material without replacing the entire document. Parts catalogs and repair manuals are often bound in ring binders.

Other methods of binding together individual sheets are plastic comb and spiral. **Plastic comb binding** uses a plastic cylinder and a special paper punch. The plastic comb is similar to the ring binder in that new sheets can be added, though not as easily as in a ring binder. **Spiral binding** uses a spiral coil made of either wire or plastic to fasten individual sheets together. Both of these methods are considered to be rather short-lived binding methods. The more that documents bound by either method are used, the greater the chance that portions of the plastic will break.

Book Binding

Larger jobs requiring more permanence need other methods of binding. One such method is adhesive binding. **Adhesive binding** fastens gathered signatures and/or individual sheets together with a hot-melt, flexible, plastic glue. After the signatures have been gathered, each book is clamped in a vice. The folded side of the signatures is roughened by a grinder, a melted plastic adhesive is applied to the roughened edges, and a single-piece front/back cover is applied. The paper cover serves to cover the glue and provides a common surface for the glued signatures. Books with paper covers are called **soft-cover** books.

Adhesive binding is used for products ranging in size from paperback books to telephone directories. The flexible adhesive allows the book to be opened to any position and to lie flat for reference. A number of magazines use this binding method as well, since it allows for the addition of individual insert pages.

In **hardcover** or **casebound** books, string is used to tie each gathered signature to the next,

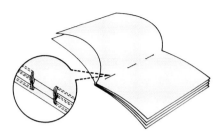

FIGURE 11–11 String forms a hinge that not only ties the signatures together, but also allows the book to lie flat when opened. *(Reprinted from UNDERSTANDING GRAPHIC ARTS by Kenneth F. Hird, © 1988 South-Western Publishing Co.)*

Figure 11–11. Sewing the signatures provides the greatest durability of all binding methods. It also allows the book to lie flat when being read. While sewn books can receive soft covers, most receive hard covers called cases, Figure 11–12. Casebound books are considered to have the highest quality binding available. Figure 11–13 shows examples of soft-cover and casebound books.

Trimming

At some time in the finishing process, the printed material must be trimmed to its final size. When this happens depends upon the type of cover and binding method to be used. **Trimming**

FIGURE 11–13 Adhesive-bound books can be opened to any position and lie flat for reference. Casebound books provide the greatest durability and the highest quality. *(Photo by Joseph Schuyler)*

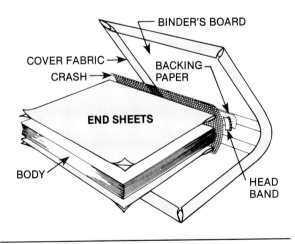

FIGURE 11–12 The cover of a hardcover book is called a case. *(Redrawn from PRINTING TECHNOLOGY, 3rd Edition by Adams, Faux, and Rieber, © 1988 Delmar Publishers Inc.)*

is the process of squaring off, by cutting three of the four sides of the printed product. The fourth side is known as the binding edge. Trimming is the last finishing operation for material that is bound using adhesive, saddle, or side stitching. Material that will be bound in a ring binder, or using spiral or plastic comb, requires trimming before being bound together. Casebound books are trimmed after sewing but before the case is applied.

Trimming may be done on a guillotine cutter or on a three-knife trimmer. The **guillotine cutter** uses a clamp to hold the paper pile in place while a knife blade is drawn across the pile, Figure 11–14. The cutter can cut only one dimension (length or width) at a time. It must be reset to make a different size cut. Computer controls can

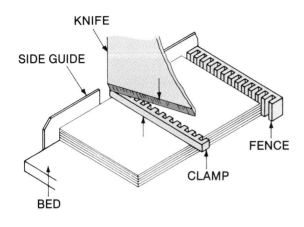

FIGURE 11–14 Guillotine cutters are powerful, hydraulic machines that can easily slice through 2,000 sheets of paper. *(Reprinted from PRINTING TECHNOLOGY, 3rd Edition by Adams, Faux, and Rieber, © 1988 Delmar Publishers Inc.)*

make fast and precise changes quickly, but the guillotine cutter does not lend itself to high-production trimming, as the operator must manually move each pile of paper into position.

Production trimming uses a cutter called a three-knife trimmer. The **three-knife trimmer** trims the front, top, and bottom sides of the book in one pass through the trimmer. The three-knife trimmer is usually one of many automatic operations set up to complete a printed product. This is called in-line finishing.

In-line finishing is used in high-production printing plants such as magazine and book printers. The printed product is loaded into inserting or gathering machines and then moved automatically through all necessary finishing operations. The printed product leaves the finishing line wrapped or tied into bundles ready for boxing, mailing, or stacking on shipping pallets.

The ultimate in-line finishing operation uses the printing press as the beginning of the automatic operation. The Cameron Belt Press is an example (see the boxed article in Chapter 10, page 161). The process starts with raw paper in the press and ends with the completed product ready for shipment without any manual handling in between.

Other Finishing Operations

Some printing jobs will require other finishing operations before the job is complete. These operations include drilling, numbering, perforating, die cutting, and scoring.

Drilling

Drilling is the process of producing holes in paper by means of a rotating, hollow, sharpened, metal tube. Drills come in various sizes, although ¼ inch is a commonly used size. Paper that is to be bound in a ring binder must be drilled with holes that match the distance between the rings. Three-ring notebook paper is an example of a printed product that is drilled after the paper has been printed. In the drilling process, the paper that is removed by the drill moves up through the center of the hollow drill. A related process, called **punching,** is used to produce holes smaller than ⅛ inch. The holes required for spiral and plastic comb binding are usually punched.

Numbering

Numbering is the process of printing consecutive numbers on individual pieces of a printed job. A raffle ticket job is an example of a printed product requiring numbering. Numbering is a relief printing process. Each time the numbering machine makes contact with the paper, the next numerical digit is advanced on the machine.

Numbering can be done on the printing press as the rest of the information is being printed. Or it can be done after the main printing is completed. In the latter case, numbering is considered a finishing operation.

Perforating

Another finishing operation required by a raffle ticket job is perforating. **Perforating** is the process of producing a series of small holes or slits. The holes or slits allow one portion of the raffle ticket to be easily and accurately separated from another portion. Postage stamps, reader information cards in magazines, and sale coupons found in the Sunday newspaper are examples of the use of perforation.

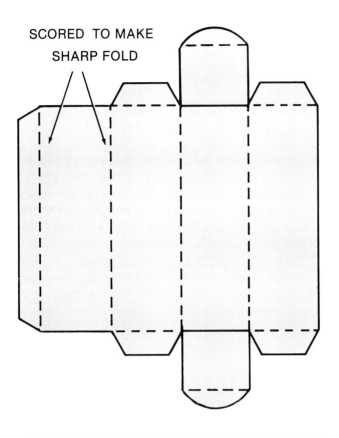

SCORED TO MAKE
SHARP FOLD

FIGURE 11–15 Boxes must be die cut and scored prior to being assembled.

Die Cutting

Not all printed material can be cut to final size in a guillotine cutter or three-knife trimmer. Irregularly shaped products, such as cartons and boxes, must be cut to final size using a process known as die cutting. **Die cutting** uses a series of thin metal edges in the shape of the item to be cut, Figure 11–15. A cookie-cutter in the shape of a star is an example of a die cutter.

Scoring

Scoring produces a narrow crease on paper in the area to be folded. Heavier paper, such as the paper used to make cartons, resists folding. Yet the box made from this paper must have sharp, neat folds to form the corners, Figure 11–16. A sharp fold can be produced if the paper is first scored. Scoring weakens the paper fibers by compressing them. This weakened area folds more easily than the adjoining areas.

Packaging

After the printed product is in its final form, it must be packaged for shipment to the customer.

Banding is the use of a two-inch to three-inch wide strip of paper taped around a convenient quantity of the printed product. For example, travel brochures, banded in packages of 100, make convenient units that can be delivered to visitor information stations along the interstate highway system.

Shrink-wrap packaging uses a clear, heat-sensitive plastic wrapping material. This material is wrapped around a bundle of printed product. The bundle then travels through a heat tunnel, where heat causes the plastic material to shrink. A tight, waterproof and dirt-proof package is produced. The clear wrapping allows the contents of the package to be visible.

Finally, the wrapped or banded bundle is placed in cardboard cartons for shipment to the customer.

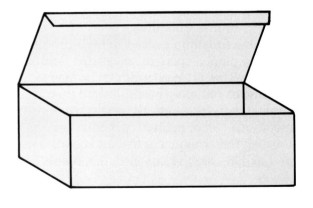

FIGURE 11–16 Scoring causes boxes and other heavy paper to form neat, sharp corners when folded.

Ink Jet Labeling

Ink jet labeling is a noncontact method of printing addresses on the front of magazines and other publications. In addition to addresses, other personalized information can be printed on inside pages of the product.

Ink jet labeling systems are generally faster than systems that require the application of a separate mailing label. This allows for the inclusion of the ink jet system within the finishing line and does not slow down the line.

Each address is created from a current computer-stored mailing list. This permits same day updates. New subscribers will begin receiving their subscriptions as soon as their names are entered in the computer.

The system can print on the inside of the product as well as the outside. This ability permits preaddressed insert order forms to be included in catalogs. This is said to encourage orders, since the customer's name is already on the order blank.

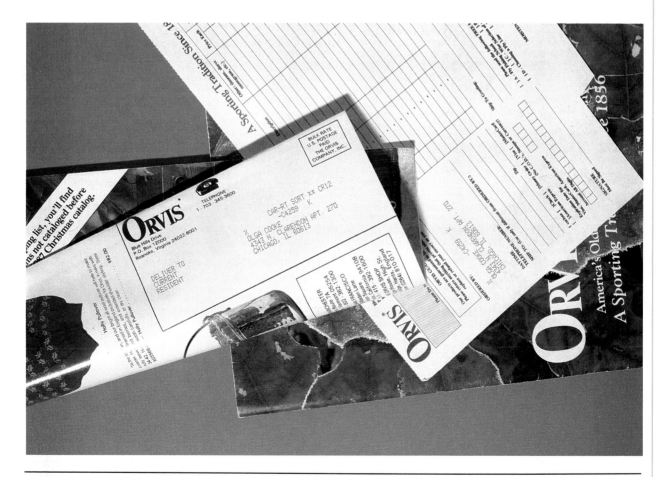

With the use of ink-jet labeling technology, addresses are as current as the client's mailing list. Labels do not have to be preprinted, but are created as needed. *(Courtesy of VIDEOJET Systems International Inc.)*

Labeling

The customer may not be the final user of the printed product. National magazines are usually mailed to the subscriber from the publication printer. In this case, a mailing label must be attached either to the product itself or to a package containing the product.

Labeling can be accomplished in one of two ways. The customer can provide a set of pre-printed address labels. The printer applies them to the package or to the product as the magazine comes off the finishing line. These labels are sorted, by the customer, by zip codes. Magazines sorted by zip codes enjoy a reduction in postal rates.

Some publication printers use a labeling method which utilizes ink jet printing. See the boxed article for more information of ink jet printing.

Shipping

The final operation in the production cycle of a printed product is shipping. Shipping is the transportation of the product to the customer or directly to the end user. Shipping may involve a simple telephone call to the customer reporting that the job is ready to be picked up. Or it may involve dispatching a fleet of trucks to deliver the product to regional distribution points across the country, Figure 11–17. Shipping could also require placing the individually labeled product into mailbags and delivering the bags to the post office.

FIGURE 11–17 Printed products move by land, sea, and air. *(Courtesy of Knight-Ridder, Inc.)*

Summary

Finishing operations are those required after a printed product comes out of a press, before it is delivered to a customer. Finishing operations are becoming increasingly automated.

Two types of folders are buckle folders and knife folders. They fold the paper in parallel folds or right-angle folds. The type of folds to be used must be considered during page imposition.

Collating is the process of combining individual printed sheets into proper page order. Inserting is the process of placing one folded signature inside another.

Once individual sheets or folded signatures are combined, they are joined together using a process called binding. Saddle stitching and side stitching both use staples to join pages or signatures. Books are bound using adhesive or case binding. Books are generally trimmed after being bound.

Other finishing operations include drilling, punching, numbering, perforating, die cutting, and scoring.

Various packaging operations include banding, shrink-wrapping, and labeling. After packaging, the final product must be shipped to the end user.

REVIEW

1. Name two types of folding machines. Under what conditions is each most likely to be used?
2. Briefly describe the processes of gathering, collating, and inserting.
3. What is the number of folds necessary for a 32-page signature? Draw a sketch of the folds.
4. Describe why the folding and binding methods used on a printed product must be considered during the stripping and page imposition stages of prep.
5. List the finishing operations likely to be used on a raffle ticket job. The job will be printed on an offset press, with 16 tickets on a single press-size sheet. First list the characteristics of a raffle ticket.
6. Describe and give an example of ring binding.
7. Describe and give an example of plastic comb binding.
8. Describe and give an example of adhesive binding.
9. Describe the ink jet process of addressing publications as they are being bound.

SECTION ACTIVITIES

SCREEN PROCESS PRINTING

OVERVIEW

This activity gives you a chance to develop your own screen and transfer the image with a screen printing process to an item of your choice. Different stencils, screens, and ink will be used for different applications in this activity. Materials will be chosen for the printing of a design on a piece of cloth, T-shirt, or sweatshirt. When you complete this activity, you will have designed your stencil, applied it to a screen, and then transferred that image onto your own project.

As you design your stencil, you need to decide what message you want to convey. Then decide how to transfer this design to a piece of paper, shirt, or some other item. Try to predict the effects of different colors and designs.

MATERIALS AND SUPPLIES

To complete this activity you need the following materials:

- screen fabric
- wooden frame materials
- cord
- tool to set the cord in groove
- water source
- liquid soap
- newspapers

- CDF-4 film
- squeegee
- scissors
- plastisol ink
- cleanup rags
- mineral spirits
- heat source for drying the ink

If the frame has not been put together yet you need wooden framing materials.

PROCEDURE

The following procedures will help you complete this activity successfully:

1. Make sure the screen has been mounted onto the frame. If this has not been done, cut a piece of screen just larger than the frame and lay it over the frame. While keeping the screen tight, press the cord into the grooves. This will hold the screen in the groove as you tighten it around. Once the screen is mounted

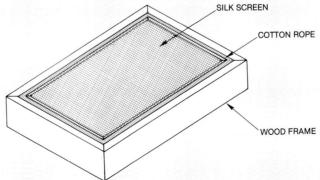

SILK SCREEN

COTTON ROPE

WOOD FRAME

tightly onto the frame, the screen must be washed well with liquid soap to remove all of the protective coating on the fabric. This also roughens the fabric so it will grip the stencil better. Leave the screen wet and go to the next step.

2. Use the roll of CDF-4 film that has been kept in the tube out of the light. Cut off a piece large enough to cover the stencil you have designed.

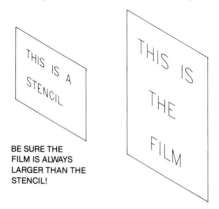

BE SURE THE
FILM IS ALWAYS
LARGER THAN THE
STENCIL!

NOTE: All activities using the CDF-4 film before exposing it need to be done in yellow light only. Take care to handle the film as little as possible to avoid fingerprints and other problems.

---

3. Cut the film to fit your positive.

4. Right away, carefully put the film back into the storage tube just as you found it.

5. Again, wet the screen. Make sure the screen is wet and move it to the table, rotating it to keep the water uniformly on the screen. The film sticks to the water.

6. Place the wet screen with the screen side up on a table lined with newspaper.

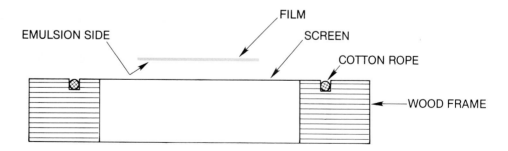

7. Place the film emulsion-side down on the screen. Turn the screen over. Run the squeegee over the film and get all the air bubbles out. This will keep the film tight to the screen.

8. Allow enough time to dry. Remember, the steps to this point need yellow or no light at all.

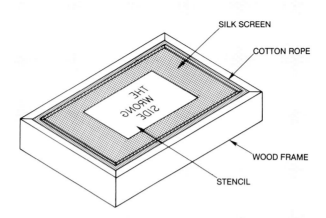

9. When the screen is all dried, remove the polyester backing from the film.

10. If you have an ortho-film positive, it is to be placed wrong side up onto the film side of the fabric and over the film.

NOTE: Wrong side up means the object is backwards.

---

11. If you need to produce an ortho-film positive, you can make one even if you do not have access to a process camera. Make two transparencies of the item you wish to use for your pattern. Place each transparency directly over the other and secure them. The image becomes opaque enough to give the desired results.

12. Place a clean plate of glass over the positive to keep it tight against the stencil during exposure.

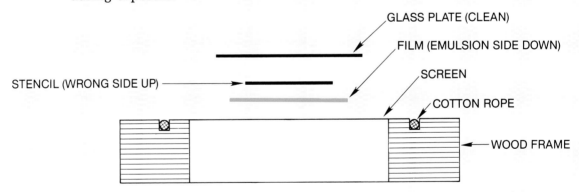

13. Place the frame into a platemaker or expose it to white light from flood lamps. You may have to experiment to find the exact amount of time needed for the best results. After exposure, wash the screen again. Areas that have not been exposed to light will wash from the screen. A good washing is needed to remove these exposed areas of film from the screen. After you are sure all of the image area is clean, blot both sides with newsprint. Allow to dry.

NOTE: Plastisol ink does not dry except when exposed to heat. This means it is easy to get all over you, your clothes, and the lab. Be very careful when handling the ink.

14. Cover all unexposed screen areas with a mask. This may be heavy paper taped to the screen, or just successive strips of tape. It is important to mask the whole area except for the image area to keep ink from coming through in the wrong places.

15. Place the screen in the desired position on your shirt, handkerchief, or other object.

16. Lay a bead of ink along the bottom of the screen below the image.

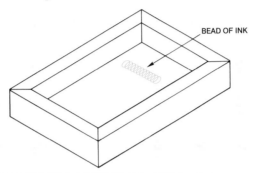

BEAD OF INK

NOTE: In some cases, it is difficult to get enough ink through the screen onto the object. This may be fixed by flooding the screen first. Flooding the screen is done by pushing ink with the squeegee across the screen, while the screen is in the up position and not on your project. Then place the screen directly on the project where you want it. Make a second pass which will force more ink through the screen and onto your shirt.

17. To make the pass, hold the squeegee with both hands. Tilt the squeegee slightly towards yourself and press down just enough to bend the rubber blade. Draw the ink on the screen across the image area. Pressure will force the ink through the screen and onto the material. Do not lift or move the screen between squeegees.

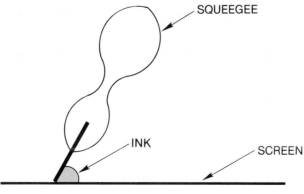

SQUEEGEE

INK

SCREEN

18. Lift the screen from the project. Carefully place the project under the heat source to set the plastisol ink.

> ► **CAUTION:** Be careful not to burn your shirt. Keep a constant watch.

Note: If there is no other heat source to use, electric hair dryers may be used for this process.

19. When you are done with the screen, the screen must be cleaned and all of the ink removed from it.
20. Put the extra ink back into the container.
21. Saturate a rag with mineral spirits and wipe down the screen. Small pieces of rag about 4 inches long seem to work best. They will be discarded as soon as they are used. Repeat the process until you are able to saturate a rag with mineral spirits and it comes away clean.
22. Clean yourself and all of the tools including the squeegee. Put all items back in their proper places.

FINDINGS AND APPLICATIONS

During this activity you have observed other students' designs of stencils for screen printing and their procedures. Note the importance of a quality design and of closely following the procedure. Straying from the procedure can result in a mess. You may want to do a second activity—use more than one color to develop a multicolor picture or pattern.

ASSIGNMENT

1. Design and develop your own pattern or stencil. Print it on an object of your choice. Because the screen process is the critical part of this activity, you can make your print on a sheet of paper and hand it in for a grade.
2. Make a list of the number of items around your home that have been printed using the screen printing method.
3. How could you use this process in a fund-raising activity for your local school club or student association?
4. If you could do this activity again, what would you do differently?

 ## PROCESS PHOTOGRAPHY AND PLATE MAKING

OVERVIEW

In this activity you will use the process camera to make line copy. The line copy will then be used to prepare a plate for offset press work. You will observe and take part in a practice that is often used in industry today. By carefully following directions you will be delighted with quality results as you prepare your materials to be used in the offset press.

When preparing materials for this activity, plan ahead to ensure a quality product. You will need to plan the location of your image on the plate.

MATERIALS AND SUPPLIES

For this activity you need the following materials:

- quality copy of item to print (laser copy or paste-up)
- process camera
- gray scale
- orthochromatic film (ortho film)
- film developing chemicals
- masking paper
- opaque solution
- brush
- plates
- plate burner
- plate processing gum
- lacquer

PROCEDURE

First, it is important to remember that the best results come from following directions carefully. Take care in handling the materials. Use caution to keep from wasting film and chemicals. To prepare your design and develop the line copy, you begin as follows:

1. Determine exactly what you want to print. Develop a quality copy of that material.
2. Set up the process camera by first making sure the lights are focused in the correct direction, the cover is off the lens of the camera, and the cover plate is clean.

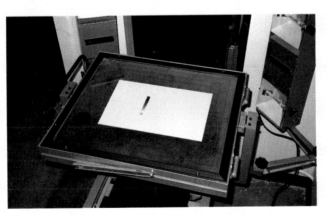

3. Place the copy and gray scale on the copy board. Close the glass over the copy board. Adjust the aperture to get the sharpest image. Usually the lens will produce the sharpest image two F-stops from wide open.
4. Swing the focus plate into position. Turn the darkroom to safe lights only.
5. Turn on the camera lights so that the image is projected onto the focus plate.
6. Check the focus.
7. Turn off the camera lights. Set the camera shutter for the correct exposure time.

8. Open the box of film and remove a single sheet. Remember only red light or no light is acceptable.

9. Place the film on the film holder, the emulsion side facing the lens of the camera.

10. Turn on the film holder vacuum switch.

11. Swing the focusing plate out of the way and close the film holder.

12. Start the shutter timer. The camera lights will turn off automatically when the exposure time is completed.

13. Take the film out of the camera. Process the film using ortho film developing chemicals.

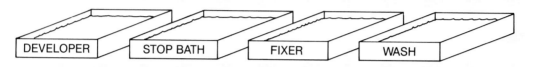

Five basic steps for developing the film are:

1. **Developing**—Place the film into the developer, emulsion side up. Rock the tray to wash the developer across the negative. Watch the gray scale and when No. 4 turns solid black, take the negative out of the developer. This can all be done in red light.

2. **Stop Bath**—Place the negative in the stop bath for 30 seconds. This neutralizes the developer and stops the development process. If the stop bath becomes dark blue in color, it is no good and should be replaced.

3. **Fixing**—Place the negative in the fixer. This dissolves the emulsion in the image areas.

4. **Washing**—Wash with clear running water for at least 15 minutes.

5. **Drying**—Remove the excess water with a clean sponge or squeegee. Never wipe across the emulsion side; it will scratch. Hang to dry in a dust-free place or dry in a drying machine. After the film has been developed you will have a negative of the material you wish to print. To make a positive for use in making the plate, repeat the process using the negative you just made as the copy that goes on the copy board with the gray scale.

Steps in preparing mask and burning the plate are as follows:

1. Attach the negative to the mask (yellow grid paper). Be sure your design is lined up in the center and correctly positioned on the mask.

2. While the mask is taped to your negative, cut a window to expose only that part of the negative that has the design you want to print.

3. Using opaque solution, opaque any pin holes that light shows through.

4. Once the opaque solution is dry, place the mask on the plate and line up the holes along the side of the mask.

5. Burn the plate for the appropriate time (expose to carbon arc or similar light source).

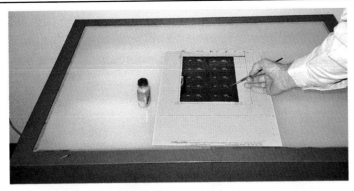

6. Remove the mask from the plate.

7. Apply the process gum to the plate, covering evenly with a thin coat.

8. Apply the developer to the plate and wipe the emulsion off.

9. Wash with clean water and a soft sponge.

FINDINGS AND APPLICATIONS

During this activity you found that you had to follow the directions exactly to get the results you wanted. You also found that it is quite simple to develop line copy which is made from a clear and black negative or positive piece of film. By using this photographic process, it is easy to get quality detail. The process results in the preparation of the offset press plates.

ASSIGNMENT

1. Design printed material to be run on the offset press. Follow the steps to prepare the negative, then the mask, and finally the offset press plate.

2. Choose a second project for this activity from the following list:

- Christmas card
- Certificate
- Note pad
- Business card
- Technology newsletter
- News flyer

COMPUTER-DESIGNED BUSINESS CARDS

OVERVIEW

As you work through this activity, you will have the chance to use the microcomputer to produce computer graphics. Your task is to design and produce a personal business card. As you do this, give attention to the kind of message you want others to get from your business card.

As you develop your business card, list five different places you may want to present it. Example: You are making a contact for your local student club. You may want to leave one of your personal business cards with the person you asked to speak at your next club meeting. Take a few minutes to list some other uses for your business card.

MATERIALS AND SUPPLIES

The equipment and supplies you need to complete this activity are as follows:

- personal computer with appropriate software
- printer
- your own storage disk
- a worksheet, pencil, and paper for layout and design

PROCEDURE

1. The first thing you will want to do is look at sample business cards. Be sure to note basic size and style. You will note that the finished size is 2″ high by 3½″ long.

2. Sketch your idea on a piece of paper (cut a card the size 2″ x 3½″ with scissors for a pattern). Remember that you need the name of the business, your name, address, zip code, and telephone number.
3. Using the computer, develop a logo that you will include on your card.
4. Have your instructor check your work. With your instructor's approval, it is time to finalize your design on the computer.

►CAUTION: Be sure to get the proper permission and instruction before you operate the computer.

5. If you have not already received it, your instructor will need to give you the proper computer training so that you can complete your card.

6. Once you have received any needed instruction, begin planning your own business card.

7. Your instructor will want to see your card when you have finished it.

8. Make sure you return all disks to their right places. The print quality of the business card is based on the type of equipment that you have available. If you can make a laser print from your computer, the quality is very good and approaches that of the typesetting processes used today in the graphic communications industry.

Generally, it is easiest to make multiple copies of your business card through the copy program on your computer so that the printout has a full page of business cards. Using this full page, you may either choose to make multiple copies on a copy machine with a lighter-weight paper or prepare a plate for use on the offset press and print directly onto a cardstock. You can also take your printout to a quick-copy store where the printout can be copied on a number of different weight papers.

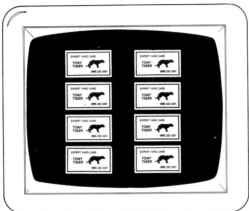

FINDINGS AND APPLICATIONS

As you developed your business card, did you find that some designs just did not work? Did you notice that type styles are different and some look better than others on your business card? You may also have become more aware of the business cards of other members of your family.

ASSIGNMENT

1. Give your instructor a copy of your personal business card attached to a sheet which lists the basic uses for business cards you listed when you read the overview.

2. Turn in a bill of materials for the supplies used in making your business card.

3. Figure the cost of each card you made.

4. If your computer could not produce graphics, how else might you have produced a business card with a logo?

ORGANIZE A COMPANY TO PRODUCE A GRAPHIC PRODUCT

OVERVIEW

Using the skills and information you have learned so far, you will organize a company, design a printed product, and produce that product for sale using the graphic reproduction processes. You will be able to apply the principles of administration, production, and marketing as you prepare your own product. This activity allows a number of students to take part in the different phases.

You must plan and predict a number of things in this activity. You should be able to select an item for production that will be marketable and that can earn money for stockholders and workers in your company. As part of the plan you may want to develop a market survey to determine whether or not your product will sell.

MATERIALS AND SUPPLIES

This activity requires all of the graphic production equipment in the lab including the photographic processes for platemaking, the plates, and other materials including paper, the press, and any materials needed for packaging.

Organizational Structures

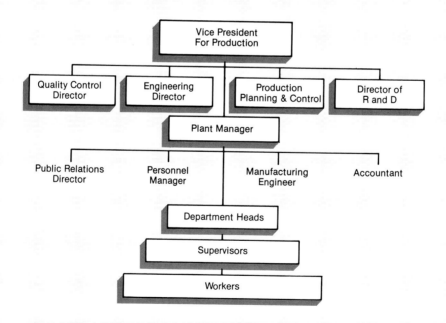

PROCEDURE

1. Organize a company, management, and production staff. This organization may include:

   - President
   - Production manager
   - Specific task workers
   - Quality control person
   - Marketing manager
   - Business manager

2. Identify a product to be produced. Possible examples include: personalized notepads for students and/or instructors, holiday greeting cards, thank-you cards, memo pads, stationery, newsletters, football programs, screen-printed T-shirts, literary magazines, or another printed object of your choice. Once you've decided on a product, your next task is to do the following preliminary tasks:

   - Design of prototype sketch
   - Market survey
   - Plans or sketches
   - Project cost and profit

3. Develop the actual production system including flowcharts, jigs, fixtures, conveyer material handling, and packaging.

4. Organize the company to secure production capital. This includes developing and selling of shares or stock in the company.

5. Make the production run including all of the steps to produce and package the product.

6. Sell or market the product.

7. Dissolve the company, paying off stockholders and employees that worked in the production.

FINDINGS AND APPLICATIONS

As you worked through this activity you found that you had to maintain high quality and standards for accuracy to ensure that the product would sell. You found you had to work with others to reach a goal. You were able to briefly look into the process of manufacturing and the activities that go into making most of the goods that we use today.

ASSIGNMENT

1. Take part in the organization, management, and operation of a company as outlined in this activity.

2. Develop and produce a product.

3. Choose a communications industry and identify the manufacturing process used within it. List places where the processes are either the same as or different from the one in which you took part.

The successful management of a company requires communication between all levels of the company's organizational structure *(Photo courtesy of Knight-Ridder, Inc.)*

SECTION FOUR
PHOTOGRAPHY AND MOTION PICTURES

Modern cameras allow us to capture events on film in many different ways. Speeding bullets can be frozen at the moment of impact with a balloon, the action on a sports field can be slowed down for analysis, a whole day's movement of a storm as seen from space can be seen in several seconds, and a birthday party can be captured and enjoyed again, almost as it happens. Photography is an interesting hobby, an important document producer, and a useful scientific and industrial tool. In all of these roles, it is a communication system.

This section describes cameras, films, and the tools needed for producing photographs and motion pictures, as well as the techniques for using them.

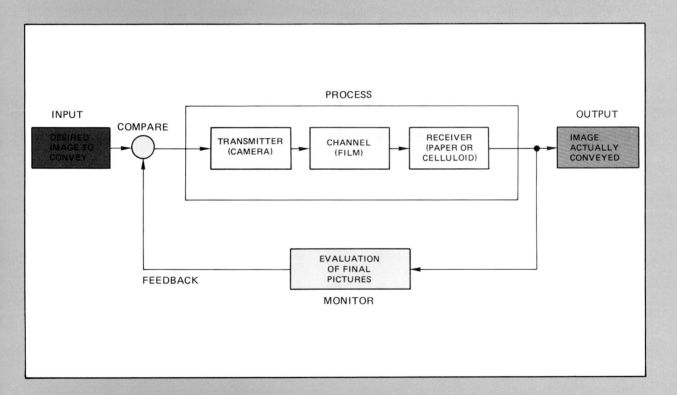

CHAPTER 12

Cameras and Film

OBJECTIVES

After completing this chapter you will know that:
- Photographs are available in black-and-white prints, color prints, and color slides.
- The first photograph was made in 1826.
- It takes light, film, chemistry, a camera, and other equipment to make photographs.
- A basic camera is a light-tight box which contains a film holder, lens, viewfinder, and shutter.
- Advanced cameras contain features that provide photographers with flexibility in picture taking.
- Microprocessors and other electronic circuits are commonly included in today's cameras.
- Five types of cameras are viewfinder, single-lens reflex, twin-lens reflex, view cameras, and instant cameras.
- Photographic film is made up of several layers.
- Film emulsions have speed ratings that commonly range between 25 (slow) to 1000 (fast).

KEY TERMS

antihalation layer	daguerreotype	parallax error
aperture	exposure modes	shutter
ASA	f-stop	single-lens reflex camera (SLR)
autoexposure	film emulsion	snapshot
autofocus	film holder	twin-lens reflex camera (TLR)
autospeed film sensing	instant photography	view camera
camera body	ISO	viewfinder
camera obscura	lens	viewfinder camera
chrome film		

Introduction to Photography

Pictures are important to people of all ages. The very young child to the oldest grandparent enjoys looking at pictures of friends and relatives, Figure 12–1.

Photography is used to record images of people, places, and events. This makes it possible

FIGURE 12–1 People enjoy looking at photographs of family and friends. *(Photo by Joseph Schuyler)*

to remember the happenings of today in the days and years ahead. Photos make accurate records so memories can be reinforced in the future.

Most pictures are in the form of photographic prints, which can be in color or black and white, and transparencies most often called slides. Prints and slides help record the events that take place in people's lives and the environment in which people live. Images of houses, schools, offices, and factories are captured in photos making it possible to show other people where we learn and work.

Historical Overview

It has taken many years for photography to advance to the high-tech level we enjoy today. Many people have given their time and talents,

so we can aim and shoot that favorite scene or event.

The first photograph was made in 1826 by a Frenchman named Joseph Nicephore Niepce. It took an eight-hour exposure to record the image, but it was a beginning. Years before that event, in the 1500s, people observed scenes of nature through the **camera obscura** (camera obscura means "darkened room" in Italian). Images of outdoor scenes were shown on the inside walls of darkened tents through tiny pinholes in the tents. The famous inventor Leonardo da Vinci wrote a description of the camera obscura in one of his well-known Notebooks. Camera obscura images were not very clear and were upside-down, but this encouraged people to keep up their efforts.

Problem Solvers in Communication

George Eastman

Until the late 1800s, most photographers were professionals who did their work in studios. The process was too complex for everyday use. George Eastman, however, turned the problem of complexity into an opportunity. He introduced a simple type of fixed focus camera that made photography available to just about anyone who wanted to try it.

The first Kodak camera cost $25. It included a roll of film that could take 100 pictures. The photographer would send the camera with the exposed film back to the Kodak Company with a $10 fee. Kodak would develop and print the pictures, and send the camera back to the photographer loaded with a new roll of one hundred pictures. It was so simple to use that the Kodak Company advertised "You press the button, we do the rest." Before long, picture taking became a fad, and almost every family owned a camera and a collection of snapshots.

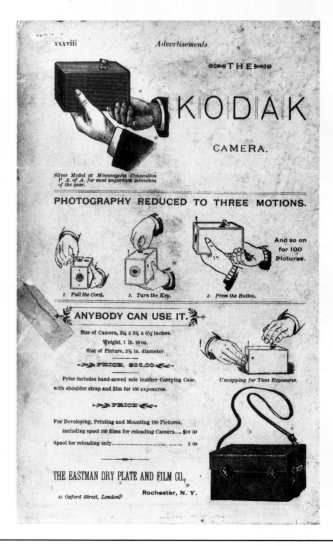

(Courtesy of Eastman Kodak Company)

FIGURE 12-2 A nineteenth-century daguerreotype of feminist Lucy Stone. *(Courtesy of the National Portrait Gallery, Smithsonian Institution, Washington, D.C.)*

Cameras with lenses were soon to follow. Also, photographic pictures made on metal plates came into use. These were called **daguerreotypes** and were named after their French inventor, Louis Jacques Mande Daguerre, Figure 12-2. Many other people made important contributions in both equipment and chemistry to bring us to the ease with which we use cameras and film of today. One such person was Dr. Edwin H. Land of the Polaroid Corporation. In 1947, he announced his invention of a camera that would make a black-and-white photographic print without a negative in less than a minute. This was a great new invention. Today people can take **instant photographs** in full color with cameras that are easy to operate, Figure 12-3.

FIGURE 12-3 Pictures of beautiful scenes can be taken quickly and with quality because of today's advanced photographic equipment. *(Photo by Kathryn Wine.)*

FIGURE 12–4 Light is necessary to photography, but quality night scenes can be taken with proper exposure. *(Courtesy of Santa Fe Southern Pacific Corporation)*

Basic Photographic Concepts

There must be light to create a photograph. The light does not need to be bright for a picture to be recorded, but it must be present, Figure 12–4.

To record the images that are illuminated by light, certain equipment and supplies are needed. These include a camera, film, processing chemistry, and lighting equipment. Enlarging equipment, photosensitive paper, plus processing chemistry and equipment are needed to make finished color prints.

Along with equipment and supplies, it takes the skills of many people to produce quality results. Photographic technology has moved ahead very quickly in recent years making it easier to take pictures. In turn, people demand higher-quality results in shorter amounts of time. It must be remembered that from beginning to end, people with great skills and dedication have designed and produced all of the equipment, tools, and supplies used in photography today. The years ahead will bring many more advancements to the imaging process made possible by light.

Basic Cameras

A camera is a light-tight box with a **film holder**, a **lens**, and a **shutter**. A **viewfinder** allows the user (photographer) to accurately aim the camera at the subject. Usually cameras include mechanisms that control the light entering the inside of the camera, and that advance the film. Basic cameras used for **snapshots** (those pictures taken quickly and with limited prior planning) contain the following parts. (The basic parts of a camera are shown in Figure 12–5.)

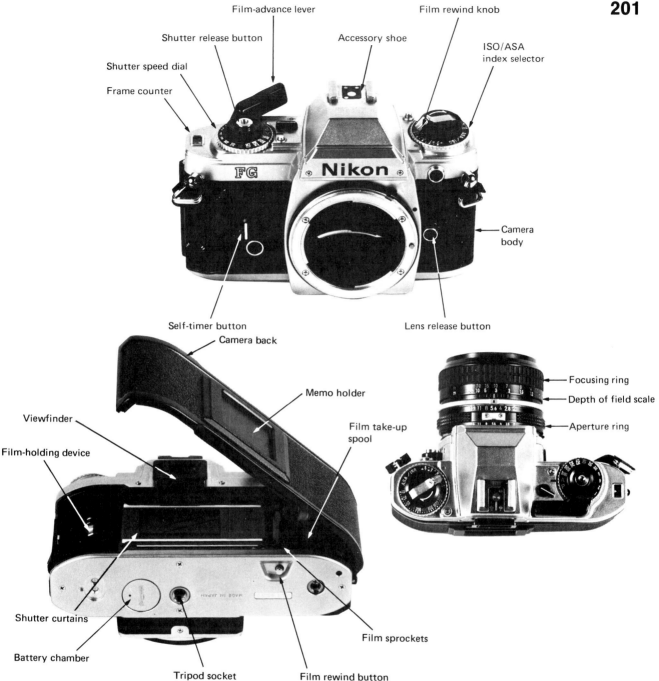

FIGURE 12–5 The many parts of a standard camera. (*Courtesy of Pentax Corporation*)

Body

This is the light-tight box that serves as the major framework of the camera. It must be large and strong enough to hold all of the other parts of a camera. Also, the design of the **camera body** should be such that it is easy to hold.

Lens

The lens serves as the "eye" of the camera. It gathers the light that forms the image and focuses it sharply on the surface of the film. Lenses can either be adjustable or fixed focus depending upon the complexity and cost of the camera.

Viewfinder

Just as the term implies, the viewfinder allows the photographer to properly aim the camera. Several types of viewfinders are used on cameras, but they all have the same basic purpose. Some viewfinders allow the subject that will be photographed to be seen through the lens. Other viewfinders use a simpler system of direct viewing.

Shutter Control

This device is a trip mechanism that triggers the opening and closing of the shutter. The shutter is a doorlike unit that when open lets light reach the film. Camera shutters can be likened to the eyelid of your eye. A shutter is either open or it is closed. The shutter control thus gives the photographer the choice of when to expose the film to light.

Film Holder

All cameras must have a place for film. Whether in the form of sheets, rolls, or discs, film must be securely held in the camera body. Because cameras are designed to use only one size of film, film holders have specific designs. This makes it impossible to use any other size and shape of film container than what is specified for the camera.

Film Advance

Usually, only one picture is taken in a certain space on a roll or disc of film. This makes it necessary for the film to be advanced a specific amount after each picture is taken. Film advance mechanisms are operated by hand, spring tension, or electric motor.

Advanced Cameras

Cameras are designed and made with nearly every feature one can imagine, Figure 12–6. This is true whether the camera is for use by the young child, adult amateur photographer, or adult professional photographer. With the current state of technology, when photographers only suggest a

FIGURE 12–6 A camera with features that allow the photographer to make numerous adjustments. *(Courtesy of Pentax Corporation)*

FIGURE 12–7 A series of aperture openings from small to large that are measured in f-stops. *(Courtesy of Pentax Corporation)*

need for a new or different camera device, manufacturers can meet the challenge.

A sampling of advanced camera features are described in the following paragraphs. Many serious amateur photographers want cameras with these options as they provide much flexibility when taking pictures. Professional photographers, on the other hand, expect these features to be included in their cameras.

Aperture Adjustments

An **aperture** is an opening of a specific size that allows light to pass through a camera lens. All camera lenses have an aperture. Some apertures are fixed (nonadjustable) while others are adjustable. Fixed apertures are often found in small, economical cameras made to be used by amateur photographers.

Apertures are identified by **f-stops** such as f/16, f/8, or f/2.8, Figure 12–7. Larger numbers represent smaller aperture openings, and smaller numbers indicate larger aperture openings.

FIGURE 12–8 A rather complete series of lenses that allows the photographer flexibility when taking pictures. *(Courtesy of Minolta Corporation)*

FIGURE 12–9 Pictures of distant subjects can be taken with a telephoto lens. *(Courtesy of Santa Fe Southern Pacific Corporation)*

Interchangeable Lenses

Serious amateur and professional photographers expect their camera system to have several available lenses. This gives the photographer the options of taking pictures near or far away, and of small or large subject areas. An example of a series of lenses for an adjustable camera is shown in Figure 12–8. Each of these lenses can cost from under $100.00 to several hundred dollars.

Camera lenses are identified by the type and size in millimeters. The standard lens for the camera shown in Figure 12–6 is 50mm. This lens size gathers in an image area similar to the human eye. Wide-angle lenses are those with a wide viewing area and have a size designation of 35 mm or smaller. Telephoto lenses are 80mm and larger. These lenses allow the photographer to take close-ups of subjects that are far away. One can see that a series of lenses is valuable to serious photographers, Figure 12–9.

Shutter Speed Adjustment

As stated before, a camera shutter is similar to a door. The shutter opens to let light expose the film and closes to keep unwanted light from reaching the film. The shutter speed adjustment allows the user to select a slow to fast shutter speed. There are up to 14 shutter speeds to select from in current model cameras, Figure 12–10. Shutter speeds generally range from 4 seconds to 1/2000 of a second. The most commonly used shutter speeds range from 1/60 second to 1/500 second.

Timer

With a timer, the photographer can prepare the camera and then have time to get into the pic-

SHUTTER SPEED	POSSIBLE PHOTO CONTENT	
1/8		SOLID AS A ROCK
1/15		
1/30		
1/60		
1/125		PERSON STANDING
1/250		DOG WALKING
1/500		HORSE RUNNING
1/1000		TRAVELING TRUCK
1/2000		SPEEDING RACE CAR

FIGURE 12–10 A standard range of shutter speeds available in adjustable cameras.

ture. Usually, the camera is mounted on a tripod when using the timer so the camera will stay still during the exposure. A timer normally provides a 10-second interval between turning it on and when it trips the shutter.

Motor Drive

Photographers who want and often need to take pictures in rapid succession (one after another) should have a motor drive. This unit is mounted on the bottom of the camera body and advances the film automatically after each exposure. It is operated by a small electrical motor powered by small, powerful batteries. As many as 3.5 pictures can be taken in one second if the photographer desires.

Electronics in Cameras

Electronics and microprocessors included in camera equipment design have greatly improved the ease of use, and the quality of cameras. Electronics have been a critical part of cameras for years. Now, computer and microelectronic technology allows the photographer to simply aim the camera and squeeze the shutter. Technological innovations have thus given more people the chance to use more advanced cameras that give greater returns in the quality of photographs.

Exposure Modes

An important feature of a quality camera is the **exposure mode** adjustment. This setting either allows the computer chip in the camera to establish the correct adjustments manually, automatically (called **autoexposure**), or lets the photographer make the selected settings before taking the picture.

Autoexposure

Two major exposure settings always had to be done by the photographer when using adjustable cameras. These were the aperture opening (or f-stop setting), and the shutter speed. With autoexposure, electronic sensors within the camera measure the amount of light entering the lens and give that information to a micropro-

cessor controller. The microprocessor then determines the correct f-stop and shutter speed based upon data preprogrammed by the camera manufacturer. Information is then passed electronically to the shutter and f-stop systems and the critical camera adjustments are made automatically.

Autofocus

This feature makes it easy for the photographer to aim and take a picture. The lens is especially designed with infrared electronic sensors that determine the distance between the camera and the subjects. The sensors control a small motor that moves the lens in or out, thus focusing the image on the film, Figure 12–11. As with autoexposure, a microcomputer chip is preprogrammed to react to the information it receives. In this case, the information comes from the infrared sensors in the lens. Photographers who take fast-action scenes like to use **autofocus** lenses because they respond quickly and accurately.

FIGURE 12–11 A 35mm SLR camera outfitted with an autofocus lens that uses electronic circuits and a microcomputer chip to accurately focus the image on the film. *(Courtesy of Vivitar Corporation)*

FIGURE 12–12 Coded information on the film cartridge is scanned by the camera-integrated contacts to set the film speed, length, and exposure range. *(Courtesy of Nikon, Inc.)*

Autospeed Film Sensing

In **autospeed film sensing**, precisely sized and positioned holes are punched into the leader (front) end of a roll of film, Figure 12–12. These holes contain coded information such as the speed rating or light sensitivity of the film. The system is often referred to with the initials DX.

Other informative data that could be included in the series of holes are the brand, type, and processing instructions. The microchip that has been preprogrammed receives the information via electronic sensors when the film is loaded into the camera. The several electronic circuits and microcomputer chips must work closely together. With this information, the film speed setting is then made so the film will be correctly exposed based upon the light that enters the lens.

Camera Types and Uses

Cameras are made in many shapes and sizes. Some are simple and some are complex. Some cameras only require that the photographer load

the film and then take pictures. Other cameras have several manual adjustments giving the photographer the chance to make settings that are right for the scene being photographed.

To help in understanding the many types of cameras, they can be placed into five major groups.

Viewfinder Cameras

The **viewfinder camera** is the most basic of the camera types, Figure 12–13. The photographer needs only to load the camera with film and begin taking pictures. Most viewfinder cameras are designed with a fixed focus lens, a fixed aperture (f-stop), and a fixed shutter speed. Some viewfinder cameras contain a choice of f-stops for the photographer. Many viewfinder cameras contain built-in electronic flash units. The electronic sensor and computer chip determine when there is not enough light for the scene, thus causing the flash to activate at just the right time.

FIGURE 12–13 This viewfinder camera is fully automatic from setting the film speed to focusing, exposure control, and film advance, and rewind. Also, it has a built-in flash unit that is controlled by a computer microchip. *(Courtesy of Nikon, Inc.)*

Single-Lens Reflex Cameras

Often referred to as SLRs, **single-lens reflex cameras** are very popular with serious and professional photographers, Figure 12–14. The

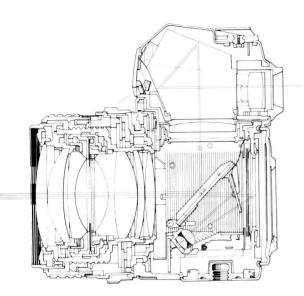

FIGURE 12–14 The viewing system of a single-lens reflex camera allows the photgrapher to see the image through the lens. *(Courtesy of Pentax Corporation)*

viewfinder allows the photographer to aim the camera by looking through the lens. This feature assures the photographer that the same image seen in the viewfinder will be captured on film.

SLR cameras are also designed to accept the many lenses shown in Figure 12–8 (page 203). A wide-angle lens can be used for one picture and a telephoto lens can be used for the next picture. Most SLR cameras contain many or all of the features as described earlier in this chapter under the heading Advanced Cameras. People with limited photographic capabilities as well as expert photojournalists, portrait, and product photographers use SLR cameras.

Twin-Lens Reflex Cameras

This type of camera was popular with news and product photographers before single-lens reflex cameras came into common use, Figure 12–15. **Twin-lens reflex (TLR) cameras** contain two separate lens systems. The top lens is used in the viewing system which has similarities to

FIGURE 12–15 A twin-lens reflex camera that has interchangeable lenses. *(Photo by Joseph Schuyler)*

SLR cameras. The photographer normally views the image through the viewing glass at the top of the camera.

The second lens system or ''taking'' lens is that which directs the image to the film. The two separate lens systems sometimes cause a viewing problem called **parallax error**. This is when the viewing lens and the taking lens do not cover the exact same image area. On most current models of TLR cameras, this problem has been corrected. Amateur as well as professional photographers find good uses for the TLR camera

FIGURE 12–16 A view camera is large and bulky, but can be used to take high-quality pictures. *(Courtesy of Zone VI Studios Inc.)*

over the SLR camera because of its larger film size. This sometimes gives a photographer greater image detail and better focus.

View Cameras

The photographer using a **view camera** is often a professional, Figure 12–16. That is because these cameras are rather expensive to buy, complicated to use, and large in size. Also, view cameras are usually expensive to operate because they use large sheets of film from 4″ × 5″ to 8″ × 10″ (20.3 × 25.4 cm).

A great advantage of the view camera is that the image is viewed through the lens before taking the picture. Also, the camera design allows the photographer to position the camera so the entire image is in focus. These features alone make it an excellent camera for taking architecture and product pictures that will later be published in catalogs, advertising literature, and magazines.

Instant Cameras

As stated earlier in this chapter, instant cameras were first introduced in 1947. Since that time, they have become very popular with amateur photographers, Figure 12–17. Almost anyone can aim and squeeze the shutter release

FIGURE 12–17 Instant cameras are very popular because they take finished pictures in less than a minute. *(Courtesy of Polaroid Corporation)*

only work necessary by the photographer is to aim the camera with the viewfinder and take the picture. Most disposable cameras even contain a flash. When the roll of film has been exposed, the entire camera is brought or sent to a commercial film processing company. The camera is cut open and the film is removed for processing.

Photographic Film

Film used in cameras described up to now in this chapter is categorized as continuous-tone film. This means that the light-sensitive film is capable of recording images from very light to very dark. The film must be made in that way so even the slightest image detail will be captured. The opposite of continuous-tone film is high-contrast film. This type of film records images as either black or white. It is mainly used in the graphics arts industry and is described in Chapter 10.

Photographic film is made up of four basic layers and two layers of adhesives which increase the layer count to six, as seen in the side view, Figure 12–19.

Emulsion Layer

Very small particles of silver halide crystals are mixed with a liquid gelatin. The silver particles turn black when exposed to light. This mixture serves as the light-sensitive layer of photographic film. It must be spread very thinly and evenly over the film base during manufacture.

FIGURE 12–18 A 35mm disposable camera with a built-in flash that is loaded with color print film at the factory. *(Photo courtesy of Fuji Photo Film U.S.A., Inc.)*

and take a good picture. These cameras feature automatic focusing and built-in electronic flash which give acceptable results nearly everytime.

Often the instant camera is referred to as a Polaroid camera. This is because the Polaroid Corporation is the only manufacturer of instant cameras.

There are many commercial uses for instant cameras. They are used by police and highway patrol officers, real estate agents, and in driver's license issuing offices. Whenever a photograph is needed quickly, an instant camera is very useful.

Disposable Cameras

The most recent camera group is that of the "disposables," Figure 12–18. These are inexpensive viewfinder cameras which are loaded with film at the factory. When purchased by the user, the camera is immediately ready to be used. The

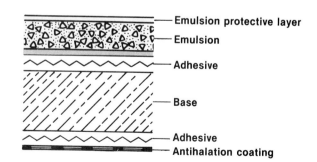

FIGURE 12–19 The basic layers of black-and-white photographic film. *(Reprinted from APPLIED PHOTOGRAPHY by Ervin A. Dennis, © 1985 Delmar Publishers Inc.)*

For black-and-white film there is one emulsion layer and for color film there are three emulsion layers.

Protective Layer

This thin layer is usually made of clear gelatin. It is placed on top of the emulsion so the light-sensitive coating will not be scratched during handling.

Base Layer

Photographic film bases must be clear, flexible, and strong. Light must be able to pass through the base layer without any change in its color. Discoloration could easily change the content of the image. The film must be flexible so it will easily bend and roll through a camera. Finally, the base is the foundation of the film, thus it must be strong. Breakage and cracking must not take place while the film is being handled in the camera and during processing. The two substances that are usually used for film bases are plastics called cellulose triacetate and polyethylene terephthalate.

Antihalation Layer

Light is absorbed by this layer. When light strikes the film during exposure, it must be controlled. Thus the **antihalation layer**, made of dyed gelatin, serves to keep the light from bouncing back to the emulsion and giving a double exposure. The word ''halation'' means the spreading of light beyond its normal boundaries. ''Anti'' then refers to keeping the light under control after it has served to expose the image in the **film emulsion**.

Adhesive Layers

A special gluelike gelatin material is used to bond the emulsion, base, and antihalation layers together. Without this agent, the emulsion and antihalation layers would peel from the base during handling. Film receives a great amount of handling from when it is packaged following manufacture through exposure, processing, and making of photographic prints.

Film Characteristics

Before the 1880s, all photographic film was made on rigid glass plates. This made it difficult to handle the film and the large cameras that were required for the light-sensitive glass plates. Also, film emulsions had to be applied by the photographer a short time before the picture was to be taken. Then, during the 1880s flexible photographic film became available for the general public. That film lacked the common characteristics which we enjoy today, but it was a beginning.

There are several important things that amateur and professional photographers must know about film before it can be accurately used.

Speed Rating

Numerical values are given to film according to how fast (sensitive) the film emulsion reacts to light. These numbers generally range from 25 to 1000. Three letters, **ISO**, often precede the numerical value printed on the film box and package. These letters are the abbreviation for International Standards Organization. This means that film is rated the same way throughout most countries of the world. Many people are still familiar with the older speed designation of the American Standards Association (**ASA**). The speed rating numbers are the same with both ISO and ASA, but world manufacturers of film products now use the common, agreed upon ISO rating system.

Most film used by amateur photographers falls into the speed rating of ISO 64 to 125. These film speeds are classified as slow to medium speed film. The other classes are listed as fast and ultrafast. Slow and medium speed films are used when the subjects are still or when there is little chance of movement. Because of the emulsion formulation, slower speed films have greater capacity for recording greater detail.

Faster speed films are most often used by serious amateur photographers. They are used when the photographer wants to take pictures where the subjects will be moving. Sporting events and stage plays are two good examples when fast films of ISO 400, ISO 1000, and higher are needed. Taking pictures from moving cars, trains, and planes also call for these films that react to light very quickly.

Black-and-White Film

Less black-and-white film is in use today than color film because most people prefer looking at color photographs. There are some important uses for black-and-white pictures in publications such as newspapers, books, magazines, and advertising literature that are printed with one color of ink. Great detail can be captured with black-and-white films because they are of the panchromatic type. This means that the emulsion is sensitive to all three primary colors — red, blue, and green. The film emulsion turns the three primary light colors to shades of gray.

Color Film

Most photographers like to use color film. This is true because most of us like to see pictures showing our natural surroundings. Color film contains the same basic layers as black-and-white film except that there are three emulsion layers, Figure 12–20. These layers are yellow, magenta, and cyan, which makes them secondary colors.

In film, the yellow emulsion layer absorbs the red and green light rays. This causes the yellow layer to record the images of the blue light rays. The magenta emulsion layer absorbs red and blue which allows it to record the green images. Finally, the cyan emulsion layer records the red light wave images. As you can see from the color wheel, Figure 12–21, cyan is next to blue and green. Because of this, the cyan emulsion absorbs the blue and green light waves.

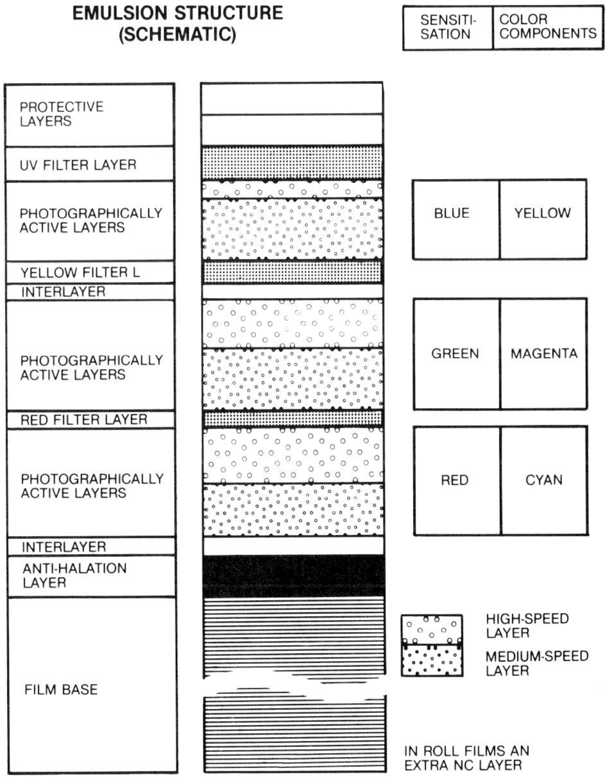

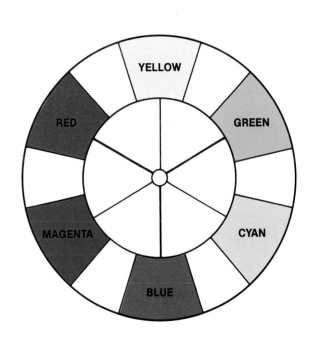

FIGURE 12–20 Color film has three emulsion layers. These are yellow, magenta, and cyan. *(Courtesy of Agfa-Gevaert, Inc.)*

FIGURE 12–21 A color wheel shows the position relationships between the primary and secondary colors. *(Reprinted from APPLIED PHOTOGRAPHY by Ervin A. Dennis, © 1985 Delmar Publishers Inc.)*

Color Photographic Film

The development of photographic film that accurately records color was a true technological breakthrough. Once black-and-white film emulsions became common during the second half of the 19th century, the inventive and creative minds of people never stopped working. Producing color photographic film was a logical next step.

Film with limited capability to record color was invented in the late 1800s, but it was impractical for marketing to the public on a mass scale. Two young musicians from New York City, Leopold Godowsky Jr. and Leopold Mannes, believed they could do better. They began their work during World War I. After considerable work and years of struggle, Godowsky and Mannes successfully prepared film that accurately recorded color. The color transparency film was first marketed during April of 1935.

The products of this film were 2″ × 2″ cardboard mounted slides. The slides could be viewed by inserting them in slide projectors and showing the pictures on a movie screen. The same procedure is still used today. Soon to follow was negative color film that could be used to make contact and enlarged prints. From this point forward, many color film emulsions have been researched and manufactured.

The years of work by Godowsky, Mannes, and many others have made it easy for people of all ages and abilities to take pictures that faithfully record the colors of the world in which we live. *(Courtesy of Gillian A. Spatz)*

FIGURE 12–22 Color film boxes tell the photographer whether the film is for slides or prints. *(Courtesy of Agfa-Gevaert, Inc.)*

Slide or Print Film

When buying color film, it is important to specify whether you want film to make slides (transparencies) or prints (photographs). The standard in the photographic industry is that **"chrome"** means slides and that "color" means prints, Figure 12–22. For example with Kodak film, slide film is identified as either Kodachrome or Ektachrome. For print film, it is labeled as Kodacolor.

There are five important specifications to give to the sales person when buying film. These are listed as follows:

■ Brand—i.e., Kodak, Fuji, 3M, Agfa
■ Size—i.e., 126, 35mm, disc
■ Speed—i.e., ISO 64, ISO 100, ISO 400
■ Slides or prints—i.e., Kodachrome, Fujicolor
■ Number of exposures—i.e., 20, 24, 36

Film Size and Packaging

Cameras are designed to accept only one size and type of film. In fact, it is impossible to use a size of film other than what is specified by the camera manufacturer. Do not try to use the wrong size or type of film because you can damage the camera.

Film is packaged in disc cartridges, roll cartridges, roll magazines, rolls, and sheets. Most amateur photographers use film that is packaged in plastic or metal containers such as cartridges and magazines. These fit the camera case perfectly because the camera and film manufacturers have worked together in their design and manufacture.

The sizes of film vary depending upon the camera specifications. The standard film sizes and packaging are 110 cartridge, 126 cartridge, and 35mm magazine. Serious amateur and professional photographers will most often use film in the forms of 35mm magazine, standard rolls with a light-protective paper covering, and sheets. You may remember from the camera description that view cameras are mainly designed to use individual sheets of film.

Summary

The first photograph was made in 1826, although images were displayed in camera obscura as early as the 1500s.

In order to produce photographs, light, film, photographic chemicals, and a camera are needed. A basic camera is a light-tight box with a film holder, lens, and shutter. A viewing system like a viewfinder allows the photographer to aim the camera.

Advanced cameras contain features that provide photographers with flexibility in picture taking. These features include aperture adjustments, interchangeable lenses, shutter speed adjustments, automatic timers, and motor drives.

Microprocessors and other electronic circuits that are commonly included in modern cameras have greatly improved the ease of use and the quality of cameras. Electronic circuits provide autoexposure, autofocus, and autospeed film sensing.

Five types of cameras are viewfinder, single-lens reflex, twin-lens reflex, view cameras, and instant cameras. Inexpensive disposable viewfinder cameras with film included are now being marketed. The entire camera (including the exposed film) is sent to a commercial film processing company.

Photographic film is made up of four basic layers: emulsion, protective, base, and antihalation. Adhesives bond the layers together. The speed rating of film is assigned a numerical ISO value. Most films range from ISO 25 (slow) to ISO 1000 (fast). Color film contains the same basic layers as black-and-white film, except that there are three emulsion layers which absorb the primary colors of light. Some standard sizes of film packaging are 110 cartridge, 126 cartridge, and 35mm magazine.

Photography is an exciting medium of communication. Through black-and-white prints, color prints, and slides, photographs reach the minds and pleasures of everyone.

REVIEW

1. Why are photographs important to people of all ages?
2. Who has been credited with making the first photograph, and when?
3. Describe a daguerreotype.
4. What essential ingredient must be present before a photograph can be taken?
5. Name five important parts of a basic camera.
6. Of what use are interchangeable lenses?
7. How have electronics been used in cameras?
8. Which camera type is not usually used by an amateur photographer? Why?
9. List the four basic layers of photographic film in the order they most often appear.
10. Would photographic film with a speed rating of ISO 64 or ISO 400 be the best to take pictures of fast-moving automobiles? Why?

Photographic Techniques

KEY TERMS

angle of view	filters	photographic composition
bracketing	flood lamps	push process
depth of field	focal-plane shutter	split-image focus
electronic flash	focus	telephoto lens
existing light	leaf shutter	wide-angle lens
exposure mode	normal lens	zoom lens
film speed	panning	

Introduction to Picture Taking

Taking pictures with a camera is exciting. One knows that this very moment will be "frozen" in time. No scene, unless it is a static display, will ever be exactly the same. This makes it important for you as a photographer to be sure the scene is accurately recorded on film.

Sometimes photographers are uneasy about the pictures they are taking. They are not sure that the lighting is appropriate or that the camera adjustments have been set correctly. These uneasy feelings are present with all amateur and professional photographers at certain times. Uncertainty can be overcome most of the time by knowing the camera and understanding how

FIGURE 13-1 Exciting photographs are taken by both amateur and professional photographers who know their equipment. *(Photo by Joseph Schuyler)*

to use it. Also, it is important to use lighting effectively and to compose the scene to make a pleasing picture, Figure 13–1.

Pictures are taken for both business and pleasure. Some people make their living taking pictures, but many more people take pictures because it is fun. Most pictures are taken of people, places, and things that serve as pleasant reminders of happy events. This is what makes picture taking so enjoyable.

Camera Preparation and Maintenance

The camera body is full of electronic circuits that help the photographer make the needed adjustments, Figure 13–2. The inside of a camera is

FIGURE 13-2 Most camera bodies contain small electronic parts and circuits including computer chips. *(Courtesy of Pentax Corporation)*

Photography in Construction

The first practical application of construction photography was demonstrated by Louis Jacques Mande Daguerre in the early 19th century. Daguerre, who was one of the first to record a permanent photographic image, called his invention the daguerreotype.

Daguerre had apprenticed with an architect and was the pupil of a scenic designer for the Paris Opera. The subjects of most of his works were buildings, houses, and Paris boulevards. Although mainly an art form, daguerreotypes also became a record of the type of building construction that was popular in Paris during the early 1800s.

Today, photography in the construction industry is used in much the same way as Daguerre's account of Paris. Photo records are kept of buildings being prepared for demolition or under construction, of bridges, tunnels, subways, and anything that needs this type of documenting.

Photos are used to show monthly building construction status of a given project, and photos are taken of work in progress. The photos are also used for public relations, brochures, annual reports, preconstruction surveys, and sales meetings.

Most shots are taken with existing light or with tungsten flood lights. Because they often must get in and out of a construction site quickly, photographers mix daylight and artificial illumination shots on the same film. An assignment may include from four to 4,000 views.

Photographs can be used to record the stages of contruction of a building. *(Photos by Robert Gorman)*

FIGURE 13–3 Neck and wrist straps are placed on cameras for the convenience of the photographer and for the safety of the camera. *(Courtesy of Sunrise Creations)*

much like a wrist watch. The parts are very small and it takes a highly qualified person to repair a damaged camera. Therefore, it must be handled with care.

Handling Cameras

Cameras are delicate. The first thing a photographer should do after picking up a camera is to place the carrying strap over the neck, Figure 13–3. Carrying straps are about three feet long and are fastened to the camera body with snap hooks. Small cameras such as the viewfinder types have short carrying straps. These cannot be placed over the neck, but they are large enough for a hand to pass through.

Straps allow photographers to carry cameras either over their necks or wrists. A strap is a good means of protection for any camera. The strap can and has saved many cameras from being dropped on a hard floor and damaged. Also, straps have helped photographers save their cameras from falling off high locations such as football stadiums, building ledges, and bridge railings. Straps are on cameras to be used, so use them every time you pick up a camera.

Care of Cameras

Generally, cameras need very little cleaning and maintenance. This is true if photographers care for their cameras properly. It is best to keep an expensive camera in a protective case. The case will help keep foreign materials such as dust, dirt, and liquids from entering the camera body or lens. Also, the case, which is usually made of leather, will help protect the camera from bumps and sharp jolts. Sometimes a camera is hit against the side of a door or car when being carried in or out. The camera case will take most of the shock.

Too much heat can damage cameras and film. It is unwise to leave a camera in the hot sun. The back window area of a car is not a good place for a camera. The heat that comes through the window glass is very high. Film does not survive well in high heat; thus, it is wise to keep cameras and the supply of film in cool places.

One of the most important points of care is the changing of the battery in a camera. Depending upon the camera brand and features, there can be one, two, or more batteries used to provide energy for the electronic and mechanical systems.

Most batteries used in cameras are about the shape and diameter of a dime but are two to four times thicker. One or more battery chambers are located in a camera body. A photographer can quickly learn the location of the batteries in a specific camera by reading the instructions in the operator's manual. The important thing to remember is to install new batteries on a regular basis. This should be done at least once a year.

Modern cameras require batteries because they need electrical energy to operate. Electrical energy is needed to operate the light metering system, the DX film sensing system, and the aperture and shutter systems on some cameras. Motorized film advancers also require electrical energy from the batteries.

Loading and Unloading Film

For many cameras, film loading is an easy task. Film that has been packaged in specially designed cartridges is easier to load. The camera back is opened after releasing the latch and the film cartridge is dropped into place. There is nearly no room for error whether the photographer is amateur or professional.

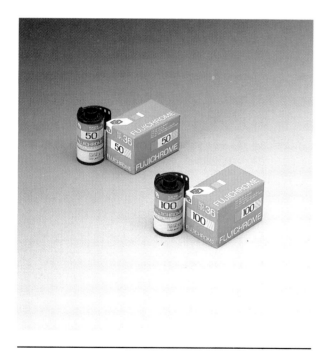

FIGURE 13–4 Film for 35mm SLR cameras is packed in magazines of 20, 24, and 36 exposures. *(Courtesy of Fuji Photo Film U.S.A., Inc.)*

Adjustable cameras such as the 35mm SLR use film packaged in magazines (sometimes called cassettes), Figure 13–4. It is best to load the film in a camera in low light conditions.

Bright sun or artificial light could fog (partially expose) some of the film while loading.

After opening the back of the camera, the film magazine is placed in its proper location. The film must then be started on the take-up spool. Upon closing the camera back, it is important to advance the film 2 to 3 exposure frames before taking the first picture. This is necessary because the first few inches of the film were exposed to light during the loading procedure. At this point it is important to remember to set the film speed adjustment to match the ISO (or ASA) number of the film.

Unloading film from a camera is just as important as loading it. After the last exposure is taken, the film must be rewound back into the film magazine. This is a must before the camera back is opened. Once the film is in the magazine, the camera back may be opened and the film magazine removed. It should then be placed in the protective container for safe keeping until processing. The camera is now ready for reloading.

Safety

Both amateur and professional photographers must remember and practice good safety habits. With camera in hand, it is sometimes easy to forget about some common safety practices. Streets can be dangerous whether the traffic is heavy or light. It is advisable to have an observer watch for vehicles if it is necessary for a photographer to stand on or near a roadway. Also, when stopping a vehicle along the side of a street or roadway, the photographer must be mindful of other cars when opening the door and leaving the car.

Safety is a concern when the photographer is working near heights. This may involve new building construction or mountains. Sometimes the best scene can be taken from the highest spot. Injury and even death can result when safety is not taken seriously.

Another example where safety is a concern for the photographer is around water and ice, which can provide excellent photographic possibilities. It is the wise photographer who

FIGURE 13–5 Lakes and streams provide excellent photographic scenes, but they also may contain dangerous situations for the safety-conscious photographer to consider. *(Courtesy of James Besha Associates)*

FIGURE 13–6 Lens millimeter designations and the angle of view in degrees. *(Redrawn with permission from Karl Heitz, Inc.)*

remembers the dangers of water as well as the beauty, Figure 13–5. The same is true for ice whether it be a coating on a walkway, a frozen lake, or a glacier.

Lens Selection

There are many lenses available for adjustable cameras such as the 35mm SLR (refer to Figure 12–8, page 203). Because lenses can be removed and replaced so easily, the photographer can take each picture with a different lens. The choice of lenses depends on the type of photographs to be taken and the amount of money the photographer wishes to pay. Lenses cost from about $100.00 to over $1,000.00 each, thus it is wise to be selective.

Lenses are divided into three groups — normal, wide angle, and telephoto. The lens groups are divided according to the millimeter specification of a lens, Figures 13–6 and 13–7.

Normal Lens

A **normal lens**, often referred to as the standard lens found on most 35mm SLR camers, is a 50mm lens. This has an **angle of view** (area of image coverage) of about 45 degrees. This amount of image area is similar to that seen by a single human eye. The millimeter specifications of normal lenses are in a narrow range between 45mm and 50mm.

Wide-Angle Lens

Wide-angle lenses gather in more image area than normal lenses. Their millimeter specifications generally range from 40mm to less than 20mm. The smaller number of millimeters of a lens means that it has a greater angle of view.

24mm

105mm

35mm

180mm

55mm

500mm

FIGURE 13–7 Camera lenses of different millimeter designations have an excellent range of angles of view. *(Photos by Joseph Schuyler)*

FIGURE 13-8 For distant scenes, cameras can be attached to telescopes. *(Courtesy of Tamron Industries, Port Washington, NY)*

Telephoto Lens

These lenses have larger millimeter specifications and usually begin around 60mm. **Telephoto lenses** have a narrow angle of view and are used to photograph distant objects, people, and scenes, Figure 13-8. Some telephoto lenses have angles of view less than 10 degrees and are in the 300mm to 1,200mm range. It is also possible to attach a camera to a telescope for taking pictures of distant objects such as planets and stars.

Zoom Lenses

Photographers often like to have flexibility built into a lens. **Zoom lenses** can be easily adjusted to change the angle of view to permit great flexibility. Typical zoom lenses may range from 35mm to 70mm, or from 80mm to 200mm, Figure 13-9. These lenses are easily adjusted during picture taking depending upon the angle of view desired by the photographer.

Camera Adjustments

Many of the cameras used by amateur photographers are automatic. This means the photographer only needs to load the camera with film, remove the lens cap, aim the camera with the

LENS ELEMENTS MOVEMENT IN ZOOM LENSES
(WITH SMC PENTAX-A ZOOM 28—135mm f/4 lens)

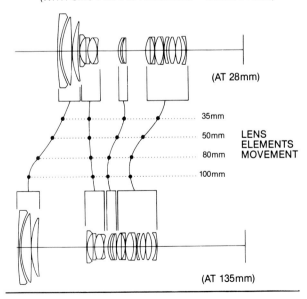

FIGURE 13-9 A zoom lens allows the photographer to select the angle of view just before the picture is taken. *(Courtesy of Pentax Corporation)*

viewfinder, and begin taking pictures. Cameras that allow convenient picture taking such as just described have many advantages. There are, though, a number of disadvantages. Completely automatic cameras take the controls away from the photographer. This takes away much of the creativity that makes photography exciting.

Manual adjustments give the photographer the chance to find the best camera settings for a scene. For instance, one scene may call for extra sharp focusing on a given object. This can be done with a manual focus lens. An autofocus lens gives an average focus for the whole scene. Whether the photographer wants automatic or manual camera adjustments is a choice that can be made just before taking each picture. That is because most of the 35mm SLR cameras today have both automatic and manual adjustment features. A brief discussion of the common camera adjustments follows.

Focusing

To **focus** a lens means to bring the picture into sharp view. The result is that the finished picture (photographic print or slide) looks just as sharp and clean as the original scene. Most camera focusing systems use a **split-image focus** technique, Figure 13–10. Within the center circle area of the viewfinder scene, there will be a double image when the image is out of focus. By

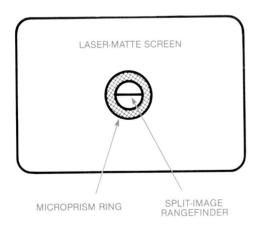

FIGURE 13–10 The split-image viewfinder is commonly used on 35mm SLR cameras.

moving the focusing ring on the lens, it is easy to see when the double image becomes only one image. Seeing one image in the viewfinder tells the photographer that the focus is acceptable.

Film Speed

As presented in Chapter 12, all film is assigned a number that indicates its sensitivity to light. This is the **film speed**. It is critical that the ISO (or ASA) film speed number be matched in the camera. All adjustable and automatic cameras have this important setting. The procedure and location of the adjustment are shown in all camera operating manuals.

The film speed setting is important because it provides basic calculating information that will be used by the camera computer system. For example, a slow film (low ISO number) will require more light to expose the film. The reverse is true for a fast film. Forgetting to set or inaccurately setting the correct film speed can cause some major disappointments upon looking at the photographic prints from a processed roll of film. Once the film speed adjustment has been made, it should not be changed until the entire roll of film has been exposed.

Sometimes the photographer wishes to alter the camera film speed setting from the film's ISO speed. An example of this is setting the camera film speed adjustment to 400 when the film ISO speed is 200. This allows the photographer to take pictures where there is less light. This procedure is called pushing the film. With the camera being told that the film speed is twice as fast as it really is, the film is receiving half as much light during each exposure. To bring the full image out of the film, it is then necessary to increase the development time. The commercial processing laboratory must be informed in advance that a film roll is to be **push processed**. This means that the development time must be increased. There are instructions in most film processing manuals for photographers who wish to push process their own film.

Aperture

The camera aperture is found in the lens. It is an opening that allows light to pass through.

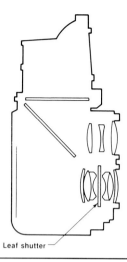

Leaf shutter

FIGURE 13–11 Leaf shutters are positioned within the lens and are often used in viewfinder and TLR cameras. *(Reprinted from Applied Photography by Ervin A. Dennis, © 1985 Delmar Publishers Inc.)*

Aperture size openings are controlled by f-stops that range in size from very small to nearly as large as the diameter of the lens glass. A typical series of f-stops range from f/22 (small) to f/1.4 (large) (refer to Figure 12–7, page 203). These various sizes of aperture openings give the photographer some choices before taking each picture. The f-stops can be set manually or automatically depending upon the camera, and how it is adjusted.

Shutter

A camera shutter controls when and for how long light will expose the film. It serves as a doorlike device that is located between the lens and the film. There are two basic shutter designs. These are the leaf shutter and focal-plane shutter. A **leaf shutter** is often located within the lens, and is made of several thin pieces of metal, Figure 13–11. Sometimes it is positioned ahead of the lens glass elements and other times it is located behind the lens glass elements.

The **focal-plane shutter** is used in most 35mm SLR cameras. It is positioned just ahead of the film, Figure 13–12. Two pieces of material called curtains make up the shutter and open and close when the shutter release button is activated. Focal-plane shutters are most often used with cameras that use interchangeable lenses.

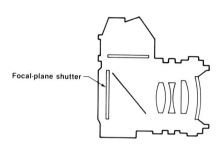

Focal-plane shutter

FIGURE 13–12 Focal-plane shutters are found in most 35mm SLR cameras and are positioned just ahead of the film. *(Reprinted from Applied Photography by Ervin A. Dennis, © 1985 Delmar Publishers Inc.)*

Typical Exposure Modes for 35mm SLR Cameras

Program: (P) The computer system selects the best aperture and shutter speed using the film speed and amount of available light. A camera with this feature is often referred to as "programmable."

Automatic: (A) The computer system selects either the best aperture or the best shutter speed depending whether the camera is equipped with **shutter priority** or **aperture priority.** Shutter priority means that the photographer selects the shutter speed for the picture; the camera computer selects the correct f-stop based upon the film speed and available light. The reverse is true for aperture priority cameras. Some 35mm SLR cameras have both priority systems.

Manual: (M) Here the photographer can override the camera computer system and personally make the f-stop and shutter speed settings. This mode is used for special effects and when the lighting might cause the computerized light meter to make incorrect camera settings. Almost anyone can take pictures when a computer selects the f-stop or shutter speed or both.

Flash: (F) This mode adjustment locks the camera into a specific shutter speed. This makes sure the shutter is open at the exact moment that the electronic flash is at full power. The shutter and flash synchronization speeds vary among cameras and flash units. Common speeds are 1/60, 1/125, and 1/250.

B: On early cameras, B was an abbreviation for bulb. This designation came into use when flash bulbs were used for artifical light instead of electronic flash. This setting allows the photographer to open and keep open the shutter as long as desired by using the shutter release. Night and low-light scenes often need long or timed exposure.

L: The abbreviation for lock. In this position the shutter cannot be activated.

FIGURE 13–13 An explanation of the six common exposure modes available with 35mm SLR cameras.

Shutter speeds (opening and closing) are measured in seconds or in fractions of a second. Most adjustable cameras have a range of shutter speeds from 4 seconds (very long) to 1/2000 second (very short). The most commonly used shutter speeds are 1/60, 1/125, and 1/250.

Exposure Mode

The photographer must be certain that the correct **exposure mode** is selected before taking a picture. A camera exposure mode engages the selected electronic system so the camera will function as the photographer desires. There are six common exposure mode settings on today's cameras, Figure 13–13. All of them but one, the lock mode, can and are used to expose film to take pictures. Each exposure mode has one or more purposes. Through use, the photographer will best determine which mode is right for each photographic situation.

Lighting the Scene

As stated before, there can be no photography without light. There are four basic lighting groups that provide enough light to take pictures. These are natural light or sunlight, electronic flash, flood lamps and existing light.

Natural Light

The source of all natural light is the sun. It is well known that without the sun there could be no human, animal, bird, or plant life. The sun directly provides light for most of the photographs taken throughout the world, Figure 13–14. In fact, film manufacturers have color balanced their films so true colors of objects illuminated by natural light will be reproduced in prints and slides. In such prints and slides grass is green not blue, apples are red not purple, and oranges are orange not yellow.

FIGURE 13–14 The sun provides light for most photographs. *(Courtesy of USDA, Forest Service)*

Electronic Flash

Flash photography has become very common. **Electronic flash** brings a burst of light on the scene just as the shutter opens. Many cameras made for and used by amateur photographers include a built-in electronic flash. The light-sensing system in these cameras determines when flash lighting is needed. This makes it easy and convenient to take pictures. In fact, it almost ensures successful pictures every time.

There are many kinds and sizes of flash units that attach to 35mm SLR cameras. Some electronic flash units are very powerful and are able to provide lighting up to 70 feet away. Most flash units, though, only provide enough lighting for scenes within 12 to 15 feet of the camera. Electronic flash units operate on electrical energy; thus, it is important to keep charged batteries installed whenever the flash will be used.

Electronic flash lighting is used by most professional photographers, Figure 13–15. Extensive control switches and transformers are needed to synchronize the flash units with the camera shutter release. This type of lighting often has low-level lamps that comfortably light the portrait subject(s) for preparation before the picture is taken. When all preparations have been made, the lamps will come to full power in microseconds when the photographer activates the shutter release.

Flood Lamps

Flood lamps are lights in fixtures attached to adjustable stands, Figure 13–16. The stands should be sturdy and be able to be extended several feet high. This gives the photographer flexibility when lighting is needed at various heights to provide light for the scene.

Incandescent light bulbs from 100 watts to 500 watts are often used for flood lamps. Another common light source is the quartz iodine lamp. This type produces a very bright light and operates much longer than incandescent bulbs.

FIGURE 13–15 Electronic flash lighting is often used by portrait photographers to get pictures like this one. *(Courtesy of Vivitar Corporation)*

FIGURE 13–16 Flood lamps are used to control the amount and direction of the lighting needed for a picture. *(Courtesy of Mail & Media)*

Photography Used by the Department of the Navy

Research in many fields is conducted by military personnel as they study outer space, military weapons, recognition systems, and reactions to metal under extreme heat conditions. The examples here show how high-resolution photography is being used to advance understandings in military science and technology.

A computer simulation of a series of 6-megaton nuclear explosions spaced 3 kilometers apart. Each blast is separated by 24 seconds. The lines represent atmospheric density increasing from red to blue, green, yellow, pink, and white, and then repeating the sequences. These photos were enlarged from 16mm negatives. *(Courtesy of the Naval Research Lab)*

False-color enhancement of an ultraviolet image of Halley's Comet. *(Courtesy of the Naval Research Lab)*

Lyman-alpha image of P/Halley 1986 March 13.46

NASA 36.017DL   NRL/UT

FIGURE 13-17 Pictures taken at night and in low light can result in excellent existing-light photographs. *(Courtesy of New York Convention and Visitor's Bureau)*

Existing Light

Existing light is light that is available from sources uncontrolled by the photographer. For example, whenever there is even the smallest amount of light, a picture can be taken. It may take a long exposure but by using a tripod to hold the camera still, the shutter can be held open long enough to expose the film, Figure 13–17. Some great photos have been taken by using existing light from street lamps. fireworks, building lights, and moonlight.

Taking the Picture

To get quality pictures, photographers must follow some basic guidelines. This is true for amateur, serious-amateur, and professional photographers alike. Once the basic guidelines are understood and used, they become part of the standard picture-taking behavior of the photographer.

Composing the Scene

The **photographic composition** of a picture is an important key to its value. Composition is described as a pleasing selection and arrangement of the elements within a picture. Photographic composition is based upon guidelines rather than firm rules, Figure 13–18.

The following guidelines and brief explanations can help photographers with composition. Each of these guidelines should be considered every time a picture is taken.

- See a photograph before it is taken. This means to look around and observe scenes that will look good as a picture. Sometimes the lighting is just right because of the time of day. Also, an event may be about to take place that has never happened or will never happen again. With camera in hand, always be ready to take a picture.

- Compose in the viewfinder. Looking at a selected scene through the camera viewfinder often helps to select the best composition. The viewfinder helps to frame the scene and to block out unwanted content. A good practice is to get as close as possible to the content of the picture scene and fill the viewfinder with the intended content.

- Create a center of interest. Select the most important feature of the scene content and make it stand out. It is much easier to look at a photograph that has less content as compared to one that has too much content. The viewers' eyes should be able to easily focus on the picture center of interest.

- Use framing techniques. Natural surroundings such as trees and buildings often serve as excellent framing items. Tree branches can be used to serve as an "in-picture" frame, thus helping to focus the viewers' eyes on the main content.

FIGURE 13–18 Most photographers will agree that composition is in the eye of the beholder. *(Photo by Elaine Fasano)*

■ Observe backgrounds. When composing the scene in the viewfinder, look closely at what could cause a picture to be unacceptable. The surrounding background is just as critical as the main subject. Trees, poles, stop signs, and the like should never appear to be growing out of the heads of people in a photograph.

■ Watch for motion. Movement of one or more subjects in a selected scene can easily be the cause of a poor photograph. If there could possibly be some movement such as by a small child or by an animal, it would be wise to use a fast shutter speed. **Panning** (moving the camera in synchronization with the subject) might be the solution to achieving acceptable results.

Depth of Field

The distance between the closest point and the farthest point that is in acceptable focus is considered **depth of field**. Sometimes a photographer will desire a large depth of field and other times a small depth of field. It is easy to control by using the aperture openings (f-stops) effectively. Small f-stops such as f/22 or f/16 will give the largest depth of field, whereas large f-stops such as f/1.4 or f/1.7 will give a shallow depth of field, Figure 13–19.

Squeezing the Shutter Release

It is very important to hold the camera still when taking a picture. Camera movement can occur at the time the shutter release is being activated. To keep the camera still, it is best to hold the camera with two hands. Upon looking through the viewfinder and keeping the camera

FIGURE 13–20 A camera should be held firmly and the shutter release should be squeezed, not pressed, when taking a picture. *(Courtesy of Pentax Corporation)*

FIGURE 13–19 A shallow depth of field results from using a large f-stop such as f/1.4. *(Courtesy of Pentax Corporation)*

firmly against your face, squeeze the shutter, Figure 13–20. This means that the hand should be positioned with fingers both on top and at the bottom of the camera. Pressing the shutter release from one side only will often cause the camera to move. This will cause some blurring of the image, especially with slow shutter speeds.

Bracketing Exposures

The **bracketing** technique is helpful in making certain that an acceptable picture will result when taking pictures of a given scene. The procedure is simple. The photographer takes a picture at the determined f-stop and shutter speed settings. Then, two more pictures are taken: one with one f-stop larger and the second with one f-stop smaller. It is also possible to leave the f-stop the same and increase the shutter speed by one for a picture and reduce the shutter speed for the final picture. This procedure gives the photographer three exposure settings for the same scene and one of them should be acceptable.

Using Filters

Filters can add interesting and exciting colors, shapes, and moods to a routine scene, Figure 13–21. Filters change colors, shapes, and images, and even add content to scenes. Many types and styles of filters are available that can be easily fastened to the camera lens. With through-the-lens light metering, the light absorbed by the filter will not cause an exposure problem. The best way to learn about filters is to experiment. Taking pictures with a variety of filters will give you the opportunity to study the results and make decisions for future pictures.

FIGURE 13–21 A special effects filter makes this beautiful scene even more striking. (Courtesy of Tiffen Manufacturing Corporation)

Summary

Taking pictures is fun, exciting, and valuable. The fun comes in knowing that the results — pictures — will give you, your family, and friends enjoyment. The excitement is in the thoughts of the unknown. Each time a picture is taken, there is a small mystery as to what the results will be. The unknown is there until the finished photograph or slide is available for viewing.

Pictures are valuable because they record a point in time. In effect, history is made every time a picture is taken because there is a preservation of a specific event or scene.

Camera cases and carrying straps are used to protect cameras during use. Cameras should also be protected from heat damage by not storing them in direct sunlight or near heaters.

Film should be loaded in low-light conditions. The ISO or ASA setting of the camera should be set to match that of the film.

When using a camera, care must be observed to ensure that the photographer does not place himself or herself in jeopardy.

Three types of lenses used include a normal lens, a wide-angle lens, and a telephoto lens. Zoom lenses are adjustable angle-of-view lenses.

Shutter speed and aperture are adjusted by the photographer or by an automatic camera to obtain the best possible photograph. The settings depend on the film speed and lighting conditions.

The photographer must compose the scene by creating a center of interest in the viewfinder, using elements of the scene as a frame. Care must be taken to compose a scene with an acceptable background, and to compensate for motion by selection of shutter speed or by panning.

To take high-quality and interesting pictures, it is important to handle the camera correctly. Lens selection, filter selection, camera adjustments, and lighting are critical to successful pictures. Every time a picture is taken, the photographer is directly involved in communication technology.

REVIEW

1. Who takes pictures and why?
2. Why is it important to use the carrying strap that is fastened to the camera body?
3. Camera batteries should be replaced at least how many times per year? Why?
4. When loading a camera with film, is it best to do so in low light or bright light?
5. When should a photographer be concerned about safety when taking pictures?
6. Name the three types of lenses that are often interchanged on 35mm SLR cameras.
7. Of the five important camera adjustments for picture taking, which one should not be changed for an entire roll of film?
8. Why are focal-plane shutters used most often in cameras with interchangeable lenses?
9. Which of four lighting groups will give the most accurate colors when taking pictures with color film? Why?
10. When taking an important picture, why is it a good idea to take bracketing exposures?

Darkroom Processes

KEY TERMS

color analyzer	fixer	processing trays
contact prints	graduates	safe attitude
darkroom	hologram	safelights
developer	photo-finishing lab	slides
drying cabinet	prints	stop bath
enlargements	print tongs	thermometer
enlarger	print washer	timer
film processing tank	processing chemistry	working solutions

Introduction

Looking at finished photographic **prints** and **slides** is both fun and informational. Prints, also called photographs or photos, are images in either black and white or color that have been made on photographic paper. Slides, or transparencies, are images generally in color that are made directly in the film. Slide images are seen by directing light with a special projector through the processed film and showing the image on a light, flat surface such as a projection

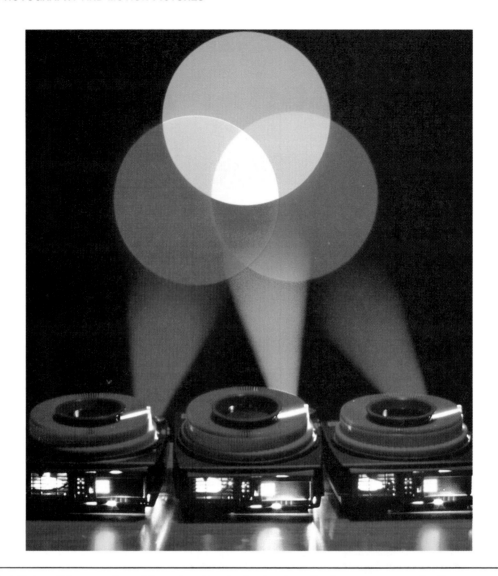

FIGURE 14–1 Color slides are viewed by using a projector to show the images on a projection screen. *(Courtesy of Eastman Kodak Company)*

screen or a light-colored wall, Figure 14–1. Slides can also be viewed with hand-held slide viewers.

Processing prints and slides must be done in an exacting manner or the scenes captured on film during picture making will be lost forever. Processing involves using specially formulated chemistry to develop black and white film and color film so the images can be seen. Processing also involves making the developed film and print images permanent so added light will not

cause added exposure to the photographic emulsions.

Film and print processing is done with specialized darkroom equipment supplies and tools. In fact, one of the most important needs for photographic processing is a light-tight space large enough to hold needed equipment, supplies, and tools. This space is called a **darkroom**. Most of the time the room contains special light that will not affect photographic print paper. Film must be handled in total darkness, but there are special light-tight processing tanks and cassettes that can be loaded easily and quickly. Because of this, darkrooms are totally dark for only short periods of time.

Darkroom Design

Several pieces of equipment are needed for processing photographic film and print paper. Each of the items must be located in the darkroom; therefore, the darkroom must be large enough to hold the equipment plus leave working room. A well-designed darkroom should have a dry and a wet side, Figure 14–2.

A light-tight door must also be included that allows the photographic darkroom worker easy access and exit. If the darkroom is designed for use by one person, a standard hinged and latched door will serve very well. For use by more than one person, there is need for an entrance and exit setup that will not allow white light to enter the darkroom by mistake. One type of door that meets these needs is a circular type, Figure 14–3.

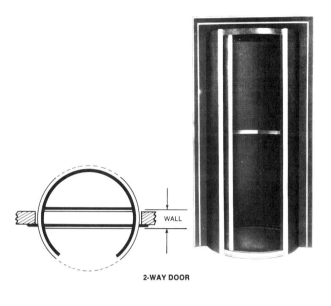

2-WAY DOOR

FIGURE 14–3 A circular door allows entrance and exit of the darkroom without letting white light pass through. *(Courtesy of Kreonite Inc., Wichita, KS)*

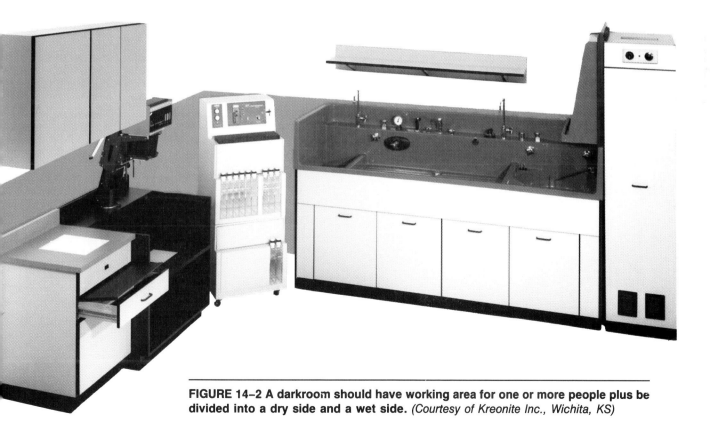

FIGURE 14–2 A darkroom should have working area for one or more people plus be divided into a dry side and a wet side. *(Courtesy of Kreonite Inc., Wichita, KS)*

This door has circular outer sides fastened to the wall with two openings. A center circular frame and sides unit is located inside. This unit has only one opening which allows a person to step inside the circular door, turn the center unit to an opposite opening, and step through.

Darkroom Equipment

The main item for most darkrooms is the sink. A darkroom sink should be large enough to fit several processing trays, one or more film processing tanks, chemical containers, and mixing tools, Figure 14–4. Temperature-controlled water should be available as film and paper processing needs quality control to give good results.

Although a common faucet assembly can be used, the best setup is to have an installed refrigeration and heating unit within the sink so the water can be kept at a constant temperature. The most commonly used temperature for photographic chemistry is 68°F (20°C). A darkroom sink must have hot and cold water. **Film processing tanks** are also needed. They are designed to hold one roll or several rolls of film. A tank is light-tight, but its design allows chemicals to be poured in and drained out with ease. Film is loaded into a tank in a 100 percent darkened room. Once the film is loaded in the tank, several processing steps may be done in a fully lighted area.

An **enlarger** is used to make photographic prints that are bigger in size than the film

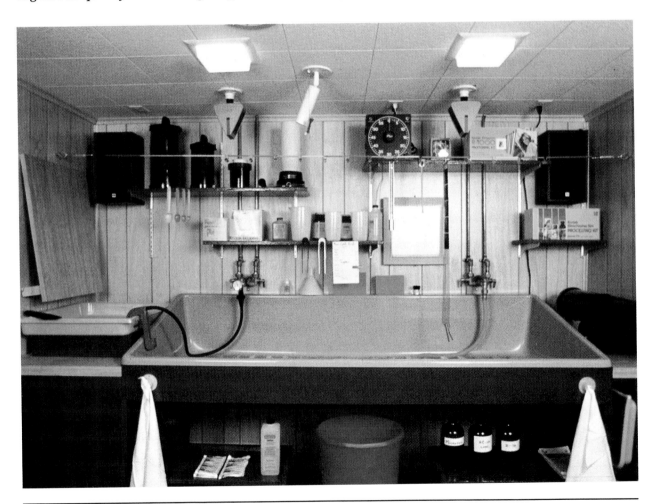

FIGURE 14–4 A darkroom sink is essential for hand processing of photographic film and print paper. (*Courtesy of Eastman Kodak Company*)

FIGURE 14–5 An enlarger is used to make photographic prints larger than the film negatives. *(Courtesy of Charles Beseler Company)*

FIGURE 14–6 Processing trays are used to develop, stop, and fix photographic prints. *(Courtesy of Eastman Kodak Company)*

negatives, Figure 14–5. A color enlarger contains a light source, a lens, and filters for making full color prints from color negative film. A typical enlarger can be used to make a print up to 11″ × 14″ in size and even larger.

Processing trays are necessary for developing, stopping, and fixing photographic prints, Figure 14–6. They are available in several standard sizes — 5″ × 7″, 8″ × 10″, 10″ × 12″, and larger. Also, they are made of white or colored plastic and stainless steel. The trays range in depth from 1″ to 3″ depending on the size.

Print washers are small sinks in which temperature-controlled water is circulated over the prints to remove unwanted processing chemicals. These units are made in different ways. One simple way is to outfit a processing tray with a small washing unit that is fastened

to a water faucet. This makes an economical but effective print washer for a photographer involved in low volumes.

To help speed processing, a film and print dryer is a useful piece of equipment. Sometimes a unit of this type is called a **drying cabinet**. A fan draws air into the cabinet through a filter. The air is heated and circulated throughout the cabinet. The heated air dries the film and prints in 15 to 30 minutes in a dust free environment. It is important to keep dust and other foreign particles from being embedded into the soft emulsions of film and paper. If this happens, the quality of the photographic products is greatly reduced.

All darkrooms should have **safelights**. A safelight is a light source that will not expose photographic film or paper but will allow the photographer to see and work comfortably in the darkroom. Safelights are assumed to be red, but yellow safelights are often used in print processing darkrooms. It is important for a darkroom worker to read and follow the safelight requirements for each photographic film or paper product used in a darkroom. An important thing to remember is that a darkroom should be as well lighted as possible. This makes it much easier for all users of this specialized area.

Darkroom Supplies and Tools

Critical to any photographic darkroom is the **processing chemistry**. This includes developer, stop bath, and fixer, Figure 14–7. Film and paper developers contain different formulations, but the main purpose of a **developer** is to make the exposed image visible. **Stop bath** is a mild mixture of acetic acid that stops or neutralizes the action of the alkaline developer. It usually takes only a few seconds for stop bath solution to do

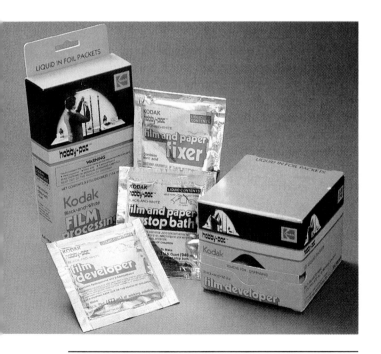

FIGURE 14–7 Developer, stop bath, and fixer are chemicals used in a photographic darkroom. *(Courtesy of Eastman Kodak Company)*

its work, but the process is important. Film and paper must be developed for specific periods of time. When the time period is completed, the developing action must be stopped quickly and completely.

Fixing solution is formulated to make the visible image permanent. **Fixer** has two major chemical mixtures, thus giving it two useful purposes. First, fixer dissolves and removes unexposed silver halide crystals from film and paper emulsions. Exposed silver halide crystals are turned dark by the developer. Unless the unexposed crystals are removed, they will also darken when further subjected to light. The second major purpose of the fixer is to harden the developed silver halide crystals. This action helps to ensure that the developed image will remain in place while the film is used in an enlarger or the photographic prints are being viewed.

Other processing chemicals include finishers to improve film washing and drying. These solutions are easy to mix and use as the specific directions are printed on the containers. Special finishers are also available for photographic papers that improve washing and help make prints lie flat after drying.

Graduates are containers used for measuring and mixing photographic chemicals. They are made in several sizes from 8 oz (250 ml), 16 oz (500 ml), to 32 oz (1 liter). It is best that the graduates be made of plastic for safety purposes although glass graduates are available. It is important that the measurement markings are easy to see while using them to make critical measures in the darkroom.

A very important tool for the darkroom worker is an accurate **timer**. A timer should be able to be set both in minutes and seconds. Also, it should be easy to see under darkroom safelight conditions. A good timer has both a large face for easy viewing and a bell or buzzer to signal the end of a pre-set timed period.

One or more **thermometers** are necessary in any darkroom. Temperatures of processing solutions are critical to quality results in photography. Thermometers should have measurement markings for both Fahrenheit and Celsius. Also, the numbers and the mercury line should be easy to read under safelight conditions.

FIGURE 14–8 The basic supplies and tools needed in a darkroom where film and paper will be processed *(Courtesy of Eastman Kodak Company)*

Some electronic thermometers have digital temperature readouts.

Basic darkroom supplies and tools are essential to film and paper processing, Figure 14–8. There are other items that a darkroom worker may wish to get after learning the basics of darkroom processing. The basic items, though, will be enough to gain a good understanding of photographic film and paper processing.

Darkroom Safety

People need to think about and practice safety every day of their lives. By thinking and acting ahead, accidents may be prevented.

Unsafe conditions can and do occur in and around photographic darkrooms. Darkroom workers must look ahead to what could happen to themselves or to someone else. There are five important safety practices that should be ob-

served by people who work in photographic darkrooms. These are listed and described as follows.

Attitude

This is the most important factor for a safe working environment. A person with a **safe attitude** is aware of the surroundings and what could happen if certain situations existed. A safe attitude makes a person able to look ahead and anticipate the unexpected. A safe attitude allows a person to think and practice safety at all times.

Eye Protection

Wearing splashproof goggles in a darkroom is a wise thing to do. This is especially true when chemical concentrates are being mixed into **working solutions**. A working solution is one that is ready to be used by the darkroom worker. Sometimes a working solution is even diluted more with water as is the case with some film developers.

Chemical concentrates, either the alkaline base used in developers or the acetic acid used in stop bath, can and will burn the soft tissue of the eye. Special care must be taken by anyone when these strong chemicals are being used. Once the concentrates are diluted with water, the injury possibilities are reduced, but this still does not remove the total danger of chemical burns to the eye. Wearing goggles in a darkroom is always appropriate. Also, a safety eye wash solution should be located near the darkroom.

Skin Protection

Because photographic chemicals contain alkalines and acids, it is important to protect your skin. When possible, fingers and hands should never be placed in photographic solutions. It is wise to wear thin rubber gloves especially if the photographer has sensitive skin. The use of **print tongs** keeps chemicals from coming into contact with fingers and hands. Print tongs, usually made of plastic, have what could be called two long fingers. The print tong fingers are squeezed together to pick up a print and relocate it in another tray. Frequent hand washing with mild soap and water helps keep photographic chemicals from irritating skin.

Body Protection

A well-dressed darkroom worker will always wear a plastic or rubber coated apron. If chemical splashes or spills occur accidentally as they sometimes do, an apron will protect both the clothes and the person. Another item that is wise to wear is a respiratory mask. This mask fits over the mouth and nose and is helpful in filtering out fumes given off by chemical concentrates.

Safe Procedures

Along with a safe attitude and thinking about safety, it is critical to practice safety. This means to keep the darkroom floor dry and free from objects that could cause a person to trip and fall. It means that bottles and containers holding photographic chemical solutions should be stored on low shelving or in floor cabinets. This is so containers do not fall from high wall shelves. Safe procedures include making sure all electrical switches and wiring are properly installed. Electrical shocks and burns can occur especially where liquids are involved. Finally, make certain there is enough ventilation in the darkroom. Fresh air is a must when people and chemicals are both in a confined area.

Processing Film

Continuous tone film processing involves several basic steps. This is true whether it is black-and-white or color film. Also, film must go through similar steps for both hand and machine processing. These major steps are preparation for processing, chemical processing, and finishing.

Preparation for Processing

The darkroom and area where film will be processed must be kept very clean. It is important to keep the floor, machines, tools, and cabinets ''hospital'' clean. Frequent washing, scrubbing, and vacuuming are very important. Cleanliness must be a natural part of photographic processing if quality results are to be expected. Therefore, it is important to take time before, during, and after processing to clean tools, equipment, and the working environment.

Preparing Chemicals

Complete mixing instructions are available on most of the photographic processing solutions. The first solution that must be prepared is the developer. For hand processing, it usually takes 8 oz (250 ml) of developer for one roll of film. If the film tank holds more than one roll, it is usually necessary to add 8 oz (250 ml) of solution for each roll. These amounts may vary; thus, specific instructions must be closely observed. The developer should be placed in a graduate ready for pouring into the film processing tank.

Color film processing may involve more than one developing step depending on whether the film is for slides or negatives. The Eastman

Kodak E–6 chemicals are mainly used for color slide film. For color negative film, the Eastman Kodak C–41 chemicals are mainly used. These two chemical solutions and procedures have been the industry standard for years. Most film and chemistry manufacturers follow these standards, but there are variations.

The next step in preparing the chemistry is the stop bath for black-and-white films and bleaches or conditioners for color films. Specific instructions by the film manufacturers must be followed for this important step. These solutions are made in such a way as to retard or stop the action of the developer.

Fixing solution must also be prepared before processing can be started. The strength of the fixer varies depending on the kind and type of film. For hand processing, position the graduate with the fixer along with the other processing solutions.

There are other solutions used to help in fixing, drying, and finishing film. These are called clearing agent, wetting agent, and stabilizer. Again, which solution is used depends on whether the film is black and white, color negative, or color positive.

Preparing the Film

After film has been properly exposed in a camera, the film must be kept from getting any more exposure to light. Also, it is important to keep exposed film from extreme heat and cold. Temperature extremes (too hot, too cold) can sometimes cause film emulsions to change structure. Exposed film should be kept in the plastic canisters and/or boxes in which it was first packaged. This will help remove the chances of damaging the film magazine or film roll, for example, light fogging the film.

Roll film, whether black and white or color, is usually placed in a processing tank for hand processing. In a 100 percent darkened room, the film is removed from the film magazine and loaded into the processing tank. After the film is in the tank and the lid has been attached, the processing tank can be brought into the lightened darkroom. In fact, the room can be totally lit with white light since the processing tank is "light tight."

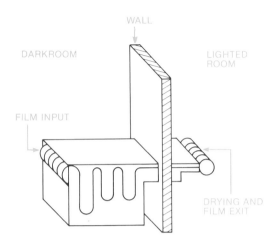

FIGURE 14–9 A common installation arrangement for a continuous tone film processor

For machine processing, the film processor is often installed in a wall within a darkroom, Figure 14–9. The drying section of the film processor can be in a lighted room. This makes it easier to remove the processed film and ready it for making prints.

Processing the Film

Film is removed from film magazines and spools and placed into the input end of a film processor. Several rolls of film can be processed continuously. The output of automatic film processors is quite fast. The processor takes the film through the specific chemistry needed to make the images visible and permanent. Depending upon the type of film, these chemical solutions are usually developer, fixer, and water wash. Also, the film is dried with heated air and ejected from the processor completely dry.

Using Holography

In order for the hologram to be visible to the naked eye, a photograph of the original hologram must be taken.

Film processing technology has led to many different developments in the field of photography. Holography is a current technology that is grounded in photographic principles.

Essentially holography is a method of creating a photographic image without using lenses to focus the light. The resulting image or photograph is called a **hologram**. The hologram picture has three visible dimensions, and is recorded on a sheet of photographic film or a photosensitive plate.

To create a hologram, a laser light beam that has a high degree of coherence (light waves that are consistent or in step with one another) and intensity must be used. The hologram itself contains all of the properties of the original image, but the recorded light waves cannot be seen by the naked eye. To make a photograph of a hologram, so the naked eye can see it, it is necessary to use a coherent laser light beam to reconstruct the original image.

Holography was invented in 1948 by a Hungarian-born scientist named Dr. Dennis Gabor. For his work, Gabor received the Nobel Prize in physics in 1971, twenty years after his invention. It simply was not immediately recognized as an invention with practical applications.

Today, holography is being used in several ways. Two major areas of use are in nondestructive testing of manufactured products and in the medical profession. Holograms can be made of physical materials such as metal, rubber, plastics, and glass without danger or damage to the material. The same is true when making X-ray holograms of human cells. Once the hologram has been made, scientists and technologists can thoroughly study the recorded information contained in the hologram without having contact with the original material.

Holography is not photography, but the photographic principles of light sensitivity are essential to the creation of a hologram. To see a hologram, it is necessary to make a photograph of the hologram image. As with many areas of science and technology, one highly technical process can be used to build upon another to create yet another useful tool.

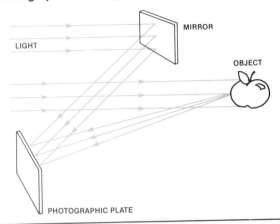

To make a hologram, light with a consistent wavelength (laser light) must be used to shine on both the mirror and actual object. The rays from both the object and the mirror are reflected on a photographic plate to form the holographic image.

Manufacturers use holography for nondestructive testing of their products. (Courtesy of Goodyear Tire & Rubber Co.)

For hand processing, the film is subjected to the prepared processing solutions of developer, stop bath, fixer, and one or two finishers. These solutions are poured in and out of the processing tank at certain times and temperatures depending on the kind and type of film. Total processing time through drying the film usually takes from one to two hours.

After processing either by hand or machine, the film is placed in protective sleeves. This is important so the film is not scratched or damaged, or becomes dirty through handling and storage. Sleeves are like envelopes and can be made of paper or plastic. Once the film is placed inside a sleeve, the processed film negatives can be viewed through the translucent paper or clear plastic. Film is usually cut into strips of 4 to 6 exposures, depending on the film size, before it is placed in sleeves. These film strips are a good length for easy handling when making prints in an enlarger.

After 35mm slide (reversal) film is processed, it is ready to be mounted into 2″ × 2″ slides. Slide mounts are most often made of cardboard

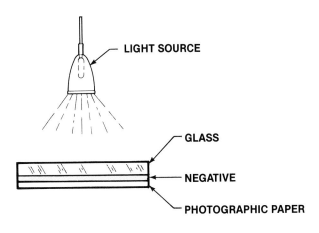

FIGURE 14–11 Contact prints are made when the negatives and light-sensitive paper are held tightly together. *(Reprinted from Applied Photography by Ervin A. Dennis, © 1985 Delmar Publishers Inc.)*

or plastic. Individual frames (pictures) of film can be mounted into frames by hand, but they are most often mounted in automatic machines, Figure 14–10. Slide mounting machines can be used to mount and label 6,000 slides per hour. Once the film is mounted into slides, the slides are packaged into boxes and shipped to the customers. At this point, the finished slides can be viewed with slide projection equipment, or on light boxes where large numbers of slides can also be stored.

Exposing Contact Prints

Photographic prints made directly from continuous tone film negatives are called **contact prints**. These prints are the same size as the film because the negatives and print paper are held tightly together while the exposure is being made, Figure 14–11.

Contact prints are very useful in determining the content of each picture. The photographer can inspect each picture for composition and image quality before spending the time, effort, and money to make enlargements. The importance of making contact prints is seen when quality results are desired.

FIGURE 14–10 Automatic slide mounting machines are used to cut film, mount, and label slides at a very fast rate. *(Courtesy of Byers Photo Equipment Company)*

FIGURE 14–12 Large photographic prints are a challenge to make but enjoyable to look at. *(Reprinted courtesy of Eastman Kodak Company)*

Exposing Enlarged Prints

Making **enlargements** is one of the most enjoyable parts of photography. Enlargements are photographic prints made from negatives that are several times larger than the original negatives, Figure 14–12. The most common sizes of enlargements are 5″ × 7″, 8″ × 10″, and 11″ × 14″. Enlargements are made in both black and white and in color.

Photographic Papers

There are two basic types of photographic paper: fiber base and resin coated. Both papers are made from wood and both have light-sensitive emulsions to accept the photographic image. The major difference in the two paper types is that resin-coated paper has layers or coatings of plastic resin on both sides of the paper. Plastic resin is a synthetic (human-made) product that resists moisture.

The two paper types can be compared by looking at side profiles, Figure 14–13. As can be

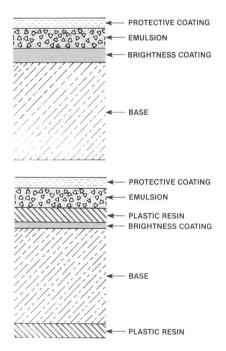

FIGURE 14–13 Profile views of the two basic photographic enlarging papers. *(Reprinted from Applied Photography by Ervin A. Dennis, © 1985 Delmar Publishers Inc.)*

seen, the major difference is in the two layers of plastic resin. The resin keeps most of the chemicals and water from soaking into the paper base during the processing cycle. Because of this, the paper stays flat while it is drying. Also, the emulsion is made in such a way so the surface will air dry without water spots. Fiber-based photographic paper must be dried with special heated drying units after processing or it will curl and have unsightly water spots.

There are several paper surfaces, but the two most popular are glossy and matte. The glossy paper surface is bright and shiny. It is used by most amateur photographers. It also gives excellent results for the printing in books, magazines, and newspapers. The matte surface has a dull, nonshiny appearance. It is often used for portrait prints.

Making an Enlargement Exposure

For black and white prints, the procedure is rather simple. The film negative is placed in the enlarger head in a special negative holder. The enlarger light is turned on and the image is shown on the base of the enlarger, Figure 14–14. An easel which holds the photographic paper in position is usually placed on the enlarger base.

Of course, the darkroom white light must be turned off and the safelight turned on when this work is being done.

The image is visually focused on a white piece of nonphotographic paper that has been placed in the easel. A small piece of photographic paper is then placed in the easel. A test print is then made by giving several areas, usually 3 to 5, of the image different exposures. The test print is processed (described later) and the results are inspected. The best exposure is selected from the test print and a full print is then exposed using that exposure. After processing, the print is ready for use.

Amateur color enlargements are made in the same basic way. The color negative is placed in the enlarger just as with the black-and-white negative. The enlarger should be designed to hold special color filters of yellow, magenta, and cyan. Color printing filters for black-and-white enlargers are available, but a manufactured color enlarger is much better to use. A **color analyzer** is then used to find the percentage of each of the three filters that must be used to create the correct colors in the print, Figure 14–15. A color

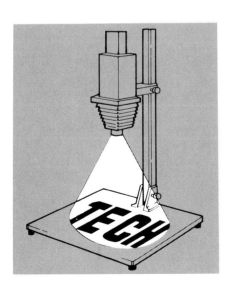

FIGURE 14–14 An image is projected onto the base of an enlarger when a negative is in position.

FIGURE 14–15 A color analyzer is used with a color enlarger to make quality color photographic prints. *(Courtesy of Durst Phototechnik GmbH)*

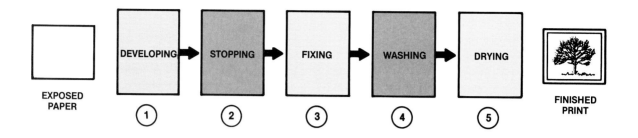

FIGURE 14–16 The five steps in processing black and white photographic prints by the tray method *(Reprinted from Applied Photography by Ervin A. Dennis, © 1985 Delmar Publishers Inc.)*

analyzer is an electronic device that is used to correctly evaluate the colors in the color film negative. It is much more accurate than the human eye. One or more test prints are exposed and processed and usable prints can then be made rapidly.

Processing Photographic Prints

There are five basic steps in processing photographic prints. These five steps are developing, stopping, fixing, washing, and drying, Figure 14–16. Each of the steps must be done in the right way so the resulting prints will be of high quality.

Black-and-White Prints

The photographic chemistry for black-and-white prints is mixed according to the instructions. Correct amounts are poured into processing trays. There are usually four trays arranged in a processing sink in a left-to-right order — developer, stop bath, fixer, and water. The developer is designed to turn the exposed image areas dark. The stop bath "stops" the action of the developer. The fixer solution makes the image permanent by dissolving the unexposed silver halide crystals and hardening those that were exposed and developed. The circulating water is used to wash away the processing chemicals before the print is dried.

Color Prints

Color print processing must be accomplished in total darkness. Thus, it is usually done in a different way than black-and-white print processing. One system used for color prints is the processing drum and motor unit, Figure 14–17. With this unit, a piece of exposed color photographic paper is placed into the drum while the dark-

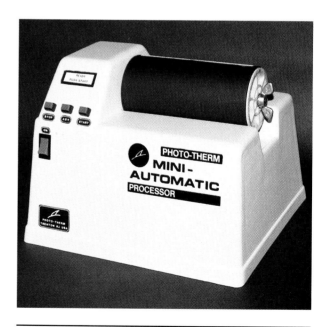

FIGURE 14–17 A color print processing system in which a processing drum and revolving motor unit are used *(Courtesy of Roman Kuzyk)*

FIGURE 14–18 A professional model exposure unit used to expose prints at a high rate of speed. This unit can make 1,898 prints in one hour. *(Courtesy of Photo Control Corporation, Minneapolis, MN)*

FIGURE 14–19 A computer-controlled color photographic print processor *(Courtesy of Colenta America Corporation)*

room is completely dark. Once the end-cap has been placed on the drum, the white lights can be turned on and the processing can be done. Each solution from the developer to the water wash is poured in and out similar to processing roll film in a tank. The motor unit turns the drum to evenly distribute the solutions over the whole print.

Commercial Printing and Processing

Specially designed equipment is used to expose and process photographic prints. An exposure unit used by professional color labs (a facility that produces the highest quality processed negatives and prints) often has several lenses and precise controls, Figure 14–18. Units such as this can be used to make hundreds of exposures per day on rolls of color photographic paper with each exposure being very precise. After the roll of paper is exposed, it is then processed in a quality-controlled print processor, Figure 14–19. Processors of this type carry the exposed photographic paper through the same basic processing solutions as described for black-and-white prints. The solutions have different mixtures that are suitable for color photographic paper, but the process is quite similar.

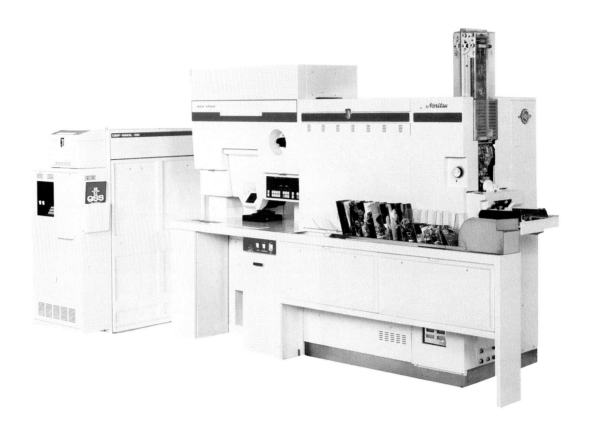

FIGURE 14–20 A photo-finishing system that is used to process color negative film and make color prints for amateur photographers. This unit can make 1,898 prints in one hour. *(Courtesy of Noritsu America Corporation)*

A **photo-finishing lab** is the type often located in or near public buildings. An example is a photographic processing laboratory located in a busy shopping center. People can bring in their exposed roll of color negative film and within a short time, usually one hour, the film is processed and prints are made. The equipment is designed for fast service and acceptable quality by the amateur photographers, Figure 14–20. It will not meet the color-control specifications of professional color labs.

Summary

Darkroom processes are important in making finished photographic prints and slides. This chapter described the equipment, tools, and supplies used by both amateur photographers and commercial labs to process film and make photos. Darkroom processing of film and prints must be done in an exact way. This is something that can be learned by anyone who can read and follow directions.

Safety is an important part of darkroom processing. Just thinking about safety is not good enough. Safety must be practiced each time work is done in a darkroom. Photographic chemicals in their concentrated state can cause serious skin and eye burns. Whenever photographic solutions are being mixed from concentrates to working solutions, splashproof goggles must be worn over the eyes. Also, it is important to protect your skin and body through use of continued safe procedures in the darkroom.

Both film and prints are processed in a similar way. The chemical solutions are made specially for film and for photographic papers, but the processing steps are similar. Developing must take place, then the developing action must be stopped with a mild acetic acid called stop bath. After this, the film and print paper are fixed to make them permanent from any added exposure when brought out into the light. Water is used to wash away the remaining chemical solutions from the film and prints. They are then dried and ready for use. Other solutions are used with some film and prints because of their special nature.

Remember that correct darkroom processing procedures for film and prints are important to successful photography. If film is not processed correctly, a photographer's work will not be successful. The same is true when photographic prints are made. Quality must be carried through from beginning to end.

REVIEW

1. Why must photographic film and prints be processed with care?
2. Name the seven pieces of equipment found in most darkrooms.
3. What is the function of the developer, stop bath, and fixer in the processing of photographic film?
4. Why is a drying cabinet used to dry a processed roll of film?
5. Why can't a red safelight be used in the darkroom when loading black-and-white and color film in a processing tank?
6. Why is it wise to wear eye protection in the form of splashproof goggles when mixing working solutions from chemical concentrates?
7. When should print tongs be used to transfer photographic prints from one processing tray to another? Why?
8. Why is film placed in protective sleeves after it has been processed?
9. What is the major difference between contact prints and enlargement prints?
10. How is a commercial photographic laboratory (a professional color lab) that produces high-quality film and print processing different from an amateur darkroom? How are they similar?

CHAPTER 15

Motion Pictures and Animation

OBJECTIVES

After completing this chapter you will know that:

- We see motion in motion pictures because of persistence of vision.
- Motion picture cameras and movie projectors normally advance film at 24 picture frames per second.
- The three standard widths of motion picture film are 8mm, 16mm, and 35mm and it is available in nearly any desired length.
- Processing motion picture film and making a finished print copy takes several exacting steps.
- Film editing takes considerable time and expert skill by one or more editors.
- There are several playback options with movie film.
- Many people are involved in making motion pictures besides the actresses and actors.
- Animation procedures have become very complex and computers are being used to produce and control the individual cels.

KEY TERMS

animation	editing	screenplay
casting	fast motion	screen writer
cel	nickelodeon	shooting
cinema	persistence of vision	slow motion
cinematography	producer	special effects
circular shutter	projector	take
director	props	tripod
dolly		

FIGURE 15–1 The Vistacruiser, a "motion control" camera that runs on tracks and whose motions are repeatable using a computer-controlled system. When used to film spaceships, the camera moves toward the stationary spaceship model. *(Courtesy of LucasFilm Ltd., TM & © LucasFilm Ltd. (LFL) 1987. All rights reserved.)*

Introduction

Most people enjoy watching motion pictures because they usually tell a story. In reality, a motion picture is a series of still pictures that are shown to the viewer in rapid order. This gives the feeling that there is movement. The continuous action of the long strip or roll of imaged film flashes hundreds of individual pictures on a screen in a short amount of time. The rapid succession of pictures makes it possible to see motion as it is happening. The individual pictures appear to be continuous because of **persistence of vision**. This is the ability of the eye to hold an observed image for a split second before the next image comes into view.

True motion pictures were made possible by the development of long strips of flexible film in 1889 by George Eastman. During the same time period, Thomas Alva Edison was busy designing a movie camera that could be used to take pictures with the new Eastman film. Several attempts by other inventors in the United States and Europe during the same time period did not meet with the success of the great inventors

Eastman and Edison. There were though, many people in the United States, England, France, and Germany who made major contributions that advanced motion pictures rapidly.

The first movie studio in the world was a tar paper-covered building in West Orange, New Jersey. It was built and used by Thomas Alva Edison. Short movies were made in this studio that were shown to the public in the first movie theaters, called **nickelodeons**, where the price of admission was five cents. The first movies were very short—less than a minute—but soon scenes were put together to make movies that were several minutes long. Nickelodeons were very popular during the first decade of the 1900s.

Movie theaters soon spread throughout the United States and feature length movies were made. It was not until 1927 that talking or sound motion pictures were made. Several inventions had to take place during the first quarter of the century to make it possible to combine sound with the moving pictures. The era of the "silent film" was over and improvements in movie making advanced faster than ever. Today movies are made with the help of computers, Figure 15–1.

In this way images never before imagined can be produced. Today motion pictures can be made that stretch the imagination of the most creative minds.

The term **cinematography** is sometimes used in place of "motion picture." Cinematography is defined as the art and science of motion picture photography. A cinematographer is a person who is involved in every stage of a film's creation except the initial script writing. The "**cinema**" in the term refers to a motion picture theater.

Motion Picture Cameras and Equipment

Motion picture cameras are complex pieces of equipment, Figure 15–2. They are made to move

long rolls of film through the process of being exposed. A movie camera can take 24 pictures every second. Each of these pictures is called a frame and a motion picture is made up of thousands of frames.

A side, cut-away view of a movie camera shows how the unexposed film travels through the camera and becomes exposed film, Figure 15–3. The lens gathers and focuses the image on the film plane in the same way as a lens on a still camera.

A major difference between a still 35mm camera and a movie 35mm camera is the shutter. A still 35mm camera has a vertical or a horizontal shutter that opens and closes to allow the light carrying the image to expose the film. A motion picture camera has a **circular shutter** that revolves as the film moves through the camera, Figure 15–3. This type of shutter spins on the center shaft letting image light pass through to expose the film. Most moving picture cameras have shutters that are adjustable, thus letting more or less light pass through.

FIGURE 15-2 A motion picture camera (Courtesy of Panavision® Incorporated)

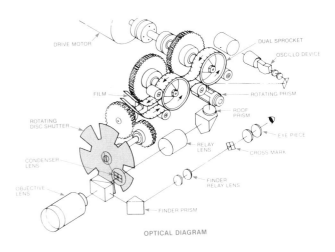

FIGURE 15-3 An optical diagram of a motion picture camera. Note the adjustable circular shutter used in a motion picture camera. (Redrawn with permission from Instrumentation Marketing Corporation)

High-Speed Motion Picture Cameras

Photography is used every day to record events that may never happen again. Look at the photographs in this article. How often does a very large plane come in for a "belly" landing? This does not happen very often, thank goodness. When it does happen, aircraft engineers, fire fighters, and medical personnel must know what to expect. Motion picture film is used to store the unusual events in the light-sensitive emulsion for later use.

What happens when two cars crash head-on? Auto manufacturers stage accidents such as the one shown here and then record the event with several high-speed motion picture cameras. Notice the cameras mounted on the sides of both cars. Cameras are positioned at several angles so a complete analysis of the crash can be studied. Based upon the findings, changes in the car design may be made and passenger/driver safety will be improved.

High-speed motion picture cameras are often used with aircraft for research purposes. For the popular movie, *Top Gun*, a high-speed camera was mounted in the cockpit area of an F-14 pointing to the rear of the plane. Now you can understand how it was possible to see the "chase" plane.

(Photos courtesy of Instrumentation Marketing Corporation)

FIGURE 15–4 The Nikonflex on the miniature mine car track for *Indiana Jones and the Temple of Doom*. The sets were made of crushed and painted aluminum sheets which could be folded over to make tunnels. *(Courtesy of LucasFilm Ltd., TM & © LucasFilm Ltd. (LFL) 1986. All rights reserved.)*

There are many makes and models of motion picture cameras. When film travels through a motion picture camera, the film stops for an instant in front of the lens to receive the exposure. If the film did not stop for a moment, the processed film would be nothing more than a blur when it was viewed with a movie film projector. The circular shutter is timed perfectly within the camera mechanism so the opaque portions of the blade are covering the lens opening when the film is being moved. In this way, the next picture can be taken. This action keeps the film from being exposed at the wrong time.

The shutter speed is 24 frames per second (fps) for regular motion pictures. To show a moving object in slow motion, the camera speed must be increased. Some motion picture cameras have adjustable shutter speeds to allow this variance in the normal operation. In sports photography, filming speeds are often increased to 70–120 fps for **slow motion** pictures. For action analysis, coaches will ask that the

filming be done at 250–500 fps. When the processed film is viewed at the standard 24 fps, the action can be seen much easier because more individual pictures were taken of the movement than usual.

The reverse of the slow motion procedure is to operate the camera slower than the standard. When viewing the processed film at 24 fps, there is the feeling of an increased tempo or **fast motion** of the scene. For example, it gives a car chase scene more excitement than usual. The reason is that more action takes place in the scene between each frame than when the action is filmed at regular speed.

Motion picture cameras are continually being adapted to create a variety of **special effects**. A 35mm Nikon camera was redesigned to operate as a motion picture camera for the mine car chase in *Indiana Jones and the Temple of Doom*, Figure 15–4. Since a smaller camera meant smaller sets, the Nikonflex camera helped to reduce production costs.

FUGURE 15–5 Underwater camera and housing used to film miniature whale models for Star Trek IV. *(Courtesy of LucasFilm Ltd., TM & © LucasFilm Ltd. (LFL) 1986. All rights reserved.)*

Motion picture cameras have also been adapted for use underwater. The underwater scenes in *Star Trek IV* would not have been possible without an underwater camera and the housing designed to keep it waterproof, Figure 15–5.

Motion Picture Film

Motion picture film is like regular photographic film except that it is in very long strips instead of short strips. For example, motion picture film is available in rolls from as short as 50 feet up to lengths of 1,200 feet. The most common lengths are 200 to 500 feet. Film for 35mm still cameras is only 3 to 4 feet long depending upon the number of exposures. These long lengths need large, light-tight film cases made of plastic or metal, Figure 15–6. Film is wound on a spool and placed in a film case so unwanted light cannot reach it.

There are three standard widths of motion picture film. These are Super 8mm, 16mm, and 35mm, Figure 15–7. For some special photography, 70mm width film is used, but this takes

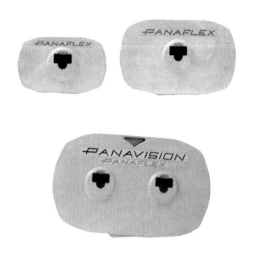

FIGURE 15–6 Light-tight film canisters are used to hold rolls of movie film both before and following exposure in a motion picture camera. *(Courtesy of Panavision® Incorporated)*

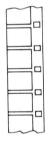

SUPER 8mm

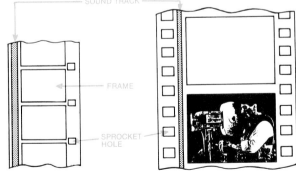

16mm WITH SOUNDTRACK   35mm WITH SOUNDTRACK

FIGURE 15–7 The three standard widths of movie film.

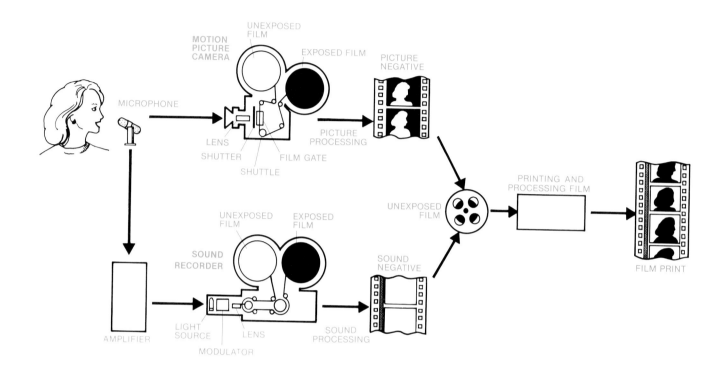

FIGURE 15–8 The basic steps in producing a motion picture print that contains recorded sound.

much larger equipment and the results do not often justify the cost. Home movies are usually photographed on Super 8mm film because of the lower cost, and the processed image is projected on small movie screens. The next larger size, 16mm, is most often used for educational films that are used by teachers in school classrooms.

Film used for motion pictures in a theater is almost always of the 35mm width. This film gives a large enough image on the film so it can be enlarged several thousand times for showing on a movie screen. The rectangular holes called perforations along the film edges are used to advance the film through the camera, the processing equipment, and the movie projector. There is space on both 16mm and 35mm films to record sound. This is an optical process. Sound waves are printed on the films as lines of varying amplitude. The motion picture projector decodes the optical signal and converts it to sound.

Motion Picture Film Processing

Film used for motion pictures is processed in much the same way as film used in still pictures.

This is true whether it is black-and-white, color, color reversal, or color negative film. It is important to use the correct photographic chemicals for the specific film being processed. Professional photographic film processing labs that handle motion picture work need only look at the label on the exposed film canister to determine the film brand and type. They can then process the film according to the procedures most often specified by the film manufacturer.

The original film footage of motion pictures that will be used commercially is never sent out to the user who will view the pictures. Duplicate films are made of the original for consumer use. In this way, hundreds or even thousands of prints can be made from the original so all of the copies are the same.

Original film footage is most often photographed on negative color film. From the processed negative film, color positive film is made through contact exposure. This not only saves the original negative film from damage, but it assures accurate exposure of the duplicates that will be used by consumers throughout the market distribution area.

Most of the time, the images and sounds of professional-quality motion pictures are recorded separately and then combined to make the positive film print, Figure 15–8. This procedure is used to ensure the quality of the sound. When the motion picture film is being exposed, there is often too much unwanted sound that should not be a part of the final prints. The actors and actresses will say their assigned lines or sing their songs while the motion picture film is being taken. Afterwards, the actors and actresses will restate or sing their sound scripts in a soundproof recording studio. This "pure" sound is recorded on the sound negative film. When the positive print is made, the picture and sound films are combined.

Sometimes the sound is recorded during the actual filming sessions. The recording is still made on separate film for later combining with the visible imaged film. This procedure also gives the film producer and director the chance to have special effects added to the sound or to the picture negative films. Special effects in the picture may be a fire or a car crash scene that could not be done with the actors and actresses present. Also, special sounds such as music or the whistle of a train not in the picture would have to be added to make the story complete.

Film Editing

Editing is often a long and slow process. It involves the complete review of all the footage that was shot on photographic film and selection of the best of each scene to be included in the final print, Figure 15–9. The editor must be a very skillful person and one who is very knowledgeable of the program script. The editor must look at every detail before making selections. For example, the editor should look at each person and determine if his/her facial expression is appropriate for the program script. Also, the editor checks the many technical aspects such as lighting and position of the **props**. Props are such things as tables, chairs, and cars that are important to the scene.

FIGURE 15–9 An effects editor, responsible for organizing all the elements in an effects shot, here aligning a series of elements using a synchronized frame counter. *(Courtesy of LucasFilm Ltd., TM & © LucasFilm Ltd. (LFL) 1986. All rights reserved.)*

FIGURE 15-10 An editing system that is completely computerized except for the human-required editorial work. *(Courtesy of CMX Corporation)*

Film editing equipment is designed to allow for content selection and then to make it convenient for a final "original" film print to be made. Completely computerized equipment used for film editing includes one or more screens to view the images, a keyboard, several special controls and switching devices, plus playback equipment for the film, Figure 15-10. This equipment makes the editing job faster and more accurate; however, editing is a human task that cannot be assigned to a computer.

Playback Options

Motion picture film that has been recorded on 16mm or 35mm film is most often shown to audiences either in the educational setting such as school classrooms and at public theaters. Specialized playback equipment called **projectors** are used for motion pictures so the images are a faithful reproduction of the original scenes

that were filmed, Figure 15-11. The film must advance through the projector at exactly 24 frames per second, which is the same speed of the movie camera. In this way all scene movements will appear natural to the viewer.

Commercial movie theaters must use larger projection equipment than that used for educational needs in school classrooms, Figure 15-12. First, the film width is 35mm and the lengths can be 4,000 feet or longer. Also, the image is enlarged several thousand times on the large movie screen. The projector equipment needs to contain a powerful lighting system as well as an excellent quality lens.

In most commercial movie theaters, a minimum of two large, heavy-duty projectors are used to show a full-length movie. Most motion pictures need two or more rolls (reels) of film to

FIGURE 15-11 Motion pictures are projected on a movie screen with a sound film projector so the viewer can see the image and hear the recorded sound. *(Photo by John F. Guthier)*

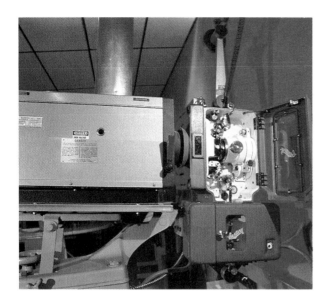

FIGURE 15–12 A large, heavy-duty projector is needed to show motion pictures on large screens at movie theaters. *(Photo by Joseph Schuyler)*

cover the entire film story. At exactly the correct moment, projector 2 will begin showing reel 2 of the movie when the final frame of reel 1 runs out on projector 1. Computerized film projectors make this switch-over look easy, but it still must be done in a smooth manner so the audience does not see the reel change.

Making Motion Pictures

Motion picture cameras must be handled very carefully when exposing film. Most of the time these cameras are set on heavy-duty tripods or dollies that hold them secure. A **tripod** is stationary and can only be moved by lifting and carrying. A **dolly** has wheels for ease in moving. Many times a dolly is large enough to hold not only a camera but the camera operator as well, Figure 15–13. An arrangement like this makes it easier to get in the correct position for taking the pictures.

FIGURE 15–13 A large dolly tripod is used to hold the movie camera and the camera operator. *(Courtesy of LucasFilm Ltd., TM & © LucasFilm Ltd. (LFL) 1986. All rights reserved.)*

Multiplex Cinemas

A recent trend in the movie industry is to have several movie theaters showing different pictures to multiple audiences, rather than one large movie theater showing only one film. Ten or more theaters in one building are not uncommon. The advantages that these multiplex theaters offer to owners are that by showing different pictures, wider audience appeal and larger average audiences are possible. Also, using new techniques, fewer projectionists can take care of a large number of projectors, reducing operating costs.

In a multiplex theater, one large projection booth houses the projectors for all the theaters. Each projector shows one movie in one theater. The film for each movie is shipped to the theater wound on small reels that contain enough film for twenty minutes. The film on the different reels (typically six reels) is spliced together so that the entire movie will show continuously. This very long length of film is wound on a large metal platter, with the beginning of the film at the middle of the disk.

The film is pulled out of the middle of the spool and through the projector, where a powerful 1600-watt Xenon lamp shines through it, projecting the image on the screen in the theater. Sound information is also extracted from the film and played in the theater. The film is then wound on another platter in the same stack as the first one. In some projection rooms, the beginning of the film is spliced to the end of the film, forming a continuous loop. A film shown from a continuous loop is ready to start again as soon as it has ended.

The use of new technology to increase productivity in the theater industry has helped to keep costs down at theaters and make more films widely available for viewing.

(Photos by Robert Barden)

It takes many people with special talents and abilities to make a motion picture film. In fact, there are about 275 different jobs in the motion picture industry. These jobs range from carpenters, painters, electricians, and computer specialists to the actors and actresses. Other important jobs are those of the hair dresser, make-up artist, costume designer, and musicians. Each of these persons plus many more have specific job descriptions so they know exactly what must be done to make a movie.

Actors and actresses are the only people seen on film, but without the large number of specialized people there would be no motion picture. Three very important positions in motion picture making are the producer, director, and screen writer. The **producer** is the key person in making a movie. This person is in charge of selecting the other important employees for positions, including the **casting** of the actors and actresses. Casting is the process of selecting the best people to fill the roles of the men, women, and children who will appear in the movie. The producer also has charge of the budget and plans the daily schedule for **shooting** or taking the movie.

The **director** is responsible directly to the producer and often helps the producer in many decision-making tasks. The director helps the actors and actresses with their rehearsing and instructs them in how the picture must be made. Directors control the events taking place on the stage, both inside the studio building and outside on a street. The director and the producer decide when a shooting can be considered a **take**. A take is a scene or footage of film that is considered acceptable quality from both the acting and technical aspects. Some scenes are filmed several times before they are approved as a take.

Screen writers are important people as they prepare the **screenplay**, the story on which the movie is based. Often a movie is based on a novel, a stage play, or a biography of an important person. A screen writer must rewrite the original manuscript so it will fit the needs of the motion picture. Often it takes several screen writers to prepare the needed script for the actors and actresses.

A motion picture shooting session can last for a few hours to many hours depending on the difficulty of the scene. A typical scene includes much equipment as well as many people. Besides the camera, there are the lighting and sound systems. A scene must have precise lighting so the film can be exposed properly. Large numbers of lights are needed similar to those used to light athletic fields. Sound equipment must be available to capture the voice and song of the actors and actresses plus the musical instruments.

Animation Procedures

The concept of animation dates back to the 17th century in China. Work with animation continued throughout the 1700s and 1800s by both English and French inventors. It was not until the invention of the movie camera by Edison in the late 1800s in the United States that the technique of animation was possible. The first animated movie was a short cartoon made in 1906. Since then, animation has reached a level never dreamed of by the early inventors and filmmakers.

Animation is a photographic and filmmaking technique that gives the viewer of a movie the illusion that objects which should not move, do in fact, appear to have movement on the screen. Animation simply stated means movement. It is a very complex process to show a cartoon character such as a rabbit (Bugs Bunny™) running, jumping, or talking. Thousands of individual scenes must be made for even a short cartoon movie that is seen at theaters or on television.

Creating an Animated Film

An animated motion picture is usually short in length — from a few minutes up to 20 or 30 minutes — such as those used for cartoons. There are, though, full-length animated motion pictures that run in length from 90 to 120 minutes. Whatever the length, there have to be graphics and a story. This calls for sketch artists, designers, and writers to create the full story that will be followed during production.

To make a standard animated film, an individual **cel** must be made for each frame exposure

FIGURE 15–14 An example of an individual cel, one of thousands needed to make an animated film. *(Courtesy of United Feature Syndicate)*

with the movie camera, Figure 15–14. A cel is an individual picture that has been drawn, painted, or made with a computer. It takes 24 cels to fill one second of time. (Remember, movie cameras and projectors run at 24 frames per second.) Thus it takes 1,440 cels for each minute of film time. Many techniques and procedures have been devised to produce the required thousands of cels, but the main ingredient is the artist, Figure 15–15. For most animated movies, several artists are employed to make the cels. These people work long hours at their computers or drawing tables to produce the images that will be viewed in but a few minutes.

After the cels have been prepared, the shooting of the film can begin. To expose an animated film, a computer-controlled movie film or video camera system is used to expose each cel, Figure 15–16. For the exposure process, each cel must be positioned correctly and quickly so the photographing can be completed in a reason-

FIGURE 15–15 This animation artist is at work preparing cels that will be used to make an animated movie film. *(Photo courtesy of Liquitex Art Materials)*

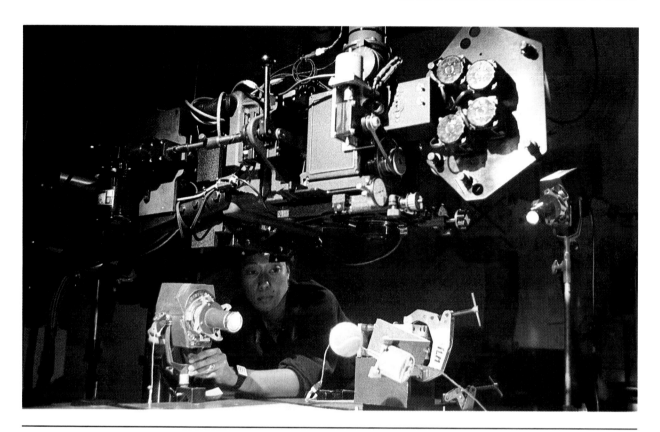

FIGURE 15–16 Animation cameras were used to film the tennis balls for *The Witches of Eastwick*, **where the actors pretended to play tennis. The actors were filmed first, then the tennis balls were shot to match the actions of the players.** *(Courtesy of LucasFilm Ltd., TM & © LucasFilm Ltd., (LFL) 1986. All rights reserved.)*

able amount of time. Even with the special computer-controlled camera systems, shooting an animated movie film can take days of work.

Following the shooting, the film editors will review the footage and determine the required editing. With animation procedures as refined as they are, there is usually very limited editing needed. Computer-assisted cel production has made animation procedures easier in recent years and has also reduced the required re-editing and/or retouching of the film footage.

Summary

Motion pictures are both enjoyable to view and very informative. They give people a great many experiences via the movie screen or the television receiver that they would not have any other way. It is much easier and less expensive to watch a movie travelogue about a country in Africa or South America than to travel there. Movie cameras can be carried and used to record events in the world in which we live. Motion picture cameras have been on the moon and in outer space, and deep below the surface of the ocean. This has shown us much about our environment.

A motion picture is a series of still pictures that are shown to the viewer in rapid order. The individual pictures appear to be continuous because of persistence of vision, the ability of the eye to hold an observed image for a split second.

Motion picture cameras take 24 pictures per second on a long, continuous roll of film. They use circular shutters that revolve as the film moves through the camera. Special cameras can operate faster or slower than 24 frames (pictures) per second, resulting in slow motion or fast motion pictures.

Motion picture film comes in three standard widths: super 8mm, 16mm, and 35mm. Generally, super 8mm is used for home movies (amateur photography), 16mm is used for educational movies, and 35mm is used for commercial movie theater production.

Motion picture film is shown using projectors that operate at 24 frames per second. The projector shines a light through the film transparency and uses a lens to focus the image on a flat viewing surface called a screen.

Hundreds of types of people can be involved in the making of a motion picture. The producer has the overall responsibility for all aspects of making the film. The director is responsible for guiding the actors, actresses, and camera crew in the actual shooting of the movie.

Animation is a photographic and filmmaking technique that gives the viewer of a movie the illusion that a drawn image has motion. Thousands of individual scenes must be made for even a short cartoon movie. One individual scene (cel) is drawn for each frame exposure of the movie film. There are 24 cels (individual drawings) for each second of animated film.

It takes creative and skilled people to produce motion pictures. Along with people dedicated to their work, it is necessary to use high-level equipment including advanced cameras, film processors, and projectors. Animation is an important part of the motion picture industry. This special process allows viewers to see imaginative programming of creatures and objects that do not have motion in real life, but seem to move through the magic of animation.

REVIEW

1. Why is persistence of vision an important part of watching a motion picture?
2. What were the first movie theaters called? Why?
3. How does the shutter in a movie camera differ from the one in a still camera?
4. How many individual frames or pictures are normally taken in a second with a movie camera? Why must this be the same in movie projectors?
5. What width of film is commonly used for commercial motion picture movies? Why?
6. What is the major difference in processing motion picture film and still camera film?
7. Why is editing a critical process for motion picture making?
8. When making a motion picture, which person has control of the whole operation? Name four other people who contribute to the making of a picture.
9. What makes animation work?

 PHOTOGRAPHIC IMAGE FORMATION

OVERVIEW

In this activity you will design a message for presentation and develop that message into a poster-bulletin board display. After deciding the layout and content, design templates to block out the light. Use these templates on photosensitive paper such as common blueprint paper from your drafting printer. Expose and develop your poster.

As you read through this activity, try to predict what happens when a mask or template is placed over the light-sensitive paper and then the paper is exposed to ammonia vapors and developed. Because most blueprints today are technically white prints, you will have a blue image on a white background.

MATERIALS AND SUPPLIES

As you work through this activity you need:

- paper, pencil, and scissors to develop mask or template
- light-sensitive paper (print paper)
- glass or clear plastic overlay (depending on the method of exposing paper)
- method for developing exposed paper

PROCEDURE

Possible ideas for your poster could include a stop drunk driving campaign poster, safety posters, or any number of other types of posters presenting class-related information.

1. First determine your message. A beginning sketch will help you proceed with less wasted materials.
2. Develop the overlays or masks to be used for your project.

NOTE: Masks can be made of a computer-generated message or graphics by taking the printed hard copy to a copy machine and making a transparency of that message. Cover the mask with the glass or plastic overlay for good results.

---

3. To keep the masks from moving around, it is best to lay something clear over the top. If you are going to expose the light-sensitive paper by using the sun or an incandescent light source, then a piece of glass works well and helps keep everything together. If the material needs to be run through the white printer, then a clear plastic sheet (carrier) can give about the same results.

4. Once you have exposed your project, you then remove all of the masks and run the light-sensitive paper through the developer portion of the white printer which contains the ammonia. Ammonia vapors will develop those areas that were masked before. If you do not have access to a white printer, a simple developer can be made using a tube of 6-inch pvc pipe with a reservoir in the bottom to hold ammonia, and a lid on top.

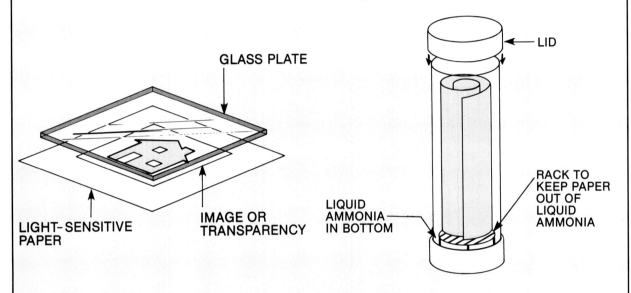

GLASS PLATE

LIGHT-SENSITIVE PAPER

IMAGE OR TRANSPARENCY

LID

RACK TO KEEP PAPER OUT OF LIQUID AMMONIA

LIQUID AMMONIA IN BOTTOM

►**CAUTION:** Adequate ventilation must be used as you work with the ammonia.

---

5. Once you have developed the image on your poster, you may now display it. White prints will fade with time. That fading process will be greatly increased if you hang it in the sun.

FINDINGS AND APPLICATIONS

In this activity you found that you could produce good results if you took time to develop quality templates and overlays. You also found the procedure quite simple yet the chances to use this process are limited only by your imagination.

ASSIGNMENT

1. Develop a poster as assigned by your instructor showing a specific activity in your technology laboratory.
2. Design a student organization poster advertising membership in your technology student club.
3. Experiment with and design a safety poster for one of the recent activities in your technology class.

EXPOSE AND DEVELOP A ROLL OF BLACK-AND-WHITE FILM

OVERVIEW

In this activity you will test your understanding of the materials presented in this section. You will compare a series of photographs following the correct procedure. You will load, expose, develop, and print a roll of black-and-white film. It is exciting to see the difference a little planning can make in the quality of the photographs you take.

Assume, for the moment, that you have been assigned as yearbook chairperson. Your task is to create a photo essay for the beginning section of the yearbook. The essay will include about 20 photos whose intent is to capture the essence of what your school is like. You may want to include sports photos, building photos, cafeteria photos, student photos, faculty photos — the choice and variety is endless and is up to you. Once you have developed a plan, you will take the photos, develop them, and then arrange them in your final photo essay.

MATERIALS AND SUPPLIES

To complete this activity you will need the following materials:

- camera and film
- scissors
- bottle-cap opener
- thermometer
- developing tank
- timer
- trays
- chemicals
- print paper
- negative carrier
- easel
- tongs
- enlarger
- tripod

PROCEDURE

1. Load the film into the camera and make the needed settings as instructed by your teacher.
2. Choose the subjects for your photographs. With the light source behind you and toward the subject, set the shutter speed (if needed) and focus the camera (if needed). Support the camera on a stationary object or hold it steady in both hands.
3. Frame the subject. Check the framed area for mergers.
4. Take the picture. Remember to depress the shutter smoothly.
5. Advance the film when ready to take another picture.
6. Repeat the process until you have exposed the entire roll of film.
7. Remove the film from the camera. Be sure you have rewound it if that process is necessary.
8. Label the film with your name and class period.

Processing Black and White Film

1. Arrange the film, developing tank, reel, scissors, and the bottle opener in the darkroom so that you can find them in the dark. As you proceed with the processing of your film, be sure to note the specific processing times for the various developing chemicals. You may get these times from your instructor.

2. Close the door of the darkroom and turn off the light. The door may be opened only after the film is on the reel and securely inside the developing tank.

3. Open the canister with the bottle cap opener and remove the film. Be careful not to touch the face of the film with your fingers. Handle it only by the edge. Pull the film out far enough to remove the camera tab (the narrow part of the film—3 to 4 inches long) on the end of the roll. When you have the tab in your hand, cut it off with the scissors so the film is square. If you do not do this, it will be hard to load your film into the film reel that goes in the developing tank.

4. Once you have cut the tab off the end of the film, unwind the entire roll of film. When you come to the end, remove the tape that holds the film to the roller. Place the film into the film reel, set the film reel in the developing tank, and put the lid on it.

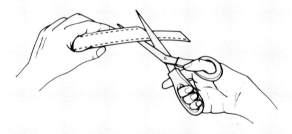

NOTE: This can be practiced in the light with a dummy roll of film until you are fairly secure with the process. It is hard to learn in the dark how to get the film onto the reel.

---

5. Fill the developing tank with developer and start the timer right away. Tap the tank to dislodge air bubbles and agitate the tank for 30 seconds. To tap the developing tank, pick it up and tap it on the counter. Watch that you do not knock the lid off or your film will be exposed. Agitate the developing tank by tilting it back and forth while moving it in a figure-eight motion. Then agitate the tank for 5 seconds every 30 seconds for the rest of the developing time. Drain the tank during the last few seconds of this step.

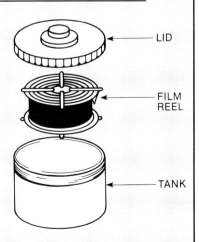

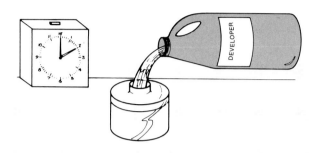

**FILL AND TAP THE FILM TANK
TO GET BUBBLES OFF THE FILM**

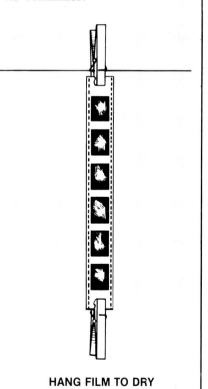

6. Pour in the stop bath. Agitate in the same manner as for the developer. When this step is complete, pour the stop bath back into its container.

7. Pour in the fixer and agitate. Continue fixing for twice the clearing time determined earlier and pour the fixer back into the container.

8. Remove the developing tank lid and wash the film under running water. The wash time is based on the particular washing aid you are using. Your instructor or the manufacturer's instructions will tell the amount of time for this and the following two steps.

9. Drain the water from the tank and pour in the washing aid. Agitate the tank for the recommended amount of time. Pour the washing aid back into its container.

10. Wash the film under running water for the recommended time. Then pour the remaining water out of the tank.

11. Pour in the wetting agent and gently agitate. After the recommended soaking time, pour the wetting agent back into its container.

12. Remove the film from the reel and hang it to dry in a dust-free place. Attach a clip such as a wooden clothespin to the bottom of the film. The clip will keep the film from curling as it dries.

HANG FILM TO DRY

Print a Picture

1. A roll of developed film should be cut into lengths of five to six negatives each. Select the negative to be developed.

2. Enlarge the negative:

   a. Remove the dust particles from the negative carrier and the negative using a soft brush.

   b. Place the negative in the negative carrier, emulsion (dull) side down.

   c. Lay a piece of plain white paper in the easel.

   d. Place the easel on the base of the enlarger.

   e. Turn off the darkroom light and turn on the safe light.

   f. Turn on the enlarger light source.

   g. Raise or lower the enlarger head until the image is the desired size. The lens of the enlarger should be wide open.

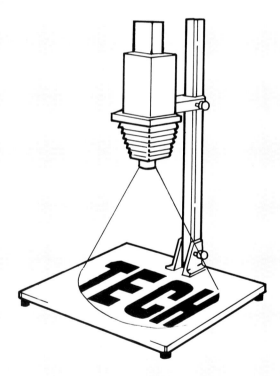

   h. Focus lens until the image has maximum sharpness. Use focusing aid if needed.

   i. Switch off the enlarger light source.

   j. Replace the plain white paper with photographic print paper.

NOTE: Print paper should be stored in a paper safe. Exposure time will differ according to paper type. Check with the instructor for exposure time.

k. Turn on the enlarger light source.
l. Set the enlarger timer and expose the print for the recommended time.
m. Turn off the enlarger light source.

NOTE: Exposure time tends to vary with the different rolls of film. Develop a test strip for best results in your printing.

3. Develop the print paper:
    a. Have four trays ready—three developing trays containing developer, stop bath, fixer, and a wash tray.

    ►CAUTION: Each tray needs tongs. Do not interchange tongs and trays. Each chemical can be contaminated by the others.

    b. Slide print paper quickly into the developer.

    NOTE: Make sure the whole sheet becomes white at the same time. Chemicals should be about 68°F; recommended developing time is one minute. Do not overdevelop or allow the print to become too dark.

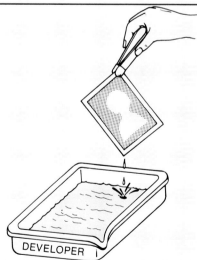

DEVELOPER

    c. Remove the print from the developer with the tongs. Hold the print above the developer tray and allow it to drain briefly.
    d. Transfer the print to the stop bath tray for five seconds.

NOTE: Do not allow the tongs from the developer to come in contact with the stop bath.

e. Transfer the print to the fixer.

NOTE: Use the second pair of tongs at this time. Check with your instructor for exact time to leave the print in the fixer bath.

f. Wash the print in running water (print washer) for instructor's recommended time.

g. Lay the print on the flat clean surface and use a squeegee to remove excess water. This promotes faster drying. Allow the print to dry completely at room temperature.

NOTE: Print quality can be improved by cropping, burning, dodging, and using filters. Your instructor will supply information on these processes.

Arrange Your Photo Essay

1. Take your final, printed photos and arrange them on pages as you would for your yearbook's photo essay.
2. Evaluate your work. Do you like how each photo turned out? Do you like the impact made by all of your photos together?
3. Ask another student to evaluate your work. What overall impression do they get from your essay?

FINDINGS AND APPLICATIONS

During this process you had the chance to design and develop quality prints of your own choosing. You found you had to follow instructions as exactly as possible for the best results. As you began this activity, you were asked to think about the uses for your pictures.

ASSIGNMENT

1. Give your four best photographs to your instructor. Include with each notes regarding the amount of time used in each of the developing processes. Describe why you feel each represents your best work.
2. Briefly list possible applications and uses for photographs.
3. If you were to do this activity again, what things would you do differently to improve your results?

COLOR SLIDE PROCESSING

OVERVIEW

In this activity, you will have the chance to develop and mount color slides. You will be surprised to find how easy the process is and will be delighted to see your own quality slides. You are going to use the slides you create and develop for a later activity in which you put together a slide show. So, do some planning now.

First, you will decide upon a theme for your slides. Here are a few suggestions:

1. Nature
2. Scenes from childhood
3. Sports

4. Colors
5. Transportation
6. Communication

The possibilities are endless. Brainstorm for a while to be sure you are allowing for your most creative ideas. Once you have decided upon a theme, begin the activity.

MATERIALS AND SUPPLIES

To develop your slides, you will need to use the following items:

- roll of exposed slide film (Ektachrome works well)
- appropriate chemicals including developer, bleach or fixer, and film stabilizer
- black changing bag
- processing tank with film reel and lid
- thermometer
- place to hang film as it is being developed
- clothespin
- bottle cap opener

PROCEDURE

This activity will work well with a small team of two students.

1. Get all the needed equipment before you begin the task. Be sure the chemicals are prepared and are at the proper temperature.

NOTE: The first time through, it is sometimes difficult to get the film loaded into the film holder. Practice with the black changing bag, the film holder and tank, and a piece of used film.

---

2. Put your materials in the bag. Place your arms into the armholes. Be sure to slide the elastic band far enough up on your arms so there is no danger of light coming in around the wrists.
3. Follow the procedure outlined in Steps 3 through 5 in the previous activity under ''Processing Black-and-White Film.''

4. At this point you can remove the developing tank from the black bag and proceed with the developing. On a separate sheet of paper, record the temperatures and times required to develop your final slides.

> **NOTE:** Developing time depends on the temperature of the solutions. Be sure you have measured the temperature correctly and determined the processing time. This will assure you quality developing results. As you begin developing your own film, predict the amount of time each chemical solution needs to be in contact with your film. Use the scale in the developing kit. What happens to the development time if the temperature is increased?

5. Tape a piece of masking tape to the sides of the solution container. As part of the activity, record on the masking tape the number of rolls of film that have been developed. This is done because the film developers become weaker with each usage. On a separate report sheet, record the temperatures and times required to develop your final slides.

> **NOTE:** The time needed to drain the tank (about ten seconds) should be included when determining the time for each processing step. In each case, start draining in time to end the processing step and start the next step on schedule. Most invertable tanks fill more easily when held at a slight angle to allow the air in the tank to escape.

6. **Processing:** The first developer is the most critical step. Pour in the first developer and start the timing. Tap the developing tank and agitate the film to eliminate any bubbles that may form on its surface. Initial and subsequent agitation are required. See the time/temperature tables for the appropriate first developer time, and remember that the development time includes the drain time.

> ►**CAUTION:** Chemicals used in this process are not good for clothes, skin, or eyes. Please use caution as you handle the chemicals. Should you get any on your hands or in your eyes, rinse them off with water right away.

7. **Rinsing:** To wash, fill the tank with water of about the same temperature as the first developer. Agitate, drain, and repeat until the film has at least four complete rinses, about 1–3 minutes total time. Use running water if available. The rinse water will be very pale yellow after the last rinse.

8. **Color developer:** See the time/temperature tables for the processing time for the temperature used. Agitate in the same way as you did with the first developer. Pour out the developer. Repeat the wash step filling the tank with water about the same temperature as the developer. Agitate, drain, and repeat until the film has had at least four complete rinses. The rinse water should be very pale blue after the last rinse.

9. **Bleach fix:** Process the film in the bleach fix for ten minutes at any temperature between 70 and 110°F (21–43°C). Longer times will not hurt the film, but will be of no benefit. Use initial and subsequent agitation.

10. **Final wash:** Wash the film continuously for at least four minutes or fill and dump the tank at least six times in four minutes. If you are using running water, tap the tank every 30 seconds to dislodge air bubbles. You can remove the lid of the processing tank for this step to ease washing. The last rinse should look clear and colorless.

11. **Stabilizer:** Process the film in the stabilizer for at least one minute at any temperature between 70 and 100°F (21–43°C). Use only initial agitation.

12. **Drying:** To dry, remove the film from the reel and hang it up to dry in a dust-free environment as you did in the previous activity. If you use heat to dry the film, do not let the temperature exceed 140°F (60°C) or the film will curl.

13. Judge the color, balance, and density of your pictures only after the film has dried. Do not squeegee unless you see excess water spots on the film. If you squeegee the film, be careful not to scratch the soft emulsion. Never squeegee more than once.

14. **Mounting:** Once the film has dried, it can be cut with scissors into individual frames or strips. Each of these frames is then mounted in an individual slide holder. Should you choose not to mount individual slides at this time, it is important that you slide short sections of the film into plastic protectors to prevent any damage to the slides.

CUT SLIDES AND MOUNT THEM IN INDIVIDUAL FRAMES

FINDINGS AND APPLICATIONS

During this activity, you found that the fundamental processes of color slide developing are basically quite simple and yet, for quality results, need very careful developing. You experienced a photographic process that has done much to improve the quality of visual displays. You also discovered that by following directions carefully you were able to get good quality slides as a result.

ASSIGNMENT

1. Turn in to your instructor the number of slides requested along with the report sheet that has the temperature and amount of time for each of the processing steps.
2. Where would you expect to see new uses for the development of color slides? How do you expect to use the slides you just developed?
3. If you could do this activity again, what would you do differently?

 SLIDE TAPE SHOW

OVERVIEW

In this activity you will create a quality audio and slide show on any topic you choose. You may be able to become a "legend in your own time!"

As you develop this activity, plan everything in advance. First, determine the amount of time needed for the slide tape show. Then select the material you wish to present. You will use the slides you took during the color slide processing activity to create a slide tape show on the theme of your choice.

MATERIALS AND SUPPLIES

To organize and develop your slide tape show, you will need:

- story board forms
- processed color slides
- tape recorder
- slide projector
- screen
- light table for organizing slides
- slide mixing unit (optional)
- second slide projector (optional)

PROCEDURE

This activity needs a good amount of planning but can give first-class results.

1. To begin, you must develop your message and determine which slides you will need to present your theme. Remember it is generally best to have extra slides to choose from as you assemble your program. Possible themes to consider might be nature, children, ecology, sports, community issues, parks, transportation, eating places, safety, etc.

BEACH

MOUNTAINS

FLOWER

2. Using the story board, identify the slides that best convey the message. These slides could include actual location pictures, or slides can be taken of pictures from a book.
3. Review the information you wish to present with the pictures. A written script will be most helpful. Enter the text for the audio on your planning story board.

FROM THE WHITE SANDY BEACHES OF SUNNY FLORIDA,

TO THE QUIET MOUNTAINS IN IDAHO, WE FIND AND ENJOY

PEACE AND OPPORTUNITY.

4. Select the slides and arrange them in order. Remember that good slides are critical to the success of your show. You may need to take more slides to have a selection.
5. Load slides into the projector tray. Prepare the audio tape.

(Photo by Ruby Gold)   *(Courtesy of Eastman Kodak Co.)*

6. If you have the equipment, it works best to develop the voice part of the audiotape first and then blend the music. You are aware of the impact that music can make in communicating an idea. This can be done with the blending of the music and the voice portion of your audiotape. The music overlay tends to get rid of dead spaces when there is no audio and will add another dimension to the quality of your show.

7. The final activity is to project the slides along with the audio track. Make any necessary changes. After the final revision, you can show the slide/tape presentation to your class.

FINDINGS AND APPLICATIONS

During this activity you have found that careful planning is important in getting a good sequence in your show. You also found that video and audio, when combined, enhance the presentation.

ASSIGNMENT

1. List two places where a slide tape show would be useful.
2. Submit the completed story board form for all slides.
3. Prepare an introductory written description for the slide tape show. Explain its purpose and topic.
4. Present the slide tape show to your class.
5. If you could do this activity again, what would you do differently?

VIBRATION DETECTION

OVERVIEW

In this activity you will find out about the stability of a surface. Vibration-free sur-faces are critical to the production of a hologram. (See *Making a Hologram* activity.) For this activity you will make a Michelson interferometer. This device will let you observe any vibration caused from a number of sources such as automobile traffic, students walking inside the building, or electric motors operating.

You will have the chance to predict what you expect the result to be from the different activities. You will also be able to predict what you think can happen and not affect the vibrations.

MATERIALS AND SUPPLIES

For this activity you will need:

- smooth surface (table top or bench top)
- laser
- front surface mirrors (at least two)
- beam splitter (piece of glass)
- beam spreader lens (concave lens)
- projection area or screen

PROCEDURE

As you work through this activity, it is important that you remember the safety rules for use of the laser. Caution must be given to control of the beam so that it does not shine in your eyes or the eyes of other students in the lab. Be very careful as you

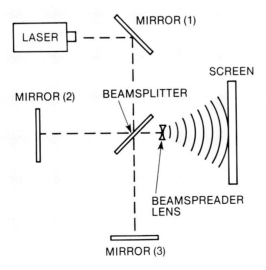

LAYOUT FOR
MICHELSON INTERFEROMETER

use mirrors which allow you to project the beam in different directions. This activity uses the following setup procedures:

1. Arrange the equipment as shown in the figure in the following sequence.
   a. Use the first mirror to reflect the laser beam. If there is enough space you do not need to have the first mirror in the sequence. Note that the mirrors are front surface mirrors. Otherwise in the angled reflection the beam will be split and reflected from both the front and back surface of the mirror.
   b. From the first mirror the beam is then directed through a beam splitter or a clear piece of glass. The glass is arranged at 45 degrees to the beam. Two beams result: one going to Mirror 2, and from Mirror 2 back to the beam splitter, and again being split with one beam going back directly into the laser; and the other beam being projected straight on through to the beam spreader. The second beam has gone straight through the beam splitter to Mirror 3, and is reflected back to the beam splitter where it is split with one beam going to the beam spreader, and the other beam going straight through back to Mirror 1 and back to the laser. This results in two beams being projected through the spreader which spreads the beam on the screen. The projected beam will appear to have lines running through it. If there is no vibration, the lines will hold still on the screen.
2. Now bump the table a little and notice what happens to the projected beam. You will also find that the room is not so stable with students walking around. Sometimes outdoor traffic will also cause vibrations that affect the projected beam.
3. Let's observe the effects of a number of things on the projected beam. Direct a fan into the area so that it causes movement in the air and see if there is any effect on that projected beam.
4. Try placing your hand directly over the beam. Be careful not to block the beam with your hand. Notice if there is any effect on the projection caused by the heat from your hand.

FINDINGS AND APPLICATIONS

During this activity you were able to observe that while most objects appear to be still, vibrations can be caused by a number of things. These vibrations that you observed in the projected beam illustrate how difficult it is to get a good hologram which necessitates that all vibration be stopped. You were also able to observe that even simple vibrations in the air can be recorded using the projected beam on the screen. In cases where absolute stillness is necessary, the interferometer can be used to determine whether or not vibrations exist.

ASSIGNMENT

1. Construct a Michelson interferometer as shown in this activity. Ask your instructor to see if it is set up correctly.
2. List five things that could cause interference and/or vibration in the laser beam as noted on the projected screen.
3. Identify any possible uses for an interferometer.

HOLOGRAPHY

OVERVIEW

Holography is a relatively new technology. It has only existed since the laser because laser light is needed to make a hologram. You will be able to set up the laser and optical equipment to view a demonstration hologram.

As you set up the experiment, try to predict the effects of the various apparatus. For example, why is it necessary to use frosted glass? Could we set the laser closer to the film?

MATERIALS AND SUPPLIES

You will need these items to complete this activity:

- helium neon laser
- beam splitting prism
- mirror or tripod stand
- clamps
- film holder
- frosted glass
- set of demonstration holograms (360° plastic tube with film)

PROCEDURE

1. Be sure you have been provided with the necessary instruction and taken a safety test dealing with use of the laser.
2. Obtain the laser and holography equipment from your instructor.
3. Set up the equipment to view a 360° hologram. To do this:
   a. Mount the laser into the clamp ring holder with the door of the apparatus open and plug it in.
   b. Mount the frosted glass in its clamp about 6 to 8 inches below the laser opening.
   c. Clamp the 360° plastic ring with its film in the place about 6 inches below the frosted glass.
   d. Turn off the room lights and move around until the image comes into clear focus. The hologram appears to be in three dimensions. You may walk around it to obtain different views.

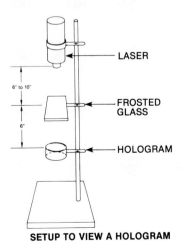

SETUP TO VIEW A HOLOGRAM

FINDINGS AND APPLICATIONS

As you set up the equipment, you found that you had to align all of the apparatus to get a clear hologram. You also found that the image was very clear and suspended (as if by magic).

ASSIGNMENT

1. List three uses for holograms.
2. How do holograms differ from photographs?
3. Where have you seen other holograms?

 MAKING A HOLOGRAM

OVERVIEW

In this activity you will make your own hologram. You will set up the equipment and expose and develop your own film.

As you work through this activity, try to predict the effects and purposes of the various procedures. Think about uses for holograms. Look for articles and other information about holograms that might be of use or interest to you or to other members of your class.

MATERIALS AND SUPPLIES

To do this activity you will need the following:

- table to absorb all vibrations
- laser
- laser mounting stand
- special filter for the laser
- front surface mirror and mirror holder
- object to photograph
- film
- film developing chemicals

PROCEDURE

1. This activity needs to be done in an area where light can be controlled and where it can be totally dark.
2. The laser table works best if it can be set up on a solid floor where no vibration can be transmitted. See the drawing of the layout. Notice that the table has foam rubber between each of the cinder blocks supporting the base, and inner

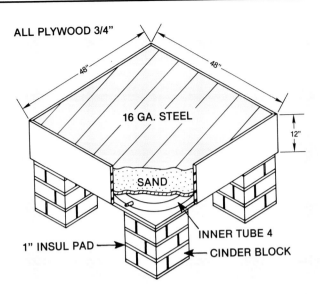

tubes between the base and the box itself. The box is filled with sand to give it weight and stability. You will also notice the sand is covered with a plate of 16-gauge metal. This seems to work best to keep the dust and other unwanted particles out of the laser and film.

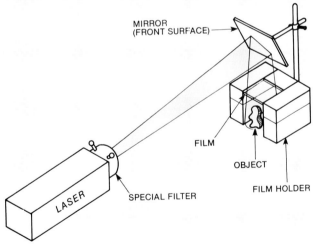

3. Set up the laser with the special filter which cleans up the light. Note the location of the mirror so that the beam is reflected onto the object. The film will be mounted between the mirror and the object. The film must be kept in total darkness until it is ready to be exposed and then only laser light may be used.

4. Practice in the dark getting a test piece of film out of the film box, loading that piece of film into the film holder, and then exposing it to the laser beam.

5. Turn on the laser and set a card in front of it to block the beam. This card will be the shutter for your laser camera.

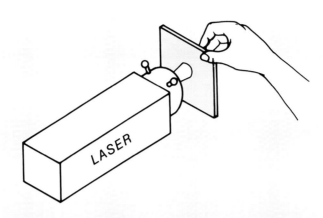

6. Turn out all lights so that everything is in complete darkness. Take out the film and place it in the film holder.
7. Allow the laser table to stabilize for one minute. If there is any vibration in the room or around the table, the hologram may not work.
8. Once the table is stabilized, very carefully move the card so that the laser beam can reflect from the mirrors through the film to the object to be photographed. Expose for the desired time. Then replace the card to block the beam.

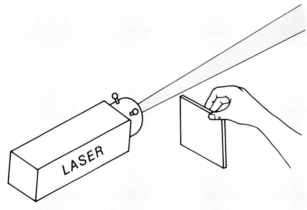

9. Turn off the laser. Carefully remove the film from the film holder. Develop the film using chemicals you prepared earlier.

FINDINGS AND APPLICATIONS

Once you have developed your hologram, notice the depth and dimension you were able to get. Holograms are now being used for identification purposes, nondestructive testing, and a number of other uses.

ASSIGNMENT

1. Complete a hologram and turn it in to your instructor.
2. Research and identify other uses of holograms. Give your notes describing your findings to your instructor.

SECTION FIVE

ELECTRONIC COMMUNICATION

All modern communication systems depend on electronics in some way. The use of electronics has completely changed the speed at which we communicate with one another, and the distance we communicate over. Electronics have made it possible and desirable for people to communicate with machines, and for machines to communicate with each other.

The use of electronics extends from improving graphic, photographic, and other forms of communication to "electronic communication," which uses an electronic channel to carry information from the sender to the receiver. Some examples of electronic channels are wires, radio waves that travel through the air, fiber-optic cables, magnetic tape, and optical disks. Electronic communication systems are the subject of this section.

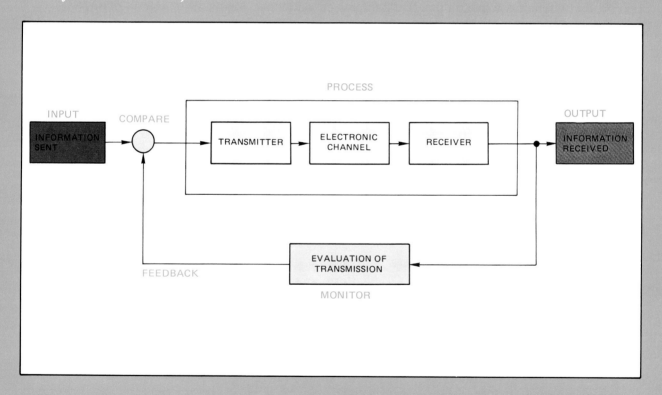

CHAPTER 16

Electricity and Electronics

OBJECTIVES

After completing this chapter, you will know that:

- The harnessing of electricity and electronics has revolutionized the way we live and communicate.
- All materials are made up of tiny particles called atoms. Atoms are made up of protons, electrons, and neutrons.
- Electric current is the flow of electrons through a material.
- Electric current will not flow unless there is a pressure, called voltage, applied to a circuit.
- Direct current (DC) does not change polarity. Alternating current (AC) changes polarity at regular intervals.
- Electronic circuits are made up of components connected together. Each component has a specific function in the circuit.
- Integrated circuits are complete electronic circuits made at one time on a piece of semiconductor material.

KEY TERMS

alternating current (AC)	frequency	proton
ampere (amp)	gallium arsenide	rectifier
amplifier	hertz	resistance
atom	insulator	semiconductor
capacitance	integrated circuit	silicon
circuit	neutron	sine wave
component	nucleus	thermistors
conductor	ohm	transformer
direct current (DC)	photoresistors	transistor
electron	printed circuit board	voltage

FIGURE 16–1 Networks of signal towers like this one in Barbados were used in many parts of the world to relay messages over great distances using flags or lights. (Courtesy of Barbados National Trust)

Introduction

The harnessing of electricity during the last hundred years has greatly changed the way we live. Electricity and electronics drive or control devices and systems that touch almost every aspect of our everyday lives. Some examples are the light we read by at night, the vehicles that transport us, our household appliances and tools, and our forms of entertainment.

The impact of electricity and electronics may be the greatest, however, in the field of communications. Before the use of electricity in communications, the fastest forms of communications were signaling from relay tower to relay tower using flags, Figure 16–1, and sending a letter by pony express.

Long-distance telegraph signaling was developed by Samuel F. B. Morse in 1843. This changed the amount of time that it took a one-word message to go one hundred miles from five hours to a few seconds. Since that time, communication speed has steadily improved. Today, fiber-optic cables and microwave radio networks carry large amounts of information from place to place in fractions of a second, Figure 16–2.

It can be seen that the role of electricity and electronics in today's communication systems is very important. Therefore, in the study of communications, one needs a good understanding of the basics of electricity and electronics.

FIGURE 16–2 The amount of information that can be sent from one place to another in one second is one way to measure the progress in communications.

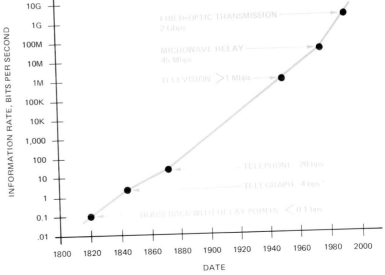

Atoms and Atomic Particles

All materials are made up of tiny particles called **atoms**. Materials that are made up of atoms of only one type are called elements. Iron and carbon are examples of elements. An atom is the smallest piece of an element that can exist and still have all the properties of that element. There are 106 known elements. Atoms are so small that you cannot see them, even with the most powerful optical microscope.

Often, atoms of different types are combined to form compounds. When a compound is formed from two or more elements, it has properties all its own. The compound may be nothing like any of the elements that make it up. For example, sodium (a metal that is poisonous if

SYMBOL	ELEMENT	PROTONS	NEUTRONS
H	HYDROGEN	1	0
He	HELIUM	2	2
C	CARBON	6	6
N	NITROGEN	7	7
O	OXYGEN	8	8
Al	ALUMINUM	13	14
Si	SILICON	14	14
Fe	IRON	26	30
Cu	COPPER	29	35
Ga	GALLIUM	31	39
As	ARSENIC	33	42
U	URANIUM	92	146

FIGURE 16–4 Some elements of interest

eaten and that reacts violently when it comes in contact with water) combines with chlorine (a poisonous gas) to form sodium chloride (table salt). Because of the many different ways elements can be combined, millions of compounds can be made from the 106 known elements.

Atoms are made up of even smaller particles. These subatomic particles do not have the properties of the element. They have their own general properties, and are the building blocks of all atoms. All atoms have a heavy center part called a **nucleus**. The nucleus is made up of **protons** and **neutrons**. Small particles called **electrons** circle the nucleus very fast, Figure 16–3. The number of protons in an atomic nucleus determines the kind of element the atom will be, Figure 16–4.

Both protons and electrons have electric charges. Protons have a positive charge. Electrons have a negative charge. The positive charge of a proton is exactly equal to, but opposite in polarity from, the negative charge of an electron.

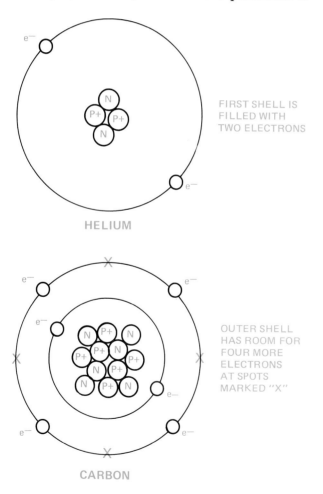

FIRST SHELL IS FILLED WITH TWO ELECTRONS

HELIUM

OUTER SHELL HAS ROOM FOR FOUR MORE ELECTRONS AT SPOTS MARKED "X"

CARBON

FIGURE 16–3 Two common atoms are the helium atom and the carbon atom.

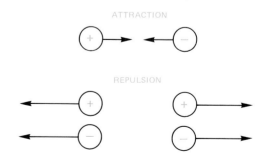

ATTRACTION

REPULSION

FIGURE 16–5 Objects with similar charges repel each other. Objects with opposite charges attract each other.

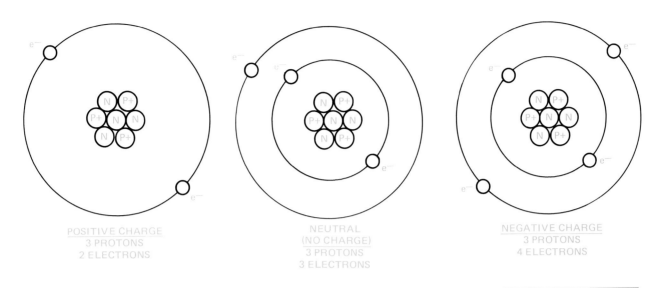

FIGURE 16–6 Atoms with more electrons than protons have a negative charge. Atoms with fewer electrons than protons have a positive charge. Atoms with the same number of protons and electrons have no charge.

Neutrons have no charge; that is, they are neutral.

Particles with opposite charges are attracted to each other. Particles with the same charges repel each other (push each other away), Figure 16–5. Negative electrons are thus attracted to the nucleus because of the protons' positive charge. However, the very fast motion of the electrons around the nucleus keeps them from crashing into the nucleus.

In most natural atoms, the number of electrons orbiting the nucleus equals the number of protons contained in the nucleus. This means that most natural atoms have no charge. The negative charges of the electrons are canceled out by the equal number of positive charges of the protons. If an atom has more electrons than protons, it is negatively charged. If an atom has fewer electrons than protons, it is positively charged, Figure 16–6.

Electric Current Flow

The electrons that orbit atoms are found in layers, or shells, around the nucleus. Shells can contain up to a fixed number of electrons. As one shell is filled, the remaining electrons fill other shells further from the nucleus, Figure 16–7.

If the outermost shell of an atom has only one electron in it, the electron is not tightly held, and may be given up easily. If the outermost shell of an atom is completely filled with electrons, those electrons are tightly held to the atom.

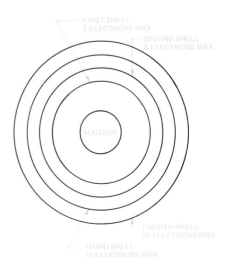

FIGURE 16–7 Electrons fill shells as they orbit the nucleus. The number of electrons in a given shell is the same from atom to atom.

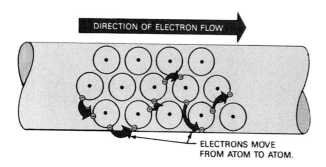

DIRECTION OF ELECTRON FLOW

ELECTRONS MOVE
FROM ATOM TO ATOM.

FIGURE 16–8 In a wire, electrons are free to move from one atom to the next. Electric current is the flow of electrons through the wire. *(Reprinted from LIVING WITH TECHNOLOGY by Hacker and Barden,* © *1988 Delmar Publishers Inc.)*

Materials whose atoms give up electrons easily are called **conductors**. Materials that hold tightly to their electrons are called **insulators**. In a conductor, electrons can move from one atom to another, Figure 16–8. An insulator's electrons do not move freely from one atom to another.

Current and Voltage

Electrons may flow through a thin wire, a solid piece of material, or even through the air (a lightning bolt is a good example of this). The measure of electric current flow is the **ampere**, or **amp**. One ampere is equal to over six billion billion (6,280,000,000,000,000,000) electrons flowing past a point in one second. Smaller units of measurement are the milliamp (one-thousandth of an amp), and the microamp (one-millionth of an amp).

To get electrons to move from atom to atom in one direction, making a current flow, an electric force has to be applied to the conductor. This electrical force is called **voltage**. Without voltage, no current will flow.

Resistance

While many materials are conductors, some are better conductors than others. For a given voltage (force), more current will flow through a good conductor than through a poor conductor. **Resistance** is the opposition to a flow of current. It is the measure of how good a conductor is. The lower the resistance, the better the con-

ductor. A material with a high resistance is a poor conductor. The unit of resistance is the **ohm**.

Voltage, current, and resistance are all related by an equation called Ohm's Law. Ohm's law states that current flow is proportional to voltage, but inversely proportional to resistance. This means that as voltage is increased, current increases correspondingly, but as resistance increases, current decreases correspondingly. Ohm's law is:

$$\text{Current (amps)} = \frac{\text{Voltage (volts)}}{\text{Resistance (ohms)}}$$

Simply stated, Ohm's law says that as the electrical pressure (voltage) across a circuit increases, the amount of current that can be pushed through the resistance also increases. Or, as the resistance increases, the amount of current that can be pushed through it decreases, if the voltage remains the same.

Alternating Current and Direct Current

If a voltage always has the same polarity, it is called **direct current**, or **DC**. While many DC voltage sources, such as batteries, produce a steady voltage, some do not. In a DC circuit, the current always flows in one direction.

In many cases, a voltage source produces a voltage that changes polarity on a regular basis. The voltage available in our homes changes polarity many times every second. This kind of voltage supply is called **alternating current**, or **AC**.

The number of times that the voltage changes polarity each second is called its **frequency**. The unit of measurement of frequency is the **hertz**, or cycle per second. Power sources that are used to drive motors and appliances around the home have low frequencies (sixty hertz in the United States and fifty hertz in Europe). Radios and televisions use voltages that change polarity millions of times per second.

AC is the current used to furnish power to homes and factories for several reasons. One is the ease with which it can be changed from high voltage to low voltage in simple devices called

AC and DC

Direct current always keeps the same polarity. It may change in strength (amplitude), but the polarity does not reverse.

Alternating current reverses polarity on a regular basis. The most common wave shape is the smooth **sine wave**. Other wave shapes, including square waves and triangle waves, are sometimes found.

Frequency is the number of times that the polarity of a voltage changes in one second. One cycle per second (polarity changes and changes back) is called one hertz.

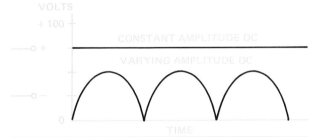

Direct Current does not change polarity, but it might change level, or amplitude.

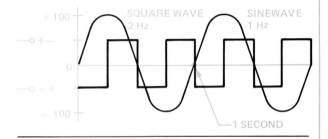

Alternating current changes polarity at regular intervals, and can be of various wave shapes.

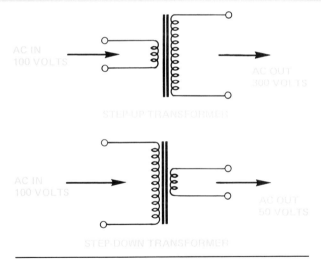

FIGURE 16–9 Transformers are used to increase or decrease the amplitude of AC voltages.

transformers. High voltage is useful for sending electric power over long distances. Lower voltages are needed to operate small motors found in appliances and machines. Transformers that increase a voltage are called step up transformers. Those that decrease voltage are called step down transformers, Figure 16–9.

Most electronic equipment such as stereos, televisions, and computers needs DC to run properly. Inside these devices, AC is converted to DC in a circuit by a **rectifier**. The power supply is made up of a rectifier and its associated circuitry for smoothing the DC and making sure it is the correct voltage, Figure 16–10.

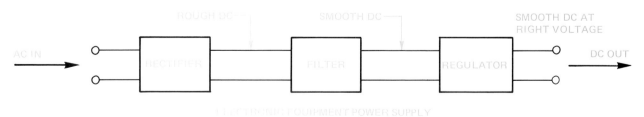

FIGURE 16–10 A rectifier changes AC to DC. Often, a filter and/or a regulator are added to the rectifier to smooth the DC produced and to control its level.

Electronic Components and Circuits

Electronic **components**, or parts, control the flow of current and perform useful functions. Some parts, such as switches or lamps, do very simple tasks. Others, such as microprocessors, do very complex and involved tasks. A group of components connected together to do a task is called a **circuit**.

Engineers and designers show how circuits are connected using special drawings called schematics. In a schematic, each part is shown by a symbol. The parts are interconnected by wires or conductors that are shown as lines on the schematic.

Resistors

While all wires have resistance, special components, called resistors, are used to give a well-defined, controlled amount of resistance in a circuit. Many resistors have a fixed value of resistance, but some, called potentiometers or variable resistors, have an adjustable amount of resistance, Figure 16–11. A potentiometer is used to control the volume or tone on a radio or stereo. Resistors are available in a wide range of values, from less than one ohm to tens of millions of ohms, Figure 16–12.

Capacitors

A capacitor is used to store an electric charge. It is made up of parallel plates, separated by a thin insulator. Sometimes, to save space, the conductor and insulator sandwich is rolled into a

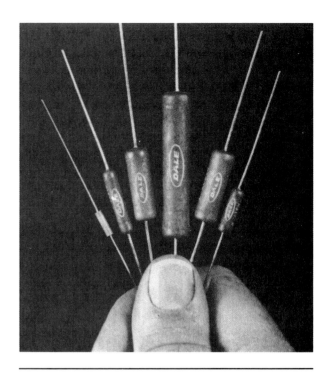

FIGURE 16–12 Resistors come in many shapes and sizes. *(Courtesy of Dale Electronics, Inc., Columbus, NE)*

cylinder. The larger the surface area of the plates that face each other, and the thinner the distance between the two, the larger the charge that can be stored. The measure of charge storage capacity is called **capacitance**, measured in farads. The more practical measures of capacitance are the microfarad (millionth of a farad) and picofarad (millionth of a microfarad).

Diodes

Some materials are neither good conductors nor good insulators. Their resistivity falls in between the two, and they are called **semiconductors** (half conductors). The most widely used semiconductor materials are **silicon** (an element) and **gallium arsenide** (a mixture of gallium and arsenic). Many different kinds of components can be made from semiconductors.

One of the simplest semiconductor components is the diode. A diode allows current to flow

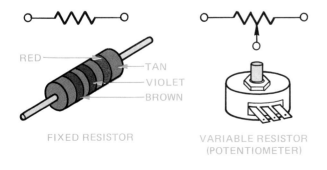

FIGURE 16–11 Resistors can be fixed in value or adjustable, such as this potentiometer.

How a Capacitor Works

If a voltage is applied to a capacitor, electrons gather on the plate that is connected to the negative voltage source. Electrons are drawn away from the plate that is connected to the positive source. Since the two plates are not connected, no current flows from one plate to another. The plate with the electron surplus is said to have a negative charge. The plate with the electron shortage has a positive charge.

Once the capacitor has been charged to the voltage applied, no more electrons will be added to or removed from the two plates, and no more current flows. The voltage source can then be removed, and the capacitor will maintain the charge until a path is found for the charge to dissipate.

Thus, current flows in a capacitor only while it is charging or discharging, or only while the

applied voltage is changing. This feature of a capacitor is used in circuits to pass the flow of alternating current, or changing current, while blocking the flow of steady or direct current.

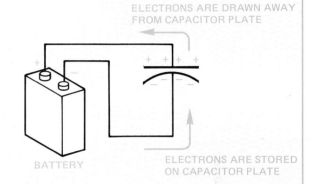

ELECTRONS ARE DRAWN AWAY FROM CAPACITOR PLATE

ELECTRONS ARE STORED ON CAPACITOR PLATE

BATTERY

(A) BATTERY FIRST CONNECTED

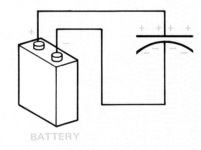

BATTERY

(B) WHEN CAPACITOR IS CHARGED TO BATTERY VOLTAGE, NO ELECTRON CURRENT FLOWS

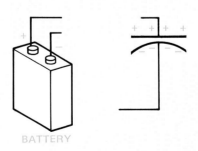

BATTERY

(C) CAPACITOR REMAINS CHARGED AFTER BATTERY IS REMOVED

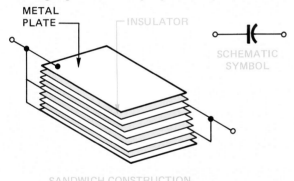

METAL PLATE

INSULATOR

SCHEMATIC SYMBOL

SANDWICH CONSTRUCTION

INSULATOR — METAL FOIL — INSULATOR — METAL FOIL ROLLED TOGETHER

ROLLED CONSTRUCTION

Capacitors are built in many different ways. Usually, multiple plates are sandwiched together to give greater capacitance. Sometimes the plates are flexible and are rolled up to keep the capacitor small.

Current flows through a capacitor only when the voltage across it is changed.

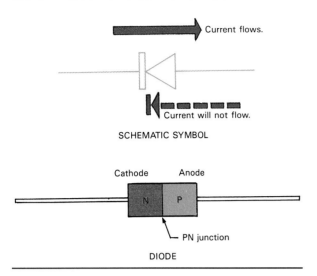

Current flows.

Current will not flow.

SCHEMATIC SYMBOL

Cathode   Anode

PN junction

DIODE

FIGURE 16–13 A diode is made when a P-type semiconductor and an N-type semiconductor meet at a PN junction. Electron current will flow across the junction in one direction, but not in the other. *(Reprinted from LIVING WITH TECHNOLOGY by Hacker and Barden, © 1988 Delmar Publishers Inc.)*

in one direction, but not in the other, Figure 16–13. This property makes diodes useful in circuits that change alternating current to direct current (rectifiers), in protection circuits, in current "steering" circuits, and for many other uses.

Transistors

One of the greatest inventions of this century was the **transistor**. It was invented in 1947 by three researchers at Bell Laboratories. The word "transistor" is short for **trans**fer re**sistor**. A transistor is a semiconductor device that allows a small amount of current into one terminal to control the flow of a much larger flow of current to another terminal.

This control ability can be used to control large amounts of current, as in an electric motor. It can also be used to control the storage of a small electric charge used to represent information, as in a computer memory circuit.

By connecting groups of transistors and other components together in different ways, many

Transistors

A transistor has two PN junctions. The transistor pictured is an NPN type. It has a thin P-type semiconductor sandwiched between two N-type semiconductors. PNP transistors are made the same way, but with an N-type semiconductor in the middle. A small amount of base current in the transistor will control a much larger collector current. Collector current can be made larger, smaller, or even turned off by a small amount of current change at the base.

Even though transistors are very small (0.01 inch x 0.01 inch is not uncommon), they are put into larger containers for protection, ease of handling, and removal of heat.

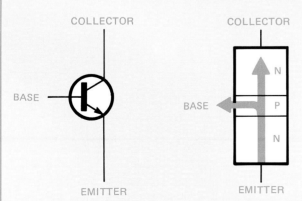

(Courtesy of Hewlett-Packard Company)

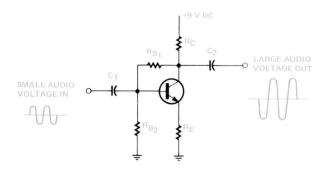

FIGURE 16–14 This simple one-transistor amplifier makes small audio signals larger.

functions can be performed. Transistors are used to make the small electrical impulses picked up from audiotapes and compact disks large enough to drive speakers. Circuits that make small currents or voltages larger are called **amplifiers**, Figure 16–14.

Transistors are also used to control the power to variable-speed motors. These circuits are often called controllers. They are very widely used in many ways to control and direct current representing information inside of a computer. These circuits are often called gates, Figure 16–15, and flip-flops.

Other Components

There are hundreds of variations of these and other types of components that are used to make circuits perform special functions. Semiconductor components used for special tasks include silicon-controlled rectifiers (SCRs), triacs, diacs, Zener diodes, and other devices used in power control circuits.

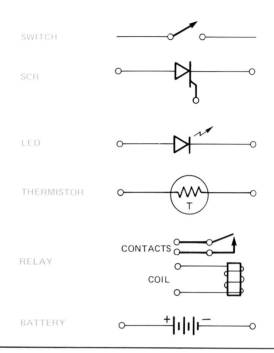

FIGURE 16–16 Common components and their schematic symbols.

Special resistors are made with resistance that changes with temperature or light. **Thermistors** change resistance with temperature change. They are used to control the temperature in ovens or to remotely sense temperature in a room. **Photoresistors** change resistance with light. They are used to automatically turn on lights when it gets dark or to measure light for photographic uses.

Other components that are used to build circuits include switches, relays (for remotely controlling a current), batteries, and LEDs (light-emitting diodes) that glow red, green, or yellow when current flows through them, Figure 16–16.

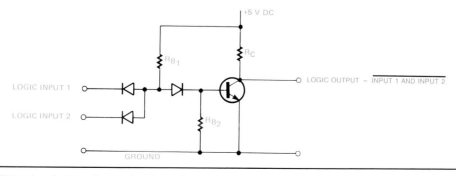

FIGURE 16–15 This simple transistor circuit performs a logic gate function needed in all computers.

Printed Circuits

Components may be connected together directly with wires. The wires are often joined together by soldering them. Soldering is a method of joining two wires together by melting a metal called solder on them. Solder is an alloy of tin and lead, with a very low melting point. The solder can thus be melted on two copper wires, joining them, without damaging the copper wires, which have a much higher melting point. Solder also has a very low resistance. It thus makes a good electrical connection between the two wires.

When wires are used to connect different components, care must be taken to ensure that the wires do not accidentally touch each other. This would cause an unintended flow of current from one part of the circuit to another (a short circuit). To prevent this, wires often have a cover of insulation on them. The insulation keeps current from accidentally flowing from one wire to another.

Early electronic circuits used large components that were connected to other components by several wires. Each wire was soldered by hand at both ends, Figure 16–17. This is often called point-to-point wiring.

Modern electronic components, however, have become quite small, with more connections to each one. It has become more difficult to hand solder wires to the components in the small spaces available. In addition, manufacturers need a method for making the same circuit over and over again, without errors, quickly, at low cost. To meet these needs, the printed circuit board was developed.

A **printed circuit board** is a thin board made of an insulating material, such as fiberglass. On one or both sides, a thin layer of a good conductor, usually copper, is plated right onto the board. Patterns etched in the copper form conducting paths, serving the same purpose as wires. Holes for mounting components are drilled in the board. The components are then soldered to the conducting paths on the board, Figure 16–18.

The conducting paths are photographically placed on the board. Therefore, many boards can be made with the same circuit appearing every time, with no mistakes or variations. After all the components are mounted on the board, Figure 16–19, they can all be soldered at one time by an automatic soldering machine. This machine produces rapid, uniform soldered connections.

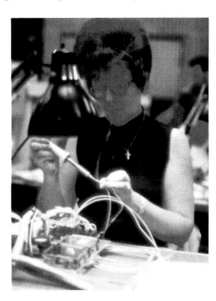

FIGURE 16–17 A soldering iron melts the solder, joining the wires to the component terminals. Soldering gives a good mechanical and electrical connection. *(Courtesy of Coopertools—Weller)*

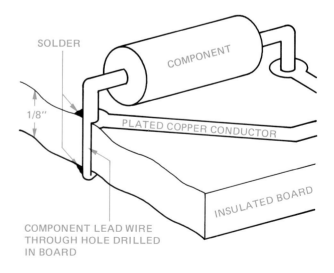

SOLDER

COMPONENT

1/8″

PLATED COPPER CONDUCTOR

INSULATED BOARD

COMPONENT LEAD WIRE THROUGH HOLE DRILLED IN BOARD

FIGURE 16–18 A typical component mounted on a printed circuit board

FIGURE 16–19 Components are inserted into holes in the printed circuit board and then soldered in place. *(Courtesy of Universal Instruments)*

Almost any type of component can be mounted on a printed circuit board, Figure 16–20.

Special types of printed circuit boards are made for certain uses. Some printed circuit boards have many layers sandwiched together for greater density. These are called multilayer boards. Others are made of flexible material that can be bent around corners into tight places. Printed circuit boards have found their way into most modern electronic assemblies.

Integrated Circuits

An invention that built on the development of the transistor was the **integrated circuit**. The integrated circuit has made possible many of our modern conveniences, including stereos, por-

table radios and televisions, autofocus cameras, fuel-efficient cars, and microprocessors.

An integrated circuit provides a complete circuit function on a single piece of semiconductor material. Integrated circuits are often less than one-tenth of an inch long by one-tenth of an inch wide. Transistors, diodes, resistors, conducting paths, and other circuit components are made on the integrated circuit, or chip, at the same time.

A chip is laid out by a circuit design engineer. The engineer makes a drawing of the chip components and paths. The drawing is several hundred times larger than the actual circuit. When the engineer is satisfied that the layout is correct, it is photographically reduced. This photographic reduction is called a mask.

The mask is used to put patterns on a thin wafer of semiconductor material by means of a process similar to making a photograph. Many tiny identical circuits are made at one time on one round wafer, which is several inches in diameter, Figure 16–21.

Integrated circuits have become more and more complex since they were invented in 1958. Many functions now are provided on a single chip, Figure 16–22. One measure of a chip's complexity is to count the number of transistors it contains. Integrated circuits are often classified in the following categories:

1. SSI—Small Scale Integration. Several dozen transistors on one chip,
2. MSI—Medium Scale Integration. Up to several hundred transistors on one chip,

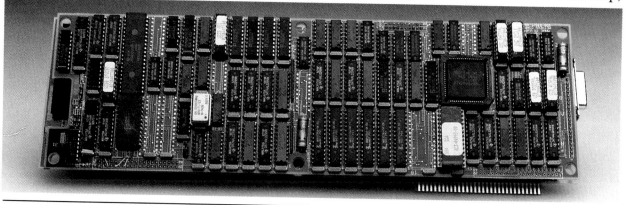

FIGURE 16–20 This printed circuit board has many components mounted in a tight space. *(Courtesy of Metheus Corporation)*

FIGURE 16–21 Each of the squares on these round wafers is a complete LSI integrated circuit containing thousands of transistors. Wafers are usually two to four inches in diameter and contain dozens or hundreds of integrated circuits. *(Courtesy of Matsushita/ Panasonic)*

FIGURE 16–22 Integrated circuits, like transistors, are packaged in larger plastic or metal containers for protection and ease of handling. Often, the more complex the integrated circuit, the more input and output connections it needs, requiring the package to be larger. *(Courtesy of Hitachi America, Ltd., Semiconductor & IC Division)*

3. LSI—Large Scale Integration. Up to several thousand transistors on one chip,
4. VLSI—Very Large Scale Integration. Up to several hundred thousand transistors on one chip.

At the same time that the integrated circuit was giving more and more functions in a smaller and smaller space, the cost per function was also coming down very fast. If similar performance gains and price decreases had occurred in the automobile industry over the last ten years, a Rolls Royce today would cost $500 and would get 1,500 miles per gallon.

Summary

The harnessing of electricity and electronics in the last hundred years has greatly affected the way we live. The biggest impact of electricity may be in the field of communications.

All materials are made out of atoms. Atoms contain smaller subatomic particles called protons, electrons, and neutrons. Protons have a positive electric charge, electrons have a negative electric charge, and neutrons have no charge at all. Particles with similar charges repel each other, while particles with opposite charges attract each other.

Materials whose atoms give up electrons easily are called conductors. Materials whose atoms hold tightly to their electrons are called insulators. Electric current flows when electrons move through a material. The force that makes electric current flow is voltage. The measure of a material's opposition to current flow is its resistance. Ohm's law relates current, voltage, and resistance.

Current that does not change polarity is called direct current (DC). Current that changes polarity periodically is called alternating current (AC). The number of times that a current changes its polarity and changes back to its original polarity in one second is called its frequency. AC voltages can be easily changed by devices called transformers. Transformers do not operate on DC. A device that changes AC to DC is called a rectifier.

Simple electronic components include resistors, diodes, transistors, and capacitors.

Components are connected together to form circuits that do certain tasks. Resistors give precise current control in circuits. Diodes allow current to flow in one direction, but not in the other. Transistors provide the control of a large amount of current by varying a small current. Capacitors are used to store electric charge, and to pass AC while blocking DC.

Some components, including diodes and transistors, are made from semiconductor materials. Integrated circuits are also made from semiconductor materials. The most commonly used semiconductors are silicon and gallium arsenide.

Components can be formed over and over into the same circuits using printed circuit boards. Printed circuit boards come in many shapes and sizes, and are found in most modern electronic assemblies.

Integrated circuits carry out whole circuit functions on a single piece of silicon, often less than $\frac{1}{4}''$ by $\frac{1}{4}''$. Integrated circuits can be very complex, containing several hundred thousand transistors, and doing many circuit functions. Many of our modern appliances are made possible through the use of integrated circuits.

REVIEW

1. Draw a picture of an atom with five protons, two neutrons, and six electrons. What electric charge does this atom have?
2. A resistor of 100 ohms is connected to a 12 volt battery. What current flows through the resistor? If the resistor is changed to a 1000 ohm resistor, what is the new current flow?
3. How did the use of electricity affect communications in the late 1800s and early 1900s? Are new developments in electronics still affecting communications? Give examples to support your answer.
4. Why is electric power delivered to your home as alternating current? What types of appliances use AC directly? What types of appliances convert AC to DC internally?
5. Describe the function of the following electronic components: resistor, capacitor, diode, transistor, thermistor.
6. Name the two most commonly used semiconductor materials.
7. What is the purpose of an amplifier?
8. Draw a sketch of a small printed circuit board.
9. What are the advantages of using a printed circuit board rather than point-to-point wiring?
10. Why was the invention of the integrated circuit important in the history of technology?
11. Name three places where integrated circuits are used today.

CHAPTER 17

Communicating by Wire: Telegraph and Telephone

KEY TERMS

cellular radio	electromagnet	relay
central office switch	Integrated Services Digital Network (ISDN)	tandem switch
cordless telephone	Morse code	telegraph
diaphragm	multiplex	telephone
dual tone multiple frequency (DTMF)	private branch exchange (PBX)	

Introduction

The electronic revolution in communications began in the 1800s with the invention and development of the telegraph. Electric signals were sent over wires for the first time to send messages from one place to another. Because electric signals travel over wires at nearly the speed of light, communication of short messages became almost instantaneous.

Improvements in the telegraph led to the invention and development of the telephone. In the more than 100 years since the invention of the telephone, there have been many changes and

Morse's first telegraph used an **electromagnet** to deflect a pencil. An electromagnet is a coil of wire that makes a magnetic field (force) when current flows through the wire. The pencil made marks on a strip of paper moving under it, Figure 17–2. It became apparent that it was just as easy for an operator to interpret the clacking of the electromagnet as it was to interpret the marks made on the paper. The telegraph became an audio rather than a visual device.

In 1837, Morse showed his device to prospective financiers. During this demonstration, he sent a message over seventeen hundred feet of wire. One of the people present agreed to back Morse's work only if his son could become Morse's assistant. The young assistant, Alfred Vail, turned out to be a mechanical genius. He helped Morse redesign the telegraph into a more practical device.

FIGURE 17–1 Claude Chappe's optical telegraph *(Reprinted from TECHNOLOGY IN YOUR WORLD by Hacker and Barden, © 1987 Delmar Publishers Inc.)*

improvements in telephone technology and service. The telephone remains the cornerstone of modern electronic communication.

The Telegraph

The word **telegraph** means to draw or make signs at a distance. The first telegraphs were optical telegraphs that signaled distant locations visually. During the French Revolution, Claude Chappe developed an optical telegraph system that used high towers located five to six miles apart. Each tower had wooden arms that could be swung into different positions to signal messages. The positions of the arms were read by a telegraph operator who stood at the top of the tower and read the signals through a telescope, Figure 17–1.

Although there were many early electric telegraphs proposed or built in the early 1800s, the first to have commercial success in the United States was one built by Samuel F. B. Morse.

FIGURE 17–2 Morse's first telegraph was not a very practical device. The sender used zinc slugs (strips) with varying spaces between them, representing letters of the alphabet. The pattern represented by the metal strips alternately connected and disconnected an electric circuit. The receiving unit used a pencil that responded to the pulses from the sender. The paper strip displayed a series of marks that corresponded to the spaces between the metal strips. *(Courtesy of the Smithsonian Institution)*

In 1843, Morse got a $30,000 grant from the United States Congress to build a telegraph system between Baltimore, MD and Washington, D.C. In 1844, he finished the system and sent the first famous message, "What hath God wrought?"

Soon newspaper reports were using the telegraph to send news releases. Railroads used the telegraph to dispatch trains and set signals. The telegraph provided the means of rapid communication needed by the railroads to allow their expansion. In 1856, the Western Union Company was started. In the ten years that followed, they strung 75,000 miles of telegraph wire.

Problem Solvers in Communication

Samuel F. B. Morse

Samuel Finley Breese Morse started his professional career as a portrait painter. He was somewhat successful at it, making $250 per week by 1817. After studying art in Europe, he returned to the United States on a ship called the *Sully* in 1832. While on board, Morse heard about European experiments in communication using electricity.

When he came back to the United States, he sought out the help of an expert in electromagnetism, Joseph Henry. Together they developed a new kind of telegraph system that was simpler to build than European systems. Henry also applied his invention of the relay to the system. This extended the distance over which the system would work.

In 1843, Congress voted on a bill to award $30,000 to Morse to build a test circuit from Baltimore to Washington, D.C. When the bill was presented, many congressmen thought that the magnetic telegraph was a joke. Morse was ready to give up. He bought a train ticket back to New York, and only had pocket change left.

The next day, the daughter of Morse's friend, the Commissioner of Patents, gave him the good news that the bill had been passed minutes before the congressional session was to end. Morse promised her that she could send the first message on the telegraph.

A year later, when the system was completed, she chose the words, "What hath God wrought?"

Samuel F. B. Morse, a self-portrait *(courtesy of the Addison Gallery of America Art, Phillips Academy, Andover, MA)*

More than one hundred years later, President John F. Kennedy used the same words when he finished the first telephone message over satellite. He was speaking to an African Prime Minister. Their voices travelled a distance of about 45,000 miles!

How a Telegraph Works

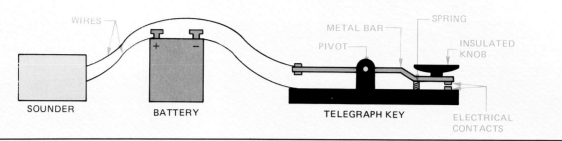

(Reprinted from LIVING WITH TECHNOLOGY by Hacker and Barden, © 1988 Delmar Publishers Inc.)

A simple telegraph system contains a telegraph key, a power source, a buzzer or sounder, and wires to connect the sending station with the receiving station. When the sending operator presses the key down, the contacts conduct electricity, sending power to the buzzer.

The sending operator holds the key down for short and long amounts of time. This forms a code in which different groups of shorts and longs (dots and dashes) stand for each letter, number, and punctuation mark. The Morse code is the most often used code of this type. It can also be used for signaling with lights.

Over very long distances, the electrical signal on the wire gets weaker and weaker. When it gets so weak that it will not operate the sounder at the far end, communications stop. To extend the useful distance of the telegraph, a **relay** is used. The relay has a coil that operates on a small amount of current. The electromagnetic coil operates a set of contacts that act as a new send-

ing key, switching the current from a new battery to send a re-energized signal along the line.

While telegraph is not widely used today, where it is used, machines send and receive the messages.

The Morse Code		
A • —	K — • —	U • • —
B — • • •	L • — • •	V • • • —
C — • — •	M — —	W • — —
D — • •	N — •	X — • • —
E •	O — — —	Y — • — —
F • • — •	P • — — •	Z — — • •
G — — •	Q — — • —	. • — • — • —
H • • • •	R • — •	/ — • • — •
I • •	S • • •	? • • — — • •
J • — — —	T —	— — • • • • —
1 • — — — —	4 • • • • —	8 — — — • •
2 • • — — —	5 • • • • •	9 — — — — •
3 • • • — —	6 — • • • •	0 — — — — —
	7 — — • • •	

The Morse Code *(Reprinted from LIVING WITH TECHNOLOGY by Hacker and Barden, © 1988 Delmar Publishers Inc.)*

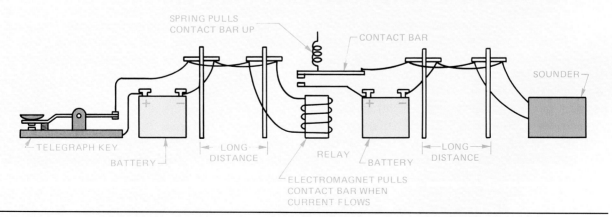

The relay is used to extend the distance of a telegraph system.

Telegraph systems grew very fast, both in Europe and the United States. Thousands of miles of telegraph cable were strung across land. Continents and islands were still isolated by the water separating them.

While some cables were laid across rivers and small lakes, the first important body of water to be spanned was the English Channel, which separates England from the rest of Europe. In 1850, a retired antique dealer and his brother laid the first cable across the channel. It provided some communication for one day, until it was pulled up and destroyed by a fisherman. A second cable, this time with an armored covering, was laid in 1851 and worked very well.

Other, larger bodies of water were then crossed by cables. These made larger and larger telegraph communications networks. The big challenge, however, was to provide instant communication across the Atlantic Ocean, from Europe to the United States. This difficult task was to take years of trying, with several failures before success was achieved.

The first cable to be successfully laid on the ocean bed from England to North America was finished in 1858 by Cyrus Field, an American entrepreneur. When the first messages were sent between England and America, people were very excited and had a celebration in New York with fireworks that set City Hall on fire. The cable was finished at great expense, and it still did not work well. It stopped working after less than one month's operation. But it proved that a transatlantic cable was possible. It also encouraged other attempts to build a reliable cable, Figure 17–3.

Finally, in 1866, another expedition led by Cyrus Field laid a new transatlantic cable that worked well. It was armored to make it stronger than the earlier cable. The armoring, however, had increased the weight to more than 5,000 pounds per mile. More than 2,000 miles of underwater cable were needed, making the weight of the cable more than any ship of the day could carry—any ship except one.

A giant ship, The *Great Eastern*, had been built several years before. The *Great Eastern* was five times larger than any other ship then afloat. It had found no practical commercial use, how-

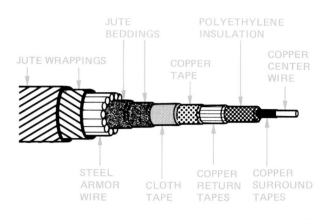

FIGURE 17–3 Cutaway view of a modern underwater cable

ever, before it was converted to a cable laying ship. It did this job very well, and was used to lay underwater cables all over the world.

Cyrus Field's transatlantic cable was the first in a long series of long distance underwater cables. These cables have linked the continents to provide instant telegraphic communication. Some parts of the early underwater cables were in use for as much as 100 years. The next challenge was to provide direct voice communication across wires.

The Telephone

The **telephone** was one of many devices that was invented by accident. Alexander Graham Bell was a speech teacher who was working on ways to send several telegraph messages at the same time over one wire. He knew that if a musical note of a certain frequency was played near a group of tuning forks, only the tuning fork tuned to the frequency of the note would vibrate.

Bell thought that he could send several musical notes over the same wire, and that several tuning forks at the other end would sort out the notes. Although Bell was never successful in making this system work, it is the same principle used in many electronic communication circuits today.

In 1875, Bell's assistant, Thomas Watson, was

hanging metal strips to make musical notes while Bell was in another room listening to a receiver. One of the metal strips accidentally got stuck and acted as a diaphragm, picking up the jingling noise of the other strips. Bell heard the jingling noises, and realized what had happened.

How a Telephone Works

The principle of the **diaphragm** is basic to the operation of the telephone. Simple telephones use a thin metal disk as a diaphragm. When you speak into the mouthpiece of a modern telephone, the diaphragm vibrates. It pushes against a container holding carbon granules. As the granules are packed more tightly by the diaphragm's pushing on them, their electrical resistance goes down. As the diaphragm relaxes the pressure on the granules, the electrical resistance goes up.

If a voltage is applied to the container holding the carbon granules, then the varying resistance changes the current flowing through the circuit. A varying electric current that exactly represents speech is thus generated.

The telephone receiver earpiece has a diaphragm made out of a metal that is attracted to a small electromagnet. The current flowing through the electromagnet is the varying current produced by the mouthpiece of the other telephone. As the current varies, the attraction of the electromagnet for the diaphragm also varies. This causes the diaphragm to vibrate in just the way that the mouthpiece diaphragm first did. The vibrating diaphragm moves the air, turning varying electric current back into sound, Figure 17–4.

The difference between the telephone and the telegraph is that the telegraph operates using on-off pulses of electric current, while the telephone uses a smoothly varying, continuous current. Telegraph signals could only be sent and received by someone who had been trained to understand the Morse code. A telephone could be used by anyone without any formal training.

The Telephone System

The growth of widespread telephone use was slow at first. This rate increased as the telephone

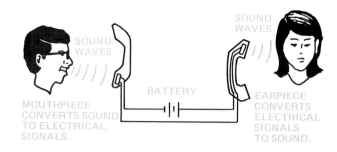

FIGURE 17–4 A telephone converts sound into electrical signals and electrical signals back into sound. *(Reprinted from LIVING WITH TECHNOLOGY by Hacker and Barden, © 1988 Delmar Publishers Inc.)*

network grew. At first, telephones were not as useful as they are today because there were only a few others that could be called. (How useful would your telephone be if it could only call two other people that you knew?) Telephone networks were started only within towns or small areas. Communication with other, distant towns was not possible.

As the number of telephones grew, networks grew to connect more people over wider areas. This made having a telephone more desirable, further increasing the number of telephones. By 1900, there were about one million telephones in use in the United States. When Bell died in 1922, all 13 million telephones in Canada and the United States were shut off for one minute in tribute.

In the United States today, there are more than 200 million telephones, and any phone can be used to communicate with any part of the world, almost instantly. There are about 1.3 billion telephone conversations each day in the United States.

Telephone Switches

One of the first problems in building today's telephone network was how to connect a telephone to any one of a very large number of telephones. To understand the problem, think of a network of three telephones. If any one phone needs to be connected to any other one, a switch panel of two positions is needed at each telephone. Two wires are needed from each house, each running directly to the other two

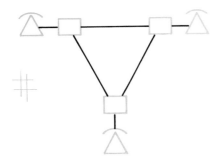

THREE–PHONE NETWORK

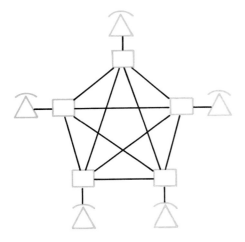

FIVE–PHONE NETWORK

△ = TELEPHONE   ☐ = SELECTOR SWITCH

NUMBER OF PHONES	TOTAL NUMBER OF WIRES NEEDED
3	3
5	10
10	45
100	ABOUT 5,000
1000	ABOUT 500,000
1 MILLION	ABOUT 500 BILLION

FIGURE 17–5 The number of wires needed in a telephone system without a central switch becomes very large as the number of telephones in the system increases.

houses. If five telephones are in the network, each house will have a four position switch, and there will be a total of ten wires in the system, Figure 17–5.

For one hundred telephones, a 99 position switch would be needed at each house. A total

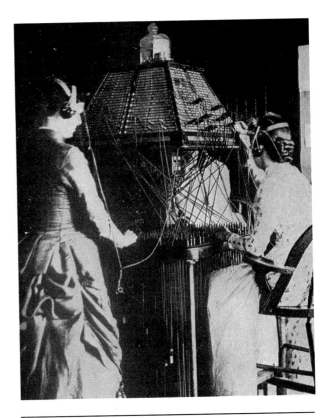

FIGURE 17–6 These operators are shown working at a small switchboard in the 1880s. *(From the collection of Tony Minichiello, Concord, NH)*

of 4,950 wires would be needed in the network. A small town with 1000 telephones would have a total of 499,500 wires, not including long distance cables. Clearly, it would not be possible to build a telephone network that could connect any two of millions of phones using this method of wiring.

This problem is overcome by the use of a central switch, rather than a switch at each telephone. The earliest telephone systems were built with a central patch board attended to by an operator. Each telephone had one line between it and the operator's patch board. A person wanting to make a call first called the operator and told her the name of the person to be connected to, Figure 17–6. (The first operators were boys, but they were found to fool around too much, and almost all operators were women for over fifty years.)

The operator would then make a connection by plugging a cord into the correct socket, joining the two wires from the two telephones, and making the connection between them. This system needed only as many wires as there were phones, but needed an operator to operate the switchboard. Early switchboards did not operate all day and night, but maintained business hours. No calls could be made if the operator was not at the switchboard, Figure 17–7.

Several attempts were made to design an automatic switch. The most successful was invented by Almon Strowger and first put into commercial service in 1892. Strowger switches are still in use today in a few places. The Strowger switch used electromagnetic stepper switches to count the pulses generated by a

The Automatic Telephone Switch

Inventions are often made by people who require their use. Almon Strowger was a funeral director in Kansas City, MO. The town's telephone operator was the wife of the owner of a competing funeral parlor. When people called the operator and asked for a funeral parlor, she connected them with her husband's funeral parlor. Mr. Strowger's business declined.

In order to save his business, Strowger designed a device that made telephone connections automatically, bypassing the operator. The switch he invented was the first automatic system to be used in a public telephone exchange. Strowger switches are still in use today in some places.

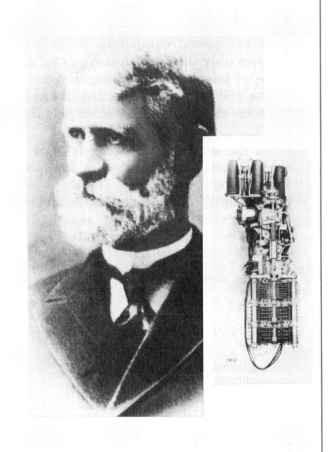

Almon Strowger and his automatic telephone switch
(Courtesy of AT&T Bell Laboratories)

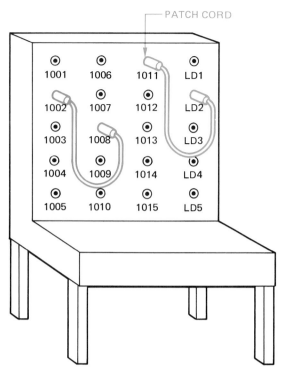

FIGURE 17-7 Connections are made in a switchboard by plugging two circuits together using a connection cord. Each phone is connected to a jack on the patch board. Patch cords make connections from one telephone to another. In this patch board, phone 1002 is connected to 1008, and phone 1011 is connected to long distance line 2.

telephone dial, automatically making a connection based on the dialed number.

Modern telephone switches are all automatic. They are actually special purpose computers built to do nothing but make and control telephone connections. Operators are only needed to help customers who are having a problem or to give information. The information operator uses a computer to locate numbers, and then an automatic, artificial voice announces the number in most modern systems, Figure 17-8.

Telephone switches are used in two basic ways. One kind of switch is used to connect individual telephones to each other and to the telephone network. When this kind of switch is located right in a customer's building, it is called a **private branch exchange**, or **PBX**, Figure 17-9. When it is located at the telephone company, it

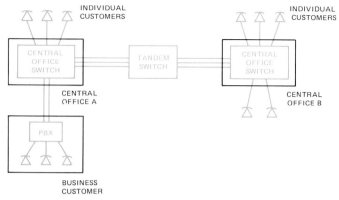

FIGURE 17-8 The modern switched network uses different kinds of switches to complete local and long-distance connections.

FIGURE 17-9 This PBX provides telephone service for several thousand people in one building. *(Photo by Robert Barden)*

is called a **central office switch**, Figure 17-10. The total number of wires needed to connect this switch is the same as the number of telephones connected to it, plus a small number of wires to connect the switch to other switches.

The second kind of switch is one that is used only to connect other switches to each other to pass calls from one local area to another. This kind of switch is called a **tandem switch**.

FIGURE 17–10 A telephone central office serves the people of a community or a neighborhood in a large city. *(Photo by Robert Barden)*

Connecting Central Offices

The telephone in your home is connected to the nearest central office by a pair of copper wires. This is often called a copper connection. The central office may be linked to other central offices by many different transmission paths, however, Figure 17–11.

It is common for two central offices that are close together to be linked by large bundles of copper pairs called trunk cables. The wires may be run on overhead telephone poles, or underground through conduits. Each pair of wires in the cable carries one conversation at a time. Therefore the total number of conversations that can be carried is set by the number of pairs in the cable.

In many areas, fiber-optic cable is replacing the large bundles of copper wires between central offices. Fiber-optic cable carries information on beams of light that are guided through a flexible glass strand that is small and lightweight. One strand of fiber-optic cable can carry thousands of simultaneous conversations through a technique called **multiplexing**. A multiplexer is an electronic device that combines many telephone or computer signals onto one wire or fiber-optic cable, Figure 17–12. At the other end of the wire or fiber-optic cable, a demultiplexer is needed to separate the signals.

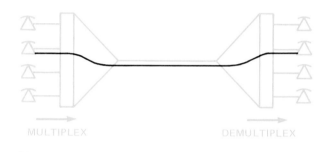

FIGURE 17–12 Multiplexers combine many different conversations so that they may be sent on the same wire.

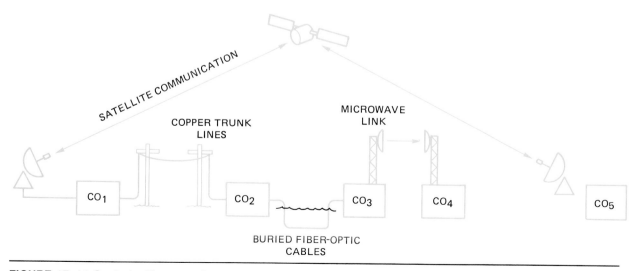

FIGURE 17–11 Central offices can be connected by many different kinds of communication links.

Central offices can also be connected by microwave radio links. Microwave radio links also take advantage of multiplexing to combine many conversations onto one radio link. It is not uncommon for more than one thousand conversations to pass over one radio link.

Telephone offices on two continents are often connected through a satellite link. Communication satellites act as relay stations in the sky for microwave radio links. Communication satellite operation is described more fully in Chapter 19.

Telephone Signaling

When you make a telephone call, you probably take for granted all the steps involved. First, you pick the handset off of its hook. The hook has a switch built into it that sends a signal to the central switch indicating that you would like to make a call. If the central switch is ready to accept your call, it sends back a dial tone. When you hear the dial tone, you start to dial.

Dialing your telephone sends a signal to the central office switch to enable it to automatically connect you with the telephone you are calling. For many years, the only method of signaling was a series of pulses generated by a rotary dial. The dial was spring loaded so that it always returned to its starting position. The number of pulses generated depended on how far around it was turned by the finger of the person dialing the number. The dial was designed to mechanically send ten pulses per second.

Modern telephones signal the central office through a series of tones. Each button that is pressed by the caller activates two tones at the same time. The combination of the two tones is sensed at the central office switch. The number that is "dialed" is read into the switch's memory. (It is interesting to note that even though numbers are selected by pushing buttons on modern telephones, it is still referred to as "dialing.")

Because this method uses two tones, it is called **dual tone multiple frequency**, or **DTMF** signaling, Figure 17–13A and B. Two tones are used to reduce the possibility of a mistake, and to allow more numbers to be signaled with fewer tones.

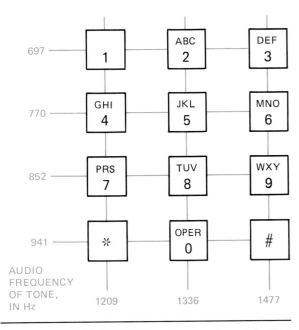

FIGURE 17–13A In DTMF signaling, each push button activates two tones that are then sent to the central office switch. Pushing the "4" button generates two tones at 770 Hz and 1209 Hz. FIGURE 17–13B Digital central office switches, like this one in Dallas, Texas, enable telephone companies to offer new services to customers. *(Courtesy of Southwestern Bell Corporation)*

DTMF signaling can also be used for communicating with computers once a call has been made. Using DTMF signaling, people can check on balances in various financial accounts or get other information at any time of the day or night, even when the bank or brokerage house being called is not open. DTMF is also used for remote control of various devices.

When you speak into the telephone, your voice is most probably converted to digital (computer-like) data at some point in its travels to the phone at the other end of your connection. Converting voice to digital form improves the quality of the voice signal at the other end. This results in less expensive transmission, and allows the telephone company to provide better services. The conversion of voice to data is described more fully in Chapter 20.

Modern Telephone Services

The adoption of new technologies to provide telephone service has produced new forms of telephone services. The integration of different technologies brings about new solutions to communication problems.

One device that has resulted from combining technologies is the **cordless telephone**. A cordless phone includes a base unit and a hand-held unit, Figure 17–14. Each has a low-power radio transmitter and receiver so that they can communicate with each other. The base unit is connected to a normal telephone line. It relays signals from the hand-held unit to the telephone line and vice-versa, Figure 17–15. The hand-held unit is powered by a battery, and has a range of several hundred feet from the base unit.

Car telephones are now very popular in many areas of the country. A relatively new technology called the **cellular radio** is responsible for the fast growth of car telephone systems. Like cordless phones, car phones transmit radio signals to a base station. Cellular phones, however, have a much wider range than cordless phones.

The base station in one cell is connected by phone lines to the base stations in adjacent cells. When a car telephone moves out of range of one base station, its call is handed off to the base

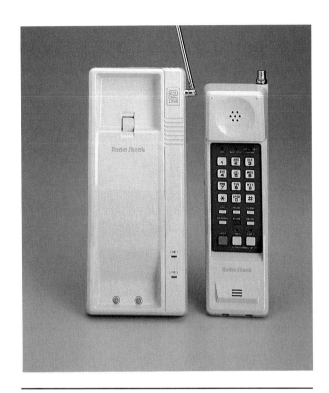

FIGURE 17–14 A cordless telephone (Courtesy of Tandy/Radio Shack)

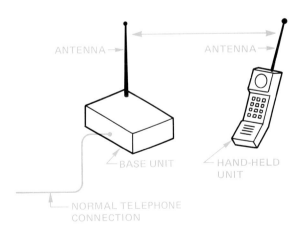

FIGURE 17–15 A cordless phone has a base unit and a hand-held unit that communicate with each other by radio transmission.

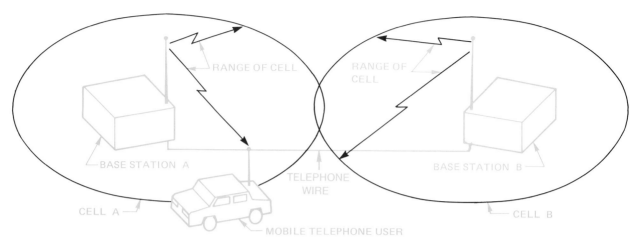

FIGURE 17–16 Cellular phones communicate with base stations within each cell. The base stations are connected to the base stations in nearby cells. As a moving phone goes from one cell to another, communication continues uninterrupted. *(Photo courtesy of Southwestern Bell Corporation)*

station in the next cell, Figure 17–16. In this way, continuous service can be given to car telephones traveling long distances, from one cell to another. Cellular phones are also available in hand-held portable form that fit into a briefcase or a pocketbook.

The introduction of digital and computer technology to the phone system has made such extra services as call waiting, call forwarding, automatic call back, speed dialing, and last number redial available to many businesses and to some homes. A large step toward making these and other services more widely available is the introduction of **ISDN** (**Integrated Services Digital Network**). It is now being tested and expected to be available in the mid 1990s.

ISDN will offer a wide variety of services to telephone subscribers, including the ability to know who is calling before answering the phone (see Chapter 22). Other services likely to be offered include accessing the telephone directory through a home terminal rather than a phone book or information operator. The use of the phone to automatically call for help in the event of a medical or security emergency will probably be more widespread than it is today.

Summary

The electronic revolution in communications began in the 1800s with the invention and development of the telegraph. The telegraph is a communication system that sends information using a code. The code is made up of a series of long and short bursts of electric current that are converted into buzzing or clicking sounds that an operator or a machine interprets as letters.

The most commonly used telegraph code is the Morse code, developed by early telegraph pioneer Samuel F. B. Morse. As telegraph networks grew, instant communication was available at very long distances for the first time. Underwater cables were put in place across lakes, rivers, and finally across oceans, giving worldwide communications by the end of the 1800s.

The invention of the telephone built on telegraph technology. Alexander Graham Bell invented the telephone while he was trying to improve telegraph transmission. A telephone mouthpiece changes sound into a varying electric current. A telephone earpiece changes a varying electric current back into sound.

The difference between a telegraph and a telephone is that the telegraph operates using on-off pulses of electric current, while the telephone uses a smoothly varying, continuous current. Telegraph signals can only be sent and received by someone who is trained to understand the Morse code. A telephone can be used by anyone without any formal training.

A telephone switch is needed to direct calls and establish circuits between people who want to communicate. The earliest switches were switchboards operated by hand. Modern switches are special purpose computers that automatically route calls and offer added features, such as call waiting and call forwarding. Switches may be located at a business customer's office and in a telephone central office.

Telephones are connected to the switch by copper wires. Switches may be connected to each other by copper wires, fiber-optic cable, microwave radio links, or satellite links. Many conversations can be sent on one link using a technique called multiplexing.

Telephones may signal a telephone switch using dial pulses or using DTMF (tone) signaling. DTMF signaling can also be used to control remote devices or communicate with computers.

The integration of new and different technologies with existing telephone technology has resulted in new services. Two examples of this are the cordless telephone and the cellular, or mobile, telephone. The adoption of ISDN will add new features to be offered by the telephone system.

REVIEW

1. Draw a diagram of a simple telegraph system.
2. What limits the distance of a simple telegraph system? Draw a telegraph system that includes a way to extend the length of the system.
3. Write the Morse code symbols for "Samuel F. B. Morse."
4. Why was the laying of the first working transatlantic cable important?
5. Draw a diagram and briefly explain the operation of a modern telephone.
6. Why did the telephone become more widespread than the telegraph?
7. Draw a telephone network with four telephones and no central switch. How many wires would be needed in such a network to allow communications from any one phone to any other phone?
8. Draw the same network as in Question 7, but with a central switch. How many wires are needed in a network with a switch?
9. Find out where the telephone central office is near your home. If possible, visit it and describe it.
10. Name four ways in which central offices can be connected to each other.
11. What is the advantage of fiber-optic cable over copper cable for carrying telephone signals from one office to another?
12. Explain how multiplexing works and why it is used.
13. Describe how the combination of new and different technologies brings about new services. Give two examples.

CHAPTER 18

Radio and Television

KEY TERMS

amplify	frequency modulation (FM)	scanning
amplitude modulation (AM)	ionosphere	share
antenna	modulation	sync
broadcast	propagation	transmitter
carrier	radio wave	video
electromagnetic field	rating	watts
electron gun	receiver	wavelength

The Early Development of Radio and Television

In the early 1850s, the Scottish physicist James Maxwell theorized that energy could travel through space. In 1887, Heinrich Hertz tried to prove Maxwell's theories in Germany. Using a high-voltage generator, Hertz produced a spark that jumped across a gap. A small distance away, he placed a metal ring with a similar gap. When a spark jumped across the original gap, a second spark appeared in the gap of the metal ring. Hertz

FIGURE 18–1 Marconi's receiving antenna for the first transatlantic radio transmission was a long wire held up by a large kite. Marconi is at the extreme left. *(Courtesy of GEC-Marconi Limited)*

thus proved the spark radiated electrical energy in the form of the radio waves.

Guglielmo Marconi, an Italian, applied Hertz's ideas to communicating across distances. He developed an antenna wire that launched energy into the air at one location and recovered it at the other. Using these antennas, he was able to send an on-off coded signal over a distance of about two miles in 1896. In 1901, he had developed his ideas to the point that he could send coded messages across the Atlantic Ocean, Figure 18–1.

Most of the early radio pioneers focused on sending telegraph signals. It was not until 1906 that voice transmission was demonstrated. On Christmas Eve, a station near New York City sent out two speeches, a song, and a violin solo. Two years later, Lee deForest, a radio pioneer and inventor, sent voice transmissions from the Eiffel Tower in Paris. In 1916, he sent news bulletins covering the presidential election. In 1920, the first commercial radio broadcast station, KDKA in Pittsburgh, began regular operation. TV was first demonstrated in 1926, but commercial TV broadcasting did not start until the late 1940s.

How Radio Works

As its early name "wireless" implies, radio provides communication between two points with-out using wires. In communication system terms, the air between the two points is used as the channel. Information is carried on radio waves that travel between the two points. The information can be voice, music, visual images, or digital data.

Electrical signals that flow through a wire set up both electric and magnetic fields around the wire. Together, these two fields are called an **electromagnetic field**, Figure 18–2. The electromagnetic field is strongest near the wire, and decreases in strength further away from the wire.

If the wire is carrying an alternating current, the electromagnetic field around the wire expands and collapses as the voltage constantly changes. You will remember from Chapter 16 that a wire placed in a changing electromagnetic field will have a current induced in it (this is how a transformer works). Using this same principle, an electromagnetic field set up by an alternating current flowing in one wire can be converted to an alternating current of the same frequency flowing in another wire some distance away, Figure 18–3.

Depending on the frequency of the AC voltage, the wire that sets up the field can be adjusted in size, shape, and orientation to make the field as large as possible in one direction. The sensing wire can also be adjusted to capture as much of the field as possible. Wires that are specially sized and shaped to make the best conversion of electric current to electromagnetic field are called **antennas**. Fields that are used for communica-

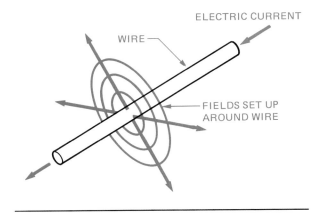

FIGURE 18–2 Current flowing through a wire causes a field to be set up around the wire.

The Sinking of the *Titanic*

One of the most dramatic events of radio history involved the great ship *Titanic*. The "unsinkable" ship was on her first and last voyage, an Atlantic crossing from England to New York. On April 14, 1912, she struck an iceberg and went down with the loss of 1,513 lives. The ship's radio operator, J. G. Phillips, stayed at his post trying to summon aid, and went down with the ship. Some of the entries from the logbook of the *Carpathia*, one of the ships that came to the *Titanic's* rescue, are as follows:

Sunday, April 14, 1912

11:20 PM   Hear Titanic calling SOS (Save Our Ship) and CQD (Come Quick—Danger). Titanic says, "Struck iceberg, come to our assistance at once. Position: Lat. 41.46 N; Long. 50.14 W:

11:30 PM   Course altered; proceeding to the scene of the disaster.

11:45 PM   Titanic says weather is clear and calm. Engine room getting flooded.

Monday, April 15, 1912

00:10 AM   Titanic calling CQD. His power appears to be greatly reduced.

00:25 AM   Calling Titanic. No response.

At daybreak, the *Carpathia* arrived at the scene of the sinking, and rescued 710 survivors. Other ships had been much closer and could have rescued many more survivors, but they either didn't have radios or didn't have their radios on because of the late hour. Earlier in the evening, the *California* had tried to warn the *Titanic* of icebergs in the area. The *Titanic* radio room was so busy sending passenger messages to Cape Cod that they did not acknowledge the *California's* warnings.

Many lessons were learned from the *Titanic* disaster. As a result, new rules for safety at sea were developed. Many of these rules are still used today, including a requirement for commercial ships to keep a radio tuned to an emergency calling frequency. The reports of the disaster sent to a waiting world from the *Carpathia* and other ships using their radios firmly established radio

The *R.M.S. Titanic (Courtesy of the Trustees of the Science Museum, London)*

A typical ship's radio room in the early 1900s *(Courtesy of the Trustees of the Science Museum, London)*

as an important communication means in the minds of the public.

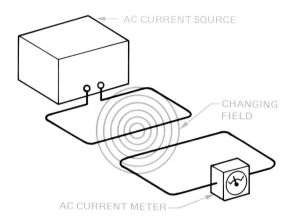

FIGURE 18–3 The changing field set up by an alternating current flowing in one wire will induce an alternating current of the same frequency to flow in a second, unconnected wire.

tion are called **radio waves**. Antennas that are used to change current to radio waves are called transmitting antennas, while antennas that are used to pick up fields and turn them into currents are called receiving antennas.

The lower the frequency of the radio wave, the larger the wavelength. The larger the wavelength, the larger the antenna must be to be effective. Antennas used for radio communication range in size from more than one hundred feet in length to less than one foot across.

As the radio wave leaves the antenna, it spreads out in different directions. Signals will radiate (spread) in all directions unless they are focused by the antenna. Focusing a signal moves some of the energy that would have gone in other directions into a defined direction, Figure 18–4. This results in a stronger signal in that direction and a weaker signal in the other directions. For a given frequency, a larger antenna can focus more of the total energy in one direction than a

FIGURE 18-5 Antennas come in many different sizes for different purposes. This small antenna is a TV antenna that receives wavelengths of one to six meters. *(Photo by Robert Barden)*

FIGURE 18–4 Antennas can send waves in all directions, or be designed to concentrate the radio energy in one direction.

smaller one. The actual size of the antenna thus depends on both the frequency and the need to focus the radio energy, Figure 18–5.

As the signal travels farther from the antenna, it becomes weaker in strength. It also becomes more likely that the signal will be affected by noise caused by man-made sources, lightning, and other natural sources. All of these combine to make it important that the receiving antenna be as efficient as possible in picking up the signal.

Transmitting the Signal

In a radio communication system, the **transmitter** generates a radio frequency signal that is sent to the antenna for transmission through the air to the receiver. The transmitter places the information to be sent onto the radio frequency signal. The radio frequency signal is often called the **carrier** or RF carrier because it carries the information from one place to another. The transmitter places the information on the carrier signal by changing the carrier in some way. This process is called **modulation**.

Frequency and Wavelength

All electromagnetic waves travel through space at the same speed. Since both light and radio waves are forms of electromagnetic waves, you may already know that this speed is 186,000 miles per second, or about 300 million meters per second. Thus, a wave sent by a radio transmitter (or a light source) is 186,000 miles away after one second. This is true no matter what the frequency of the wave.

During that same second, the radio transmitter generates the number of waves defined by its frequency. Thus, a 10 MHz transmitter will generate 10 million cycles in one second. These 10 million cycles cover 300 million meters (186,000 miles) in one second. Therefore, the **wavelength**, or length of one cycle (wave), is:

$$\text{Wavelength} = \frac{300 \text{ million meters per second}}{10 \text{ million cycles per second}}$$

$$\text{Wavelength} = 30 \text{ meters}$$

In general, the wavelength of a radio or light wave in free space is:

$$\text{Wavelength} = \frac{\text{speed of light in meters per second}}{\text{frequency in Hertz}}$$

$$\text{Wavelength} = \frac{300,000,000 \text{ meters per second}}{f, \text{ in Hz}}$$

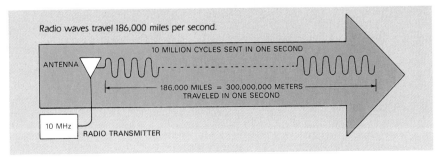

Radio waves travel 186,000 miles per second.
10 MILLION CYCLES SENT IN ONE SECOND
ANTENNA
186,000 MILES = 300,000,000 METERS TRAVELED IN ONE SECOND
10 MHz
RADIO TRANSMITTER

(Reprinted from LIVING WITH TECHNOLOGY by Hacker and Barden, © 1988 Delmar Publishers Inc.)

Modulation

To "modulate" means to change. In radio systems, the information (message) to be transmitted is placed on the carrier signal through a process of modulation. The carrier signal has three characteristics that can be changed to carry information: its size (amplitude), its frequency, or its phase.

In **amplitude modulation (AM)**, the strength of the carrier signal is changed by the varying size of the information signal. AM is very easy to pick up with a simple receiver. However, the information may be distorted when the received signal is small or when there is a lot of noise nearby (such as lightning or electric drill noise).

In **frequency modulation (FM)**, the frequency of the carrier signal is changed by the varying size of the information signal. FM receivers are somewhat more complex than AM receivers. FM gives good quality messages when the received signal is small or when there is a great deal of noise nearby.

Phase modulation (PM) is now used mainly for digital data communication. In PM, the phase, or position of the carrier signal at any time, is changed by the information signal. When phase modulation is used for data communication, it is often called phase-shift keying, or PSK.

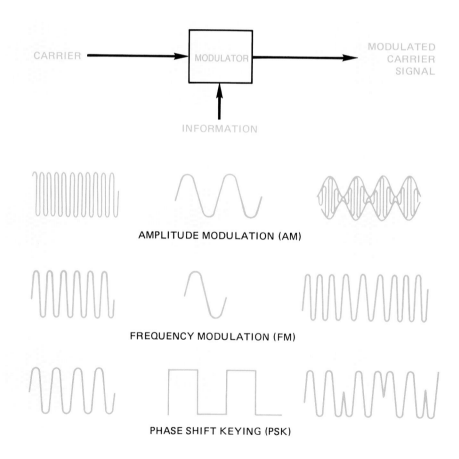

CARRIER → MODULATOR → MODULATED CARRIER SIGNAL

INFORMATION

AMPLITUDE MODULATION (AM)

FREQUENCY MODULATION (FM)

PHASE SHIFT KEYING (PSK)

As noted before in the section on antennas, the further a signal travels from the transmitting antenna, the weaker it gets. One way to compensate for this is to build a larger antenna that focuses the signal more tightly at the receiving antenna. Another way to offset this is to build a transmitter that sends a larger carrier signal, Figure 18–6. The size of the carrier signal is measured in **watts** of power. Transmitters used in radio communication systems range in size from less than one watt in systems that use highly focused antennas, to one hundred thousand or more watts in broadcast systems that radiate signals in all directions.

Receiving the Signal

When the carrier signal reaches the receiving antenna, it must be separated from other carrier signals present in the air. The information (message) must be separated from it and returned to original form. These are the jobs of the **receiver**.

The radio frequency carrier signal collected by the receiving antenna is generally very small. The first function of a receiver is to build up, or **amplify**, the size of this signal. The receiver must also separate the desired signal from other, unwanted carriers at different frequencies. This is done in a part of the receiver called the tuner, Figure 18–7.

FIGURE 18–6 This mobile telephone uses a relatively low power transmitter in a small box in the trunk of the car. *(Courtesy of Motorola, Inc.)*

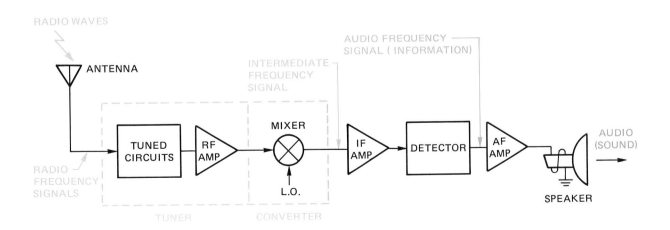

FIGURE 18–7 Block diagram of a radio receiver

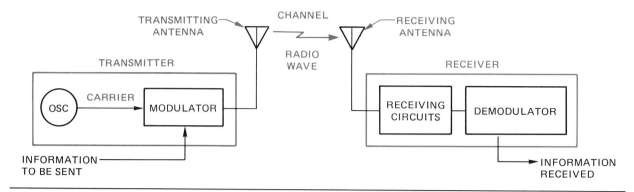

FIGURE 18–8 A radio communication system consists of a transmitter, transmitting antenna, channel, receiving antenna, and receiver.

In most receivers, the carrier signal is shifted to another, lower, frequency that is easier for the receiver to amplify and process. This new frequency is called an intermediate frequency (IF). The process of changing the RF signal to an IF signal is called conversion.

Once the signal has been changed to an intermediate frequency, it is amplified further in an IF amplifier. The information is then separated from the carrier by a part of the receiver called the demodulator or the detector. The type of demodulator used depends on the type of signal expected by the receiver. Demodulator circuits for FM are quite different from those for AM, though they both perform the task of removing the information from the carrier signal.

In summary, information is sent over a radio communication channel by a transmitter that adds the information signal to a carrier signal through a process called modulation. The modulated carrier signal is sent to a transmitting antenna that converts the electrical signal to a radio wave in the air. The radio wave then spreads, or **propagates**, in a direction determined by the design of the antenna.

A receiving antenna intercepts the radio wave and turns it back into an electrical signal which is sent to a receiver. The receiver separates the wanted signal from other signals picked up by the antenna, makes the signal larger, and separates the information signal from the carrier signal, Figure 18–8.

Types of Radio Communication

Radio communication can be classified in several ways, giving many combinations of types. Radio communication can be one-way or two-way. In one-way communication, one site always transmits the message, and the other location(s) always receive(s) the message. Entertainment received from an AM or FM station is one example of one-way radio transmission.

In two-way communication, both locations have both a transmitter and a receiver, and can send and receive messages. CB radio, car radio telephones, and walkie-talkies are examples of two-way radios.

Other types of radio communications are point-to-point, broadcast, or narrowcast, Figure 18–9. In point-to-point communication, two stations exchange information only between themselves. Point-to-point communication can be one-way or two-way. One-way point-to-point communication is called simplex. Two-way point-to-point communication is called duplex.

One transmitter sending signals to an unlimited number of receivers is called **broadcast** communication. Commercial AM and FM stations are broadcast stations. If the number of possible receivers is limited in some way, the communication is called narrowcast. A video network sending educational programs to doctors and nurses at several hospitals is an example of narrowcasting. Most broadcast and narrowcast communication is one-way. Two-way communication is sometimes used as well.

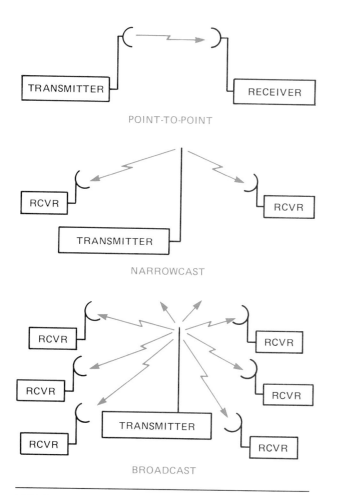

FIGURE 18-9 Radio communication systems can be point-to-point, broadcast, or narrowcast.

Choosing the Right Frequency

Radios generally operate with a carrier frequency somewhere between 100 kHz and 100 GHz (100,000,000 kHz). The operating frequency chosen for a radio depends on many factors. One of the most important factors is that radio waves of different frequencies travel (propagate) differently in the atmosphere.

Between 50 and 250 miles above the earth, the atmosphere has a group of layers that are called the **ionosphere**. The ionosphere is made up of many small particles that are charged by radiation from the sun and other heavenly bodies. The charge of the ionosphere changes from day to night, as the sun shines on it and then is blocked by the earth at night. Radio signals of different frequencies react differently to the charge of the ionosphere.

Waves in the region of 100 kHz to several MHz are generally not reflected by the ionosphere in the daytime, but do reflect off it at night. This is why you can often hear AM radio stations from other cities at night but not during the day. AM radio broadcast frequencies fall between 0.54 and 1.6 MHz (540 and 1600 kHz).

Waves in the shortwave region, between several MHz and about 30 MHz, are reflected by the ionosphere at some times during the day, and are not reflected at other times. The amount of reflection, or skip, also depends on the frequency, and changes from day-to-day and year-

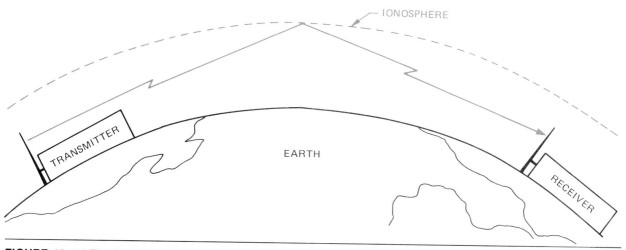

FIGURE 18-10 The ionosphere reflects (skips) radio waves of some frequencies back to the earth, giving the waves coverage beyond the horizon.

FIGURE 18–11 This shortwave amateur (ham) radio station uses the skip provided by the ionosphere to communicate with other stations all around the world. *(Photo courtesy of American Radio Relay League)*

to-year, although the reflection at some frequencies is relatively predictable. Shortwave broadcast stations use the reflective properties of the ionosphere to transmit programs around the world, Figures 18–10 and 18–11.

Above 30 MHz, and up to several hundred MHz, waves are generally not reflected by the ionosphere except under unusual conditions of weather or meteor showers. FM broadcast and TV broadcast stations use these frequencies for predictable, reliable transmission to listeners. You may have noticed that during the spring and fall, TV reception is sometimes disturbed by stations from other cities. This is due to the long-distance propagation caused by unusual weather conditions or meteor showers that sometimes plague radio waves at these frequencies.

Above 1 GHz (1,000 MHz), radio waves pass straight through the ionosphere. These frequencies are used for communicating with space vehicles and for satellite communications. Weather can also affect the propagation of radio waves at these frequencies. Rain and snow decrease the wave strength. The higher the frequency, the greater the decrease due to rain or snow. This is one major reason why microwave communication is confined largely to the lower microwave frequencies (1–20 GHz). (See also Chapter 19)

How Television Works

Television is a special kind of radio in which the transmitted information is visual. The word "television" comes from the Greek words "tele," meaning "far," and "vision," meaning "to see." Television lets us see things from a long distance away.

A device called a video camera converts visual images into electrical signals. Electrical signals that represent visual images are called **video** or video signals. When transmitting video over a television system, the video signals are fed to a television transmitter. In addition to the video, audio is transmitted at the same time. In North America standard television, the visual image is transmitted using a form of amplitude modulation. The sound is sent using FM transmission.

The amount of information that is in a full-motion, color video signal is far greater than the information in an audio signal. Using standard TV video, there is about 1,000 times as much information in the video portion of the signal as there is in the audio signal. This means that the portion of the frequency band needed to transmit the video signal is much larger than the portion of the band needed to transmit an audio signal, such as an AM or FM broadcast signal, or the audio portion of a television signal. (See a more complete discussion of bandwidth, the measure of information, in Chapter 19.)

Broadcast television signals are sent at two separate frequency bands: Very High Frequency (VHF) and Ultra High Frequency (UHF), Figure 18–12. Each television signal is called a channel. Each channel uses 6 MHz of the transmission band. Channels 2 through 13 transmit at 54 MHz to 216 MHz and are in the VHF band. Channels 14 through 83 use 470 MHz through 890 MHz and are in the UHF band. (Cable TV channels 14 and above do not correspond to on-the-air channels 14 through 83.)

Cameras

Television, like animated cartoons and motion pictures, relies on the ability of the human eye to blend a rapidly changing series of

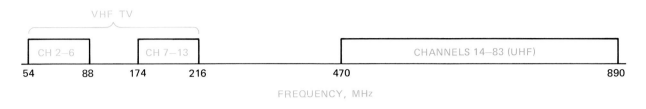

FIGURE 18–12 TV channels use frequencies from 54 MHz to 890 MHz.

fixed pictures into a seemingly continuous moving picture (see Chapter 15). In motion pictures, there are 24 still pictures, or frames, per second. A television camera creates 30 frames per second. At this rate, human eyes do not sense any flicker, but rather a continuous, fluid motion from one frame to the next, Figure 18–13.

Two basic types of video cameras are in current use: the vacuum tube vidicon and the solid state charged coupled device, or CCD. CCD cameras are smaller, lighter weight, and less expensive than vidicon cameras. CCD cameras are often used now in home video cameras and camcorders (camera-recorders). Large commercial cameras often use the older style but very high quality vidicon tubes.

In a vidicon television camera, the scene to be changed into video information is focused onto a plate by a series of lenses, just as in any camera. The plate is a light-sensitive area that creates an electric charge. The charge corresponds to the amount of light falling on each part of the plate. The plate is then scanned by a beam of electrons shot by a device called an **electron gun**, Figure 18–14.

Areas of the plate that are very bright allow a lot of current to flow. Areas of the plate that are dark allow no current to flow. The amount of current flowing thus shows the brightness of the light at any point on the focused image. The varying current is changed to a voltage. This voltage is called the video signal.

The electron beam is swept across the plate in a horizontal line from left to right and then quickly returned to the beginning edge. It is then swept across the plate again, a small amount

FIGURE 18–13 TV uses 30 still frames per second to create the illusion of a smoothly flowing motion picture. *(Courtesy of Westinghouse/Dana Duke Photography)*

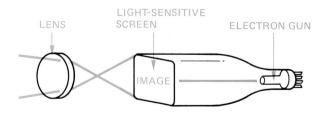

FIGURE 18–14 A vidicon tube is used to convert visual images to electrical signals in a video camera.

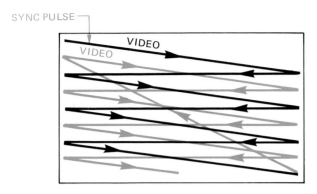

SYNC PULSE

VIDEO

VIDEO

FIGURE 18–15 The scanning process converts the visual image into a series of electrical signals. 262 Scan lines are used to make each frame. The next frame places its scan lines between the previous frames, giving better resolution of the picture. Thirty frames are made each second.

lower than the previous sweep across the plate, Figure 18–15. This process is called **scanning**. In North American television, there are 525 lines in each full frame, or 15,750 scan lines per second.

A timing pulse generated by an electronic circuit called the sync generator starts each horizontal scan. This synchronizing, or **sync**, pulse is also sent along with the video so that the receiving end will be able to correctly recreate the video image. The sync pulse controls both the start of each horizontal scan (horizontal sync) and the shift back to the top of the screen when the horizontal scans reach the bottom (vertical sync).

A CCD solid-state semiconductor camera works in a similar way to the vidicon camera, Figure 18–16. However the light-sensitive plate is made of semiconductor elements that are sensitive to light, allowing current to flow when light strikes them, and stopping light from flowing when they are dark. The light-sensitive elements are arranged in a grid pattern across the plate, Figure 18–17. The current flow in each one is sampled in a pattern similar to the scanning pattern created in the vidicontube.

Color cameras take advantage of the fact that any color can be represented as a combination of the primary colors red, green, and blue. The camera splits images into their primary colors. It then either sends three different signals (one for each primary color), or a single signal made up of a brightness part and a color part that indicates how much of each of the primary colors is being represented.

FIGURE 18–16 CCD cameras can be made quite small. *(Courtesy of Canon, U.S.A., Inc.)*

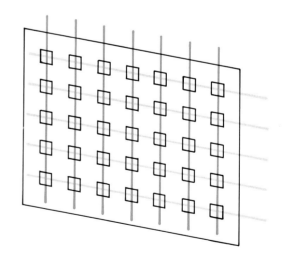

FIGURE 18–17 A solid state charge coupled display (CCD) can also be used to convert visual images into electrical video signals. Each element in the matrix is scanned in the same pattern used in a vidicon camera, giving similar video output signals.

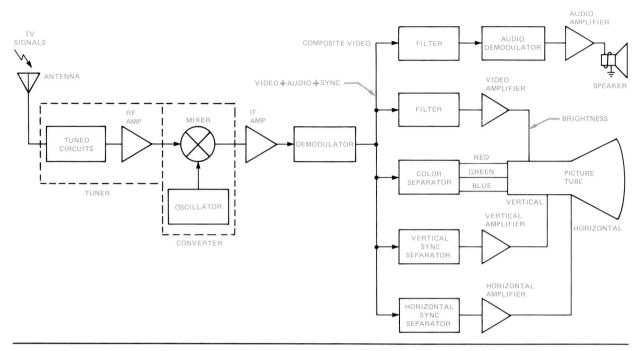

FIGURE 18–18 Block diagram of a TV receiver

Television Receivers

A television receiver is very similar to a radio receiver. It does, however, have to decode a more complex signal. This complex signal is made up of the video brightness signal, the video color signal, the sync signal, and the audio signal.

The signals collected by the antenna are fed to a circuit that amplifies the weak signals, and separates the desired signals from the unwanted interfering signals, just as in a radio. In a television receiver, this circuit is called the tuner. The desired signal is then changed to an intermediate frequency (IF) and is amplified further. The audio carrier is then separated from the video portion of the signal. The audio carrier is separately demodulated to remove the sound part of the TV signal, which is then amplified to drive a speaker, Figure 18–18.

The sync signal is removed from the video signal by a circuit called the sync separator. The sync signal is then used to make sure that the television receiver is working in lock step with the incoming signal. The remaining portion of the video signal is changed into a visual image by a cathode ray tube (CRT), also called a picture tube.

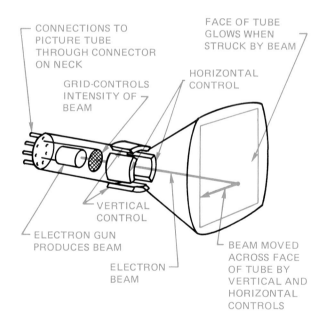

FIGURE 18–19 A TV picture tube changes electrical video signals back into visual images.

In a picture tube, an electron gun generates a stream of electrons that travel through the tube and strike the face of the tube. The face is made of a phosphor material that glows when struck by electrons. The face of the tube is scanned from side to side and from top to bottom at exactly the same rate as the scanning in the camera at the transmitting end, Figure 18–19. The intensity of the electron beam is varied by the amplitude of the video signal. Because the video and scanning are locked together by the sync signal, the picture that is produced on the picture tube is an exact copy of the scene that was viewed by the camera at the transmitting end of the system.

Color televisions have an added circuit that detects the color information on the signal and provides red, green, and blue signals to the picture tube. The color picture tube has three electron guns, one each for red, green, and blue.

Types of Television Communication

As in radio, television can be used in several ways. The way that is most familiar is as a broadcast medium, in which a central transmitting station sends a signal that is received by thousands or millions of receivers located in individuals' homes, Figure 18–20.

Television can also be used in point-to-point or narrowcast communications. Television is used to provide educational programs to small audiences spread over large metropolitan areas, or even across a continent with satellite transmission (see Chapter 19). Video transmission has also been used in point-to-point communication to provide a face-to-face meeting for people located in cities far apart. This kind of communication is called videoconferencing (described in detail in Chapter 19).

Although broadcast radio and television are thought of as one-way, or simplex, transmission, it should be recognized that there is a feedback communication channel that viewers can use to communicate back to the originator of the transmission (the TV or radio station). Viewers can (and often do) call the station or write letters, expressing pleasure or displeasure over programs they have seen or heard. Surveys are also often taken by the station management to find out how the programming is being received, and whether it should be changed. While these are not radio or TV reverse channels, they are certainly alternate communication channels, providing a complete feedback communication system, Figure 18–21.

FIGURE 18–20 This TV transmitting antenna provides coverage to more than ten million people in the New York metropolitan area. *(Courtesy of The Port Authority of New York and New Jersey)*

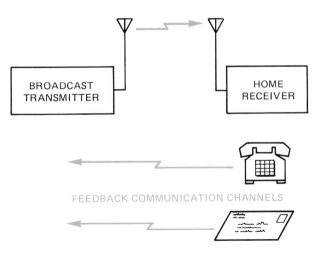

FIGURE 18–21 You can provide feedback to radio and TV stations even when you do not have a radio or TV transmitter.

Managing a Broadcast Station

Broadcast radio and TV are both part of the entertainment industry. In the movie industry, the customer (the movie-goer) pays each time a picture is viewed. Broadcast radio and TV stations usually get their income from someone other than the viewer or listener. In commercial broadcasting, the one who pays the station is the advertiser. The advertiser pays for the chance to promote a product to the listening (or viewing) audience.

The amount that an advertiser has to pay depends on the amount of air time that the advertisement takes and the total audience that is listening or watching at that time. The total audience is found through surveys and monitoring done by ratings organizations. Because the amount of income actually depends on the number of people in the audience, it is very important for a broadcaster to offer attractive, popular programming so that the audience will be as large as possible.

In order to attract a large or targeted audience, the managers of the station must select the type of entertainment, or format, they want to provide. Format ranges from album-oriented rock to religious to all talk. It could include beautiful music, country, all news, ethnic, classical, and many others. By selecting the format, the station management hopes to attract a segment of the listening audience and develop a devoted following of listeners.

Each format fits within a pattern. Patterns include all music, all news, all talk, or full service, which is made up of 70% music, 10% news, 5% sports or specials, and 15% features. A pattern gives a unified image to the station's programming. The image is designed to appeal to the audience segment that the station has targeted.

Many broadcast stations have similar management structures to help them get the most market share. The general manager, or station manager, is responsible for all aspects of station operation. The station manager directs the activities of all the station personnel and makes sure their activities are directed at the station's goals.

The program director is responsible for every sound or picture that goes out over the air. The program director selects the records that will be played, the videotapes that will be shown, and the network programming that will be aired. The traffic manager organizes the minute-by-minute schedule of what will go on the air, including the scheduling of the ads, records, movies, and news. Each minute of each day is carefully planned in advance and recorded in a log. Any on-the-air changes from the planned schedule are recorded in the log.

Radio and TV Ratings

The area over which a station can be heard is called its coverage. The total number of people who listen to a station during a given period (week, month, etc.) is called the station's cumulative audience, or cume. The percentage of a station's listeners compared to the total number who could be listening is called the station's **rating**. The percentage of listeners compared to the total number of people who are actually listening to radios (or watching TVs) at that time is called the station's **share**.

In a city that has 1,000,000 potential TV viewers, and 300,000 TV viewers on Friday night at 8:00, a TV show that has 100,000 viewers would have a 10 rating and a 33 share. The 10 rating comes from 100,000/1,000,000, multiplied by 100 to convert to percent. The 33 share comes from 100,000/300,000, multiplied by 100 to convert to percent.

The ratings are compiled by independent research companies. Ratings are very important to the stations. The amount of money they can charge for advertising time depends on the number of listeners they can attract.

FIGURE 18–22 Staff talent members—news, sports, and weather reporters—at work during a live news broadcast. The director (not shown) selects which of the camera shots to use on the air at each moment. *(Courtesy of WRGB—Newscenter 6)*

Staff talent includes disk jockeys, announcers, news reporters, and other ''personalities'' who appear or who are heard on the air, Figure 18–22. Their activities are guided by producers and directors. Directors are responsible for the artistic appearance (sound) of a program. Producers are responsible for the overall success of a program.

Other station management includes the general sales manager, who is responsible for selling air time to advertisers; the chief engineer, who is responsible for the reliable operation and improvement of all station equipment; and the office manager, who oversees all the administrative tasks, Figure 18–23. At larger stations, each of these managers has a staff of people to carry out the day-to-day activities of the department.

Summary

The earliest radios sent telegraph signals from one place to another using large antennas. One of the first uses of radio was to communicate with ships at sea. The first regularly scheduled commercial broadcasts began at KDKA in Pittsburgh in 1920. Television was first demonstrated in 1926. Commercial TV broadcasts did not begin until the late 1940s.

Radio provides communication between two points without using wires. A current flowing in a wire sets up an electromagnetic field around the wire. A changing field set up by alternating current flowing in one wire can induce an alternating current to flow in a second wire that is in the field. When the field is used to carry information, the field is called a radio wave. A wire that is designed to create the largest possible field for a given current is called an antenna.

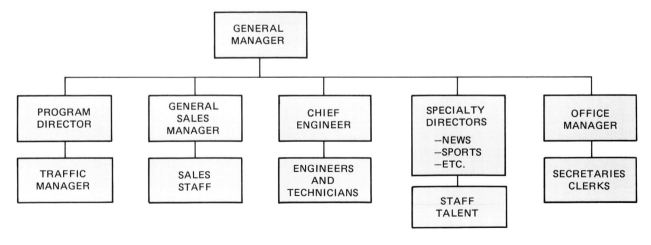

FIGURE 18–23 An organizational chart for a typical broadcast station

The higher the frequency of a radio wave, the shorter its wavelength. The shorter the wavelength, the smaller the antenna needs to be.

A radio transmitter puts an information signal onto a radio frequency carrier signal using a process called modulation. Modulation means changing of the carrier signal by the information signal. Three types of modulation commonly used are amplitude modulation (AM), frequency modulation (FM), and phase modulation (PM). Phase modulation is often called phase shift keying (PSK) when the information signal is digital data.

Receivers separate the desired carrier signal from other carrier signals collected by the receiving antenna. The receivers then change the carrier signal frequency to another frequency, called an intermediate frequency (IF). The information signal is removed from the carrier signal in a process called demodulation or detection.

Radio communication can be classified as one-way or two-way. In one-way systems, one end uses a transmitter, and the other end uses a receiver. In two-way systems, each end has both a transmitter and a receiver. Transmissions can also be classified as either broadcast, point-to-point, or narrowcast.

A group of layers in the earth's atmosphere called the ionosphere is made up of small particles that are charged by the sun's radiation. The ionosphere affects radio waves of different frequencies in different ways at different times of the day. Different types of radio links are designed taking advantage of the ionosphere's effects on radio waves. This makes around-the-world communication possible on some frequencies, and satellite communication possible on other frequencies.

Television is a special kind of radio in which the information signal represents visual images. A video camera changes visual images into an electrical signal called video through a process called scanning. In scanning, sequences of fixed images, called frames, are changed into video with a synchronizing (sync) pulse defining the beginning of each frame. In TV, 30 frames per second are created.

A TV receiver separates the sync pulse from the video. The sync pulse and the video are used to recreate the visual image in a picture tube. The video and sync pulse are sent using amplitude modulation. The color information, if any, is sent using phase modulation. The sound is sent along with the picture signal using frequency modulation.

Radio and TV broadcasters make money by charging advertisers for air time. The amount that a broadcaster can charge depends on the number of people in the listening/viewing audience. The number of people in the audience is found by independent polling agencies.

Broadcast station management is often made up of a general manager, program director, traffic manager, sales manager, staff talent, chief engineer, and office manager. They and their staffs are responsible for all aspects of operating the station.

REVIEW

1. How was early radio different from early telegraph systems?
2. What was the significance of the sinking of the *Titanic* in terms of radio's history?
3. Describe how radio signals are carried from a transmitting antenna to a receiving antenna.
4. What are the wavelengths of the following radio waves?
   a) 1 MHz (1,000,000 Hz) AM broadcast
   b) 30 MHz (30,000,000 Hz) shortwave broadcast
   c) 150 MHz (150,000,000 Hz) TV broadcast
5. What can you say about the sizes of the antennas needed for each of the radio waves in question 4?
6. What are the frequencies of the following radio waves?
   a) 15 meters
   b) 2 meters
   c) 100 meters
7. Why has FM become more widely used for music broadcasts and AM used more for news and all-talk broadcasts?
8. Draw a diagram of a radio communication system. Label all important parts.
9. What are two ways that a radio communication system might be improved if the received signal is too weak?
10. How are you able to provide feedback to a TV broadcast station, even if you do not have a TV transmitter to send a signal back to it?
11. Explain how the frequencies for satellite communication were chosen.
12. Why do you sometimes hear AM broadcast stations from other cities at night, while you do not hear them during the daytime?
13. How is TV different from radio? How are they similar?
14. Draw a diagram of how a visual image is scanned in a TV camera. Why is a sync pulse needed?
15. Why are a station's ratings important to the station management?

CHAPTER 19

Microwaves, Satellites, and Fiber Optics

OBJECTIVES

After reading this chapter, you will know that:
- Bandwidth is a measure of information carrying capacity.
- A large increase in long-distance telephone traffic, video circuits, and corporate data communication has created a need for new high-bandwidth communication links.
- Microwave links, satellite links, and fiber-optic cables are three types of high-bandwidth links used in modern communication networks.
- Microwave, satellite, and light signals are all part of the same electromagnetic spectrum. Light is much higher in frequency than microwave signals. Microwave signals are higher in frequency than FM or AM radio signals.
- Microwave links are made up of high-frequency radio transmitters and receivers whose antennas have a line of sight path between them.
- Satellite links are microwave links with a repeater (transponder) located in a satellite.
- Communication by light can carry far more information than microwave or radio links because light is at a much higher frequency.
- Light can be guided from a transmitter to a receiver by a flexible fiber-optic cable.
- Designers choose microwave, satellite, or fiber-optic links to meet the unique requirements of each application.

KEY TERMS

antenna farm	laser	satellite communication
bandwidth	light emiting diodes (LEDs)	single mode
coherent	line of sight	spectrum
downlink	microwave	telecommunications
earth station	multimode	teleport
Federal Communications Commission (FCC)	multiplexing	transponder
fiber-optic cable	reflection	TV receive only (TVRO)
geosynchronous	refraction	uplink

Introduction

The word **telecommunications** means "communicating at a distance." Our modern society depends on telecommunications for conducting business and for conducting our daily lives. Microwave, satellite, and fiber-optic systems have been developed to respond to the need for larger and larger amounts of electronic communication channels. These technologies carry telephone, video, and data communications over long paths that make telecommunications economically possible today.

Microwave radio grew out of the development of radar during World War II. Radar (**RA**dio **D**etection **A**nd **R**anging) sends out a pulse of microwave power that bounces off a plane or other target. The bounced signal, or radio echo, is then received by the radar receiver, which measures the delay time between sending the pulse and receiving the echo, and calculates the distance to the target, Figure 19–1. Some radars also calculate the speed of the target's movement toward or away from the radar. During the war, radar technology was Top Secret. After the war, however, some of the parts developed for radar became available for commercial uses. Two-way radios were built that operated at radar's very high (microwave) frequencies. Thus, commercial microwave radio development began.

Satellite communication is really microwave radio communication using a radio relay station located in a satellite in a very high orbit. The original idea for satellite communication as we know it today was proposed more than 40 years ago by Arthur C. Clarke, the noted science writer and science fiction author (*2001: A Space Odyssey* is one of his works).

Fiber-optic communication makes use of the fact that light is electromagnetic radiation, the same as radio waves, but at a much higher frequency. Because it is at a much higher frequency, it has the potential of carrying a very large amount of information. Fiber-optic systems carry more information than radio systems. There is still, however, much development work to be done in fiber optics to reach its full potential.

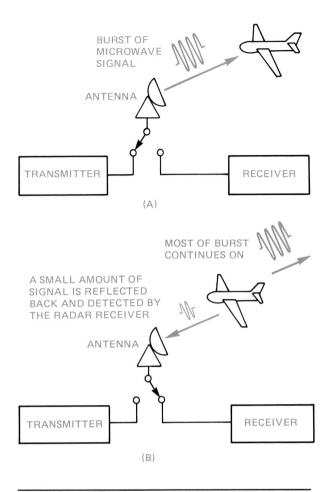

FIGURE 19–1 In a radar system, the transmitter sends out a short pulse of microwave energy. Some of the microwave signal that strikes a plane or other target is reflected back toward the radar station. A sensitive radar receiver receives the radio echo. Measurement of the time between the transmitted pulse and the received echo determines the distance between the target and the radar station.

Bandwidth: Measuring Information Flow

In Chapter 18, you learned that video signals need a larger portion of the frequency band than the audio signals that are sent with them. The information contained in a video signal is much greater than the information contained in an audio signal. **Bandwidth** is a measure of the information carrying capacity of a communication channel. Bandwidth is measured in cycles per second (Hz), the same units used to measure frequency.

A standard television video signal uses a bandwidth of about 4.5 MHz, while a telephone transmission uses a bandwidth of about 4.5 kHz. Thus, the video signal uses about 1,000 times the bandwidth of the audio signal. If many telephone signals are multiplexed (see Chapter 17) together into one signal, the combined signal will need many times the bandwidth of any one signal.

As telephone traffic grew after World War II, and long distance calls became more routine, greater and greater bandwidth was needed between telephone central offices. During the 1980s, there has been a great expansion in the business need for high capacity data communication and video circuits.

To a large extent, this increased need for bandwidth was met by the installation of more and more copper wires. More recently, three technologies have been prominent in providing the needed high bandwidth circuits: microwave radios, satellites, and fiber-optic cables.

All three of these technologies use electromagnetic waves as carriers (see Chapter 18). These electromagnetic waves are higher in frequency than the radio or television stations described in Chapter 18. In general, higher frequency carriers can transmit higher bandwidth signals. For example, a radio frequency carrier at 10 MHz (10,000 kHz) can carry 1000 times the

bandwidth of a 10 kHz carrier, using the same modulation technique, Figure 19–2.

The Electromagnetic Spectrum

AM radio, FM radio, and television signals all occupy their own frequencies. This allows us to select one station or another even though many different signals are in the air at one time. The total range of frequencies available for communication is called the electromagnetic **spectrum**. Electromagnetic signals produce both electric fields and magnetic fields. A field is produced by current flowing through an antenna (see Chapter 18).

The electromagnetic spectrum extends from audio frequency signals (not to be confused with sound waves, which are mechanical compressions of the air at audio frequencies) to frequencies above lightwaves. Between these two extremes are radio signals, radar signals, microwave and satellite transmission, and many other kinds of signals, Figure 19–3.

By taking advantage of the characteristics of each part of the spectrum, different services can be offered to the public. The electromagnetic spectrum has been recognized as a natural resource in the information age. This spectrum has been regulated and controlled by international organizations of which the United States is a member, and by the **Federal Communications Commission** (**FCC**) inside the United States. Every industrialized country has an agency that regulates the use of the electromagnetic spectrum by its people.

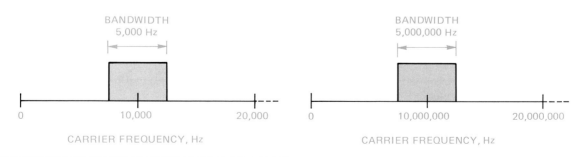

FIGURE 19–2 Higher frequency carriers can carry more information because they support larger bandwidths. Using the same modulation technique, a 10,000,000 Hz carrier can carry 1000 time more information than a 10,000 Hz carrier.

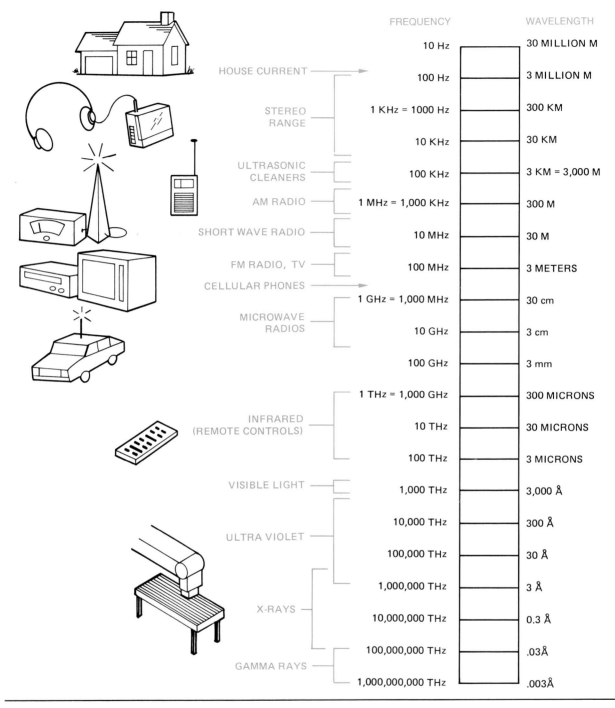

	FREQUENCY	WAVELENGTH
	10 Hz	30 MILLION M
HOUSE CURRENT	100 Hz	3 MILLION M
STEREO RANGE	1 KHz = 1000 Hz	300 KM
	10 KHz	30 KM
ULTRASONIC CLEANERS	100 KHz	3 KM = 3,000 M
AM RADIO	1 MHz = 1,000 KHz	300 M
SHORT WAVE RADIO	10 MHz	30 M
FM RADIO, TV	100 MHz	3 METERS
CELLULAR PHONES		
MICROWAVE RADIOS	1 GHz = 1,000 MHz	30 cm
	10 GHz	3 cm
	100 GHz	3 mm
INFRARED (REMOTE CONTROLS)	1 THz = 1,000 GHz	300 MICRONS
	10 THz	30 MICRONS
	100 THz	3 MICRONS
VISIBLE LIGHT	1,000 THz	3,000 Å
ULTRA VIOLET	10,000 THz	300 Å
	100,000 THz	30 Å
X-RAYS	1,000,000 THz	3 Å
	10,000,000 THz	0.3 Å
	100,000,000 THz	.03Å
GAMMA RAYS	1,000,000,000 THz	.003Å

FIGURE 19-3 The electromagnetic spectrum

Microwave Radio Communication

Radio frequency communication systems come in many sizes and types and are built to serve many uses. They range from national and inter-national broadcast networks that provide news and entertainment to millions of people, to mobile radio telephones, to satellite TV receivers, to amateur radio stations used for experimenting and for conversing with people all over the world.

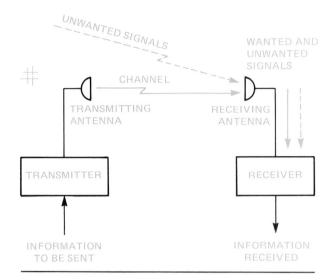

FIGURE 19–4 All radio communication systems have a transmitter, a transmitting antenna, a receiving antenna, and a receiver. The receiver separates the desired signal from the other, unwanted signals in the air.

All radio communication systems have some common elements, Figure 19–4. Each has a transmitter, a transmitting antenna, a receiving antenna, and a receiver. The transmitter encodes the information to be sent, places it onto a radio frequency (RF) carrier, and sends the RF carrier to the antenna. The transmitting antenna sends the signal into the air, where it is picked up by the receiving antenna some distance away. The receiver processes the signals from the antenna, separating the unwanted signals from the desired signal, and removes the information from the carrier. The part of the electromagnetic spectrum in the air between the transmitting antenna and the receiving antenna is the communication channel in this system.

The type of information may vary from system to system and the frequency of the carrier may be quite different, but all radio communication systems work in this way.

Microwave Communication

The **microwave** portion of the electromagnetic spectrum includes the frequencies from 1 GHz (GigaHertz, or one billion cycles per second) to above 100 GHz. The most often used frequen-

cies for commercial communication are between 1 GHz and 23 GHz. The wavelength at these frequencies ranges from about 12 inches to ½ inch; hence the name microwaves, or ''little waves.'' (Contrast this with AM radio wavelengths of about 1000 feet, short wave wavelengths of about 120 feet, and FM radio wavelengths of about 10 feet.)

At microwave frequencies, waves travel in straight lines and are not reflected by the ionosphere. Microwave radio links are therefore usually **line of sight**. This means a link can be

FIGURE 19–5 Microwave antennas are often placed on tall buildings or towers to give a line of sight to the next antenna, clear of obstructions and the curvature of the earth. *(Photo by Robert Barden)*

built between two points only if one point can be seen from the other. Because of the curvature of the earth, microwave links are limited in distance.

As frequency increases, the size of the antenna needed for effective communications gets smaller. At microwave frequencies, relatively small antennas can point the microwave signal into a very narrow beam. This concentrates all of the transmitter's power toward the receiving antenna. Microwave radio transmitters are usually very low in power.

Microwave antenna sizes range from less than two feet to about twelve feet across. They are usually placed high on towers or buildings to overcome the curvature of the earth and surrounding obstacles to achieve a line-of-sight path with the antenna at the other end of the link, Figure 19–5.

Microwave communication systems are often built with relay stations to extend the total distance of coverage, Figure 19–6. These relay stations are spaced at twenty to thirty miles along the route, depending on the frequency of the radios and the expected weather conditions. Weather is important because heavy rain can greatly affect the operation of microwave links at some frequencies. This is more important at higher frequencies than at the lower microwave frequencies, but must be considered when designing a microwave link at any frequency.

Because of their high frequency of operation, microwave links offer larger bandwidths than radio or television broadcast stations. Links are often used by telephone companies to provide

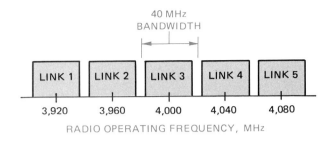

FIGURE 19–7 Each radio link is assigned a frequency slightly different from others in the same general area. In this way, many different radio links can operate in the same town without interfering with each other.

communication between central offices, or along a major pathway between cities. Microwave links used in these ways carry more than 1000 telephone conversations per link. A typical microwave radio operating at 4 GHz (4000 MHz) provides a bandwidth of 20 to 40 MHz. Many radios can be operated on slightly different frequencies. This allows simultaneous operation of radios serving different points, owned by different organizations, in the same general geographic area, Figure 19–7.

Other uses of microwave links include the relaying of television signals from cities to their surrounding suburbs. This is often done to feed cable TV systems which provide TV coverage to communities beyond the range of the TV station. Microwave links are also used as relay paths from TV studios to satellite central feedpoints or from portable news-gathering units back to a TV studio, Figure 19–8.

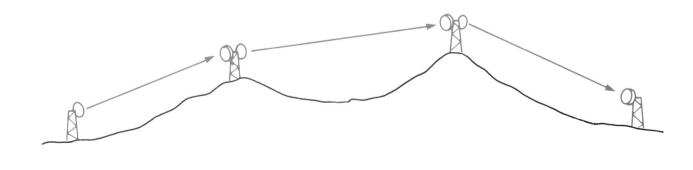

FIGURE 19–6 A typical microwave relay network has intermediate repeater stations every twenty to forty miles.

FIGURE 19-8 The antenna on top of this remote video newsvan is sending live pictures to the broadcast station located beneath Tokyo Tower in the background. *(Photo by Robert Barden)*

Satellite Communication

Satellite communication is a special application of microwave communication. In satellite communication, a microwave relay station, called a **transponder**, is built into a satellite and placed in a special orbit high above the earth. Signals from an **earth station** are sent to the satellite on one frequency. The satellite transponder shifts the signal from the **uplink** frequency to a different **downlink** frequency. The satellite sends the shifted signal back to earth to the general area of the expected receiving station. Some satellites use spot antennas to direct the signal to a relatively small area, such as the eastern part of the United States, Figure 19–9. Other satellites use antennas that give coverage of much larger areas, for example, the whole United States.

Satellites in Orbit

When a satellite orbits the earth, many different forces act on it. One force is the force of momentum, which tends to keep it going in the same direction that it is traveling. This force tends to keep the satellite in orbit above the earth,

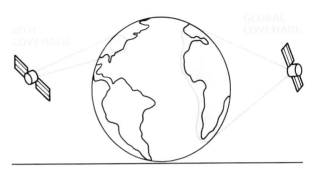

FIGURE 19-9 Satellites can beam signals down to large areas of the earth (global coverage) or to smaller, targeted areas (spot beams).

and even to increase its altitude. A second force is the force of gravity, which tends to pull the satellite back to earth. If the force of momentum exactly balances the force of gravity, then the satellite stays in orbit, circling the earth, Figure 19–10.

As the satellite orbits the earth faster and faster, its momentum increases, producing a greater outward force. A satellite in a low earth orbit has a greater force of gravity acting on it than one in high earth orbit. To counteract the increased force of gravity, satellites in low earth orbit must move faster than satellites in high earth orbit. A satellite in orbit at an altitude of about 100 miles needs to circle the earth every 90 minutes in order to stay in orbit.

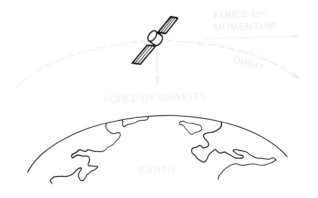

FIGURE 19-10 The momentum created by the satellite's motion around the earth fights against gravity, which tries to pull the satellite back down to earth. When the two forces are exactly equal, the satellite stays in orbit, circling the earth at the same altitude.

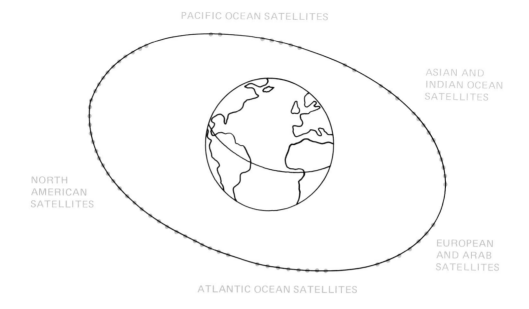

PACIFIC OCEAN SATELLITES

ASIAN AND INDIAN OCEAN SATELLITES

NORTH AMERICAN SATELLITES

EUROPEAN AND ARAB SATELLITES

ATLANTIC OCEAN SATELLITES

FIGURE 19–11 Communication satellites are assigned exact positions over the equator, called orbital slots. This ring of satellites is often called the Clarke Belt.

A satellite that orbits the earth once every 24 hours needs to be about 22,500 miles high. This rate is the same as the rate that the earth spins. If a satellite is placed in orbit at this altitude over the equator, it will appear to be stationary in the sky to an observer looking up from the earth. Because it appears to be stationary, an antenna pointed at it does not have to move to keep communications with the satellite. This kind of orbit is called a **geosynchronous**, or geostationary, orbit ("geo" means earth, and "synchronous" means locked together).

Each satellite is assigned a certain spot in the sky over the equator, called an orbital slot, Figure 19–11. Slots cannot be too close together, or antennas on the earth would not be able to pick out the signal of one satellite from the signals of those in adjacent slots. In total, there are more than 3000 satellites in orbit today. Only a small fraction of these are communications satellites in geosynchronous orbit.

Satellite Frequencies

Satellite transponders are made to work at different frequencies. One of the most widely used frequency bands for satellite communication is 4 GHz uplink and 6 GHz downlink. This band gives 500 MHz of total bandwidth in the transponder, which is usually broken up into 24 equal-sized channels, so that different people can use the same satellite transponder.

Another commonly used frequency band is 14 GHz uplink and 12 GHz downlink. Other satellite frequencies are used for military communications and for special purpose communications.

Satellite TV Broadcasts

Satellites are used in broadcast television in two ways. The first is to relay reports and programs from one location to another so that they may be broadcast at the second location. An example of this is a news report that starts in Europe and is sent by satellite link to a television studio in New York where it is either broadcast live or taped for later use, Figure 19–12.

A second way that satellites are used for television is called satellite broadcast. In satellite broadcasting, a program transmitted to a satellite from a ground station is rebroadcast by the

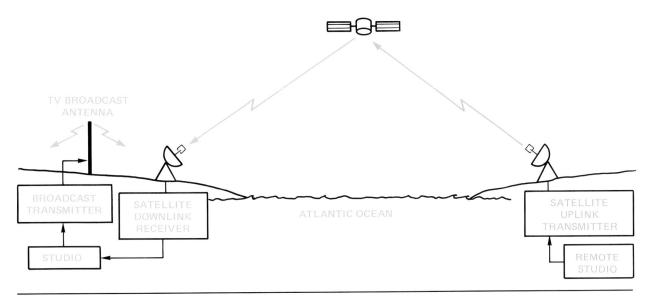

FIGURE 19–12 Network television news gathering often uses satellites. In this case, a live report is being sent from Europe to New York by satellite. In New York, the report is sent to a transmitter and broadcast to millions of homes.

satellite to a wide area of the earth, where it is directly received by many other **TV receive only (TVRO)** sites. These TVRO sites can be TV stations that are part of a network that broadcasts the same programming in many different cities, cable TV networks that distribute many TV signals to customers in a relatively small area, or people with satellite receiving antennas in their yards.

Much of the programming that is available from satellites today is intended for a relatively small number of receivers, rather than for direct home use. Newer, direct-to-home satellite broadcasts will be made on the higher frequency microwave band (12 GHz), where the home antenna can be much smaller than it needs to be for 4 GHz satellite broadcasts.

In one-way video transmission, the delay introduced by the signal's long trip from the earth to the satellite and back again is not important. It does become important, however, if two-way communication (video or other) is used. You may have seen this delay in news reports in which a person in the United States interviews another person on another continent via live satellite TV. There is a small but perceptible delay between the asking of a question and its answer.

The delay introduced by the satellite link is due to the 22,500 mile altitude of the satellite. The microwave signal travels at the speed of light from the earth to the satellite. The signal then travels at the same speed of light from the satellite back to the earth. The total one-way trip is thus about 45,000 miles. Since the speed of light is about 186,000 miles per second, the time for a transmission to complete its path from the sending station to the receiving station is about 45,000 miles/186,000 miles per second = 1/4 second. It then takes the reply another 1/4 second to reach the first station. This results in a round trip delay of about 1/2 second, Figure 19–13.

Satellite Telephone Communication

Most of today's intercontinental telephone traffic is now carried by satellite links. International satellite telephone service began in 1962. It spread very quickly to include over 100 countries on all continents. Today, it is also used for some long distance calls within a country.

Satellite telephone traffic is usually carried in digital form. Many conversations are combined into very high data rate streams of traffic. This multiplexing (see Chapter 17) makes better use of the satellite transponder.

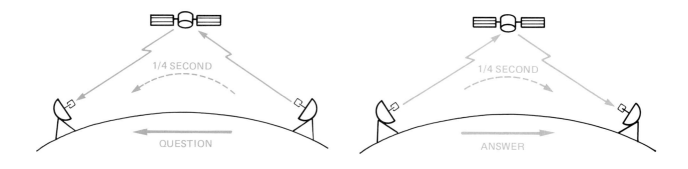

FIGURE 19–13 Because geosynchronous satellites are so high, it takes a long time for the signals to travel from one earth location to another. A question takes 1/4 second to reach the listener. The answer takes another 1/4 second to reach the original asker. The total delay is about 1/2 second.

The round-trip delay of signals through a satellite is annoying to some people. For this reason, other high data rate transmission lines are often used instead of satellites for domestic long distance service. On international calls, however, satellites will be dominant until a new network of very high capacity fiber-optic undersea cables can be completed on the ocean bed.

Fiber-optic Links

Light, like radio waves and microwave transmission, is a form of electromagnetic radiation. Light's frequency is very high, however (its wavelength is very short). It is detectable by our eyes, while radio waves and microwaves are not detectable by any of our natural senses.

Because light is of such a high frequency, it has a very large bandwidth. Light can carry very large amounts of information. Simple light links have been used for years aboard ships, as one ship signalled by blinking a light in Morse code to another ship nearby.

The explosion in communication by light began when thin, flexible guides for light, called optical fibers, were developed in the 1970s. Optical fibers are encased in protective wrappings, and then put with other similarly protected optical fibers to make fiber-optic cable, Figure 19–14.

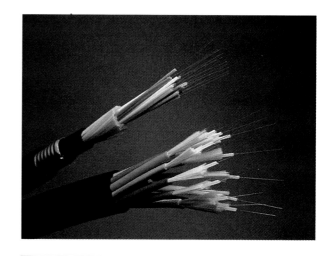

FIGURE 19–14 Very large amounts of information can be carried by light guided through very thin optical fibers. (Courtesy of Optical Cable Corporation)

The Reflection and Refraction of Light

Light, like microwaves, travels in a straight line through air. When it strikes a polished surface, its direction changes as it bounces off the surface. This bouncing is called **reflection**. The angle that the light leaves the surface (the angle of reflection) is equal to the angle that the light struck the surface (the angle of incidence), Figure 19–15.

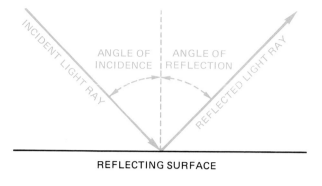

FIGURE 19-15 When a ray of light is reflected at a flat surface, the angle of reflection is the same as the angle of incidence (the angle the ray strikes the surface).

When light passes from a thick substance, such as water, into a thin substance, such as air, the light speeds up. If the light passes from the thick substance to the thin substance at an angle, one part of the wave will speed up before the other. This bends the light ray as it passes from the water to the air. This bending of the light wave is called **refraction**, Figure 19-16.

The amount of refraction of the light will depend on how much thicker one substance is than the other. If the light passing from the thicker material into the thinner material strikes the boundary between the two materials at an angle greater than a certain critical angle, all of the light will be reflected back into the thicker medium, Figure 19-17.

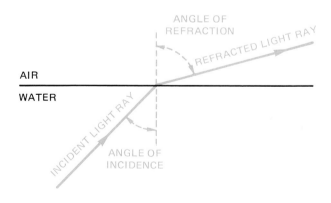

FIGURE 19-16 When light passes at an angle from water into air, it bends, or refracts, at the surface. This is because light travels faster in the air (nearly the speed of light, 186,000 miles per second).

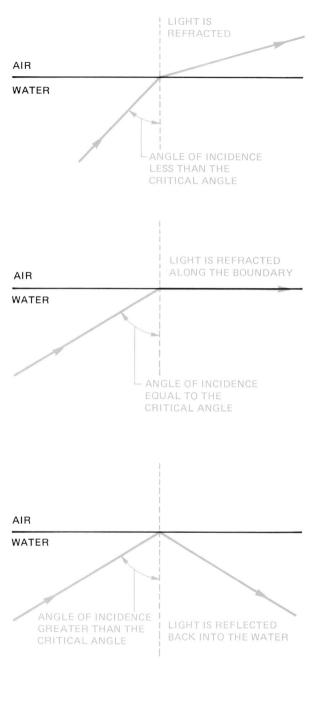

FIGURE 19-17 At a certain critical angle, light rays striking the boundary between one material and another will not enter the second material, but will skim along the surface. Light rays that strike at any angle greater than this critical angle will be reflected back into the first material.

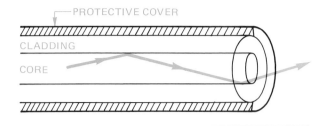

FIGURE 19–18 An optical fiber is made of two materials with slightly different indexes of refraction. Light rays traveling down the fiber that strike the boundary between the core and the cladding material at an angle greater than the critical angle are bent back into the core, ensuring that they will travel all the way to the other end of the fiber.

Optical Fibers

Optical fibers are flexible strands of glass spun into long lengths. They are made of a thin circular core material that transmits light at low loss. They are covered by a cladding material that transmits light at low loss, but that is slightly thinner (light travels through it faster). Any light that is passing down the fiber will strike the boundary between the two different materials. Any of the light entering the end of the fiber within a certain aperture angle, or acceptance angle, will always be internally reflected back into the core fiber because it always strikes the boundary at an angle greater than the critical angle, Figure 19–18. This light is thus totally contained within the fiber, and follows the path of the fiber, even though the fiber might be bent or twisted.

Optical fibers come in different sizes, each made for a different use, Figure 19–19. In commercial communications, diameters range from 6 to 100 microns, or millionths of a meter (this is about 1/4 to 4 thousandths of an inch). The smaller in diameter the fiber, the more difficult it is to work with, but the better the performance generally is. As in many systems, a trade-off is made by the system designer between performance and difficulty of installation and maintenance (cost).

In general, very long or very high capacity **fiber-optic cables** that carry signals from one telephone central office to another are very thin

FIGURE 19–19 Optical fibers used in communications are very small. Each point of light in this picture is the end of an optical fiber. *(Courtesy of United Telecom)*

and very high performance. These are called **single mode**. Optical fibers used to carry data transmission around a single building or on a college campus are often the larger diameter fibers. These are called **multimode**.

Light Transmitters and Receivers

While optical fibers can be tested by shining a flashlight down them, or holding them up to sunlight, special light sources have been developed for communicating over fiber-optic cables. These generally fall into two categories: **light-emitting diodes (LEDs)** and **lasers**.

Light-emitting diodes are the less expensive of the two. LEDs are semiconductor devices (see Chapter 16) that give off light when current flows through them. The semiconductor material that a diode is made of determines what color light it emits. Commonly available LEDs are red, green, and yellow. LEDs are small and inexpen-

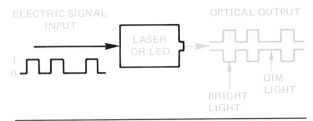

FIGURE 19–21 A transmitter in an optical communication system converts electronic signals (usually digital) to light signals. Often, digital signals are represented by bright and dim levels of light for 1's and 0's.

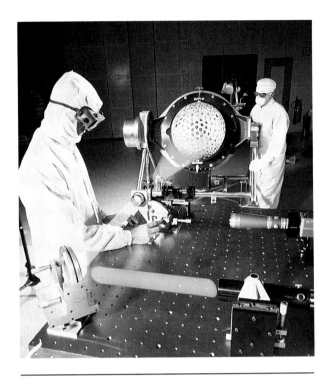

FIGURE 19–20 Lasers emit very pure, narrowly focused beams of light. *(Courtesy of NASA)*

sive, take small amounts of power, and give off small amounts of light. They are used in short distance fiber-optic systems.

Lasers are available in several forms for different jobs. Lasers can be gas lasers, solid lasers, or semiconductor lasers. Gas lasers and solid lasers are often used for high-power tasks, such as etching, cutting, and heating. While some gas lasers (such as helium-neon) are used in communication labs for experimental systems, most lasers in commercially operating laser communication systems are semiconductor lasers, Figure 19–20.

A laser is a device that emits a very intense, narrowly focused, single color (monochromatic) beam of light. It is often called a **coherent** light source because of the purity of color of the light beam. The light beam is so narrow and intense that a person should never look into it, because eye damage is very likely to result. (Lasers are used by doctors under very carefully controlled conditions to perform eye and other surgery.)

The light output from both lasers and semiconductor LEDs can be changed, or modulated, by changing the voltage supplied to them. This is useful in communication systems. The light beam can be used as a carrier, just as in a radio system, with information modulated onto it. The most commonly used way of modulating the light source is to turn it on and off or bright and dim, to represent digital data. A bright light may be used to indicate a "1" while a dim light or no light is used to indicate a "0." Each "0" or "1" represents one bit of information, Figure 19–21.

Because of the very high frequency of light, it has a very high bandwidth. A great deal of information can be modulated onto it. One average page of typed information can be represented by 14,400 bits. A laser can be modulated at over one billion bits per second, allowing it to send the contents of about 100,000 typed pages per second.

Commonly used high capacity fiber-optic links carry 560 million bits per second. Some links that carry more than one billion bits per second are in commercial use.

Light Receivers

At the receiving end of an optical link, special semiconductors called photodiodes or photodetectors produce an electrical output when light strikes them. These are used to detect the presence or absence of light sent over the fiber from a laser or LED. The rapid on-off light pulses are thus converted to on-off electrical pulses, Figure 19–22. These can be interpreted directly as computer data (in data communications), or converted back to analog telephone signals (in telephone trunk circuits).

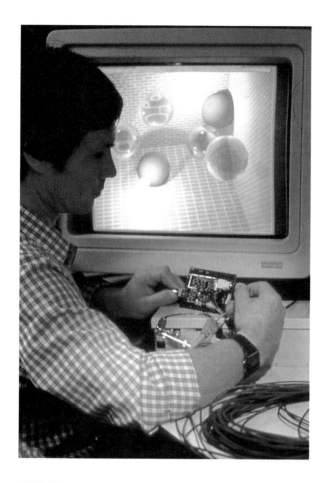

FIGURE 19–23 Optical fibers are encased in layers of protective materials to form fiber-optic cables. The cables are then installed on telephone poles or buried underground, as shown here. The cable plow must maintain a proper amount of tension on the cable to keep it from breaking as it is buried. *(Courtesy of PETROFLEX INC. Gainesville, TX)*

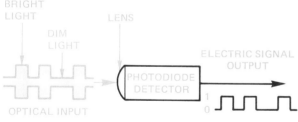

FIGURE 19–22 A receiver in an optical communication system converts light pulses back into electrical signals. The optical receiver in this system is carrying digital data that produces the image on the screen in the background. *(Photo courtesy of Motorola, Inc.)*

Communicating Using Fiber Optics

Fiber-optic cables are strung on telephone poles or buried underground, just as copper wires have been for years. A small number of fibers, however, can carry the same number of telephone conversations as a very large bundle of copper cables. Fiber-optic cables are generally used between two points that have a very large amount of telephone or data traffic between them, such as two telephone central offices, Figure 19–23.

Just as with copper cable and microwave links, repeaters must be installed periodically along a long path. This is because the amount of light power decreases as the length of the cable is made longer. Repeaters ensure that enough light reaches the receiving end to give high quality communications. In modern systems, repeaters are located from 30 to 100 miles apart.

Large numbers of telephone conversations or data communication circuits are combined into a single, high data rate digital signal. This signal is put onto one fiber-optic pair (one fiber is needed to send the signal, the other to receive the answering signal from the far end). At the receiving end, the high data rate signal is broken down into the individual signals that first made it up. The process of combining many signals into one is called **multiplexing**. The process of breaking down the high data rate signal into the original signals is called demultiplexing (see Chapter 17).

Using multiplexing, a single fiber can carry a large number of telephone circuits. Typical fiber-optic transmission systems use many pairs of fibers, each one carrying many signals multiplexed together. The total number of telephone conversations carried by a multiple fiber cable is the product of the number of signals multiplexed on each pair times the number of pairs. This can result in tens or even hundreds of thousands of conversations.

Selecting a Transmission System

Microwave, satellite, and fiber-optic systems do similar jobs. Each carries a large number of telephone or data circuits from one point to another. Yet all three are used in modern communication systems. When selecting a system to be used in a particular place, the system designers must consider the length of the path, the terrain along the path, and the costs of installing each system.

Microwave links need to have line-of-sight paths between relay stations. No poles or underground conduits need to be placed between the relay stations. While the operator of the communication system needs to buy or lease the land to place the relay stations on, there is no need to buy or lease land between the stations. In addition, natural obstacles, such as rivers, lakes, and cliffs, do not present problems for the microwave link, as long as there is line of sight between relay points. For this reason, many relay stations are on the tops of hills or mountains, so that a clear line of sight can be maintained.

Fiber-optic transmission systems can carry heavier communication traffic (more voice conversations or data signals) than typical microwave links. The quality of signal reception at the fiber-optic receiver is very good and very constant. Microwave radio links can also deliver very good quality signals, but signal quality can suffer during snowstorms, rainstorms, and other weather conditions.

FIGURE 19–24 Video conferences carried over microwave or satellite circuits allow people at many locations to conduct face to face meetings without traveling. *(Courtesy of Compression Labs, Inc.)*

Fiber-optic systems require the owner to purchase or lease land to bury the cable under, or use existing poles along the transmission route. Sometimes this is very expensive. It may even be economically impractical due to natural obstructions. If the path is important enough, and enough money is available for the project, most natural obstacles can be overcome. Fiber optics can be placed across rivers, buried in lakebeds, and made to scale cliffs.

The choice of fiber-optic or microwave links between two points is usually a matter of evaluating the traffic capacity (bandwidth) needed, the existing physical paths and poles or underground conduits, and the relative costs of the two systems along that route.

Satellite links are generally used only for the longest of paths, such as intercontinental links or transcontinental links. They are also used for links in which many users need to be attached at the same time, such as in videoconferencing, Figure 19–24. Some large corporations routinely use satellite links to conduct meetings of people located in company offices in many locations.

Teleports

Part of Teleport's control center *(Courtesy of Teleport Communications)*

New York Teleport's antenna farm *(Photo by Robert Barden)*

Several cities throughout the world have established **teleports**. These are communication hubs used by businesses throughout the city for reliable, high capacity communication with other locations throughout the world. Just as some cities are noted for their good seaports or airports, cities are now becoming known for their ability to handle information flow through their teleports.

Teleport Communications in New York City has a large space for satellite antennas (often called an **antenna farm**) protected from outside interfering signals by a high wall. It also has a building that contains supporting electronic equipment and an emergency power system that will keep the communication links operating if normal power fails.

Connections from the satellite communication site to businesses throughout the city are provided by fiber-optic cable that is run through underground conduits and through subway tunnels.

Summary

Bandwidth is a measure of information carrying capacity. The recent growth in telephone, video, and data communication traffic has led to a need for increased bandwidth communication links. This need was first met by the installation of more and more copper wires. More recently, microwave radios, satellite links, and fiber-optic cables have been supplying these needs.

All three technologies are examples of electromagnetic signals, carrying information from one place to another. Microwave radio signals are at a higher frequency than FM or AM radio signals. Satellite links are microwave radio links with a repeater located in a satellite. Light communication over fiber-optic cable takes place at a frequency much higher than microwave communication. The higher the frequency, the more information a signal can carry (the wider its available bandwidth).

Microwave radio systems, like all radio systems, contain transmitters, antennas, and receivers. Microwave frequencies used for communication range from one to twenty-three GigaHertz, with wavelengths of four inches to one-quarter of an inch. Because of the small wavelengths, antennas used for microwave radios can be quite small.

Microwave radios need to have a line-of-sight path between them, because microwave signals travel in straight lines and are not reflected by the ionosphere. To ensure that an antenna has a clear path with no obstructions, microwave antennas are often placed high on towers on hills or tall buildings.

Transmitters used in microwave links use low power and have a range of up to thirty miles, depending on frequency and antenna size. To build long microwave systems, relay stations are built along the paths to receive and resend the signal.

Microwave links are used to send large numbers of voice conversations from one central office to another. They are also used to relay video signals to cable TV systems and for data communications between businesses.

In satellite communication, a special kind of microwave relay station called a transponder is placed in a satellite orbiting the earth. The satellite orbits the earth over the equator once every twenty-four hours. This is the same rate that the earth spins. The satellite thus appears to be stationary in the sky, allowing antennas pointed at it to remain focused on it.

High-power transmitters using very large antennas send signals from the earth to the satellite on an uplink frequency. The transponder in the satellite shifts the signal to a downlink frequency and sends it back to the earth. Receiving stations on the earth pick up the signal on the downlink frequency.

Most signals carried by satellite today are intended for one or a small number of commercial receiving stations. Newer satellite services will include direct broadcasts of TV signals to homes equipped with very small antennas.

Light is a form of electromagnetic radiation with a very short wavelength (very high frequency). Because the frequency of light is so high, it has the potential of carrying very large amounts of information (has a very large bandwidth).

Light can be carried from one place to another through flexible light guides called optical fibers. Optical fibers packaged into bundles with protective coverings are called fiber-optic cables.

Fiber-optic cables are installed on telephone poles or buried in the ground. These cables carry tens of thousands or hundreds of thousands of voice conversations from one major metropolitan center to another. As in microwave links, repeaters are needed periodically along the path of fiber-optic cables. Repeater spacing in modern fiber-optic systems is from thirty to one hundred miles.

In fiber-optic systems, the light is generated by lasers or LEDs. Information is placed on the light stream through a process called modulation, or changing of the light intensity. Most light links use digital modulation. The pulsing light is received and converted back into electrical signals by a photodiode.

Microwave, satellite, and fiber-optic systems are used by communication designers where each is appropriate. The decision of which to use is based on the distance between the two points, the terrain between them, and the costs of each for that installation.

REVIEW

1. What need developed that high-bandwidth communication systems, such as microwave, satellite, and fiber-optic links, satisfy?
2. In what way(s) are microwave, satellite, and fiber-optic links alike? In what way(s) are they different?
3. Why are light links capable of carrying more information than microwave or satellite links?
4. Draw a diagram of a microwave radio link. Label the main parts of the equipment at each end, and show the channel.
5. What is the wavelength of a microwave link operating at 3.0 GHz (3000 MHz)? (Refer to Chapter 18 if you need to review the equation for frequency and wavelength.)
6. Why are relay stations needed in some microwave radio systems?
7. Why does a geosynchronous satellite appear not to move in the sky?
8. Describe the operation of a satellite communication system.
9. Why is there a delay in an answer to a question over a satellite link? Is this delay important in all types of communication? Why or why not? What is the delay introduced in a microwave link of 22.5 miles length?
10. Light strikes a reflective surface at an angle of 30 degrees. At what angle will the reflected light ray leave the surface? Draw a diagram showing this reflector.
11. Draw a diagram of a fiber-optic link. Label all the parts.
12. Digital voice conversations are grouped so that 24 voice circuits make up a multiplexed signal at 1.5 MHz. How many voice conversations can be carried on an optical fiber pair operating at 1.5 GHz (1500 MHz)?
13. How many voice conversations can be carried on a fiber-optic cable made up of ten pairs of optical fibers described in Question 12?
14. Name the kind of communication system you think would be appropriate for each of the following. Give a reason for each answer.
    a. 48 voice circuits, London to New York
    b. one telephone circuit from your house to the nearest telephone central office
    c. 24 voice circuits, across a swift river
    d. 24,000 voice circuits, Boston to Washington, D.C.
    e. 120 voice circuits, San Francisco to Los Angeles.

CHAPTER 20

Our Digital World

OBJECTIVES

After completing this chapter, you will know that:

- Analog electrical signals are replicas of natural, smoothly varying signals, such as voices or graphic images.
- Digital, or on-off, codes can be used to represent information, including analog information.
- One principal advantage of digital transmission over analog transmission is that digital transmission is not as susceptible to transmission noise as analog transmission.
- The more bits contained in a digital character, the more information it contains.
- A communication protocol is a set of rules governing how a connection is made, how data is sent from a transmitter to a receiver during the connection, and how the connection is ended.
- A modem is used to send digital data over an analog telephone line, usually at data rates of 9.6 kbps or less.
- Special transmission lines, often using fiber optics, satellites, and microwave radios, carry high data rate transmissions.
- Local area networks provide very high data rate communication within an office for attaching desktop computers, printers, modems, and other office automation devices.

KEY TERMS

A/D converter	D/A converter	protocol
analog	data rate	resolution
asynchronous	digital	sampling rate
ASCII	EIA RS–232C	serial communication
bit	local area networks (LANs)	synchronous
byte	modem	timing
clocking	parallel communication	transducer
code		

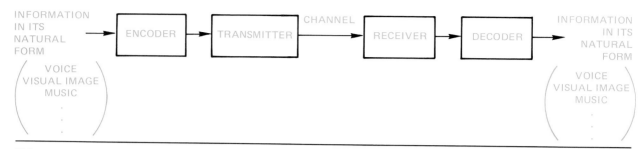

FIGURE 20–1 Like other communication systems, electronic communication systems have an encoder, a transmitter, a channel, a receiver, and a decoder.

Representing Information Electronically

Information in audio, graphic, or machine readable form can be represented by a series of electrical signals. The earliest telephones sent information by electronic means by sending voice signals directly over wires. In order to send information electronically, it must be converted from its natural form into electronic signals. The device that does this conversion is sometimes called a **transducer**. A transducer is a device that converts one form of energy (audio, optical, motion) into another (electrical).

In communications, transducers are also called encoders at the sending end and decoders at the receiving end. Thus an electronic communication system, like other communication systems, is made up of an encoder, a transmitter, a channel, a receiver, and a decoder, Figure 20–1.

Analog and Digital Signals

In early telephone transmission, and in voice transmission for nearly a century to follow, the voice signal was represented by a continuously varying **analog** voltage. As the voice grew louder, the signal grew larger. As the frequency (pitch) of the voice rose and fell, the frequency of the electrical signal rose and fell. The electrical signal was thus an analog, or copy, of the original voice, Figure 20–2.

One disadvantage of using analog electrical signals for representing information is that the

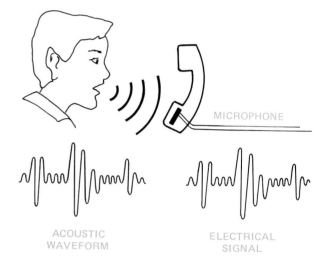

FIGURE 20–2 The microphone of a telephone handset produces a voltage that is an analog, or copy, of the speaker's voice.

noise present in all transmission systems distorts the received signal, making the recovered information imperfect. This imperfection shows itself as static in an AM radio or snow on a television set. Also, if an electrical signal is relayed from station to station in a long-distance communication system, the noise added over each segment of the link makes the signal worse and worse as it travels along the path, Figure 20–3. In some cases, the signal at the far end of a long communication path can be unrecognizable as the original signal.

Another way of sending information electronically is by **digital** transmission. The earliest widespread example of this was the on-off

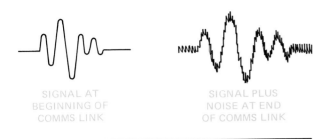

SIGNAL AT
BEGINNING OF
COMMS LINK

SIGNAL PLUS
NOISE AT END
OF COMMS LINK

FIGURE 20–3 All communication systems add noise to the signal. The more noise in a system, the harder it is for the receiver to reproduce the original message accurately.

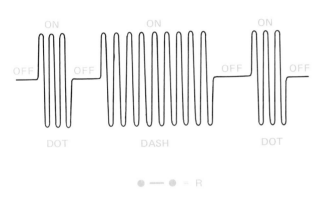

ON   ON   ON

OFF   OFF   OFF   OFF

DOT   DASH   DOT

● — ● = R

FIGURE 20–4 The telegraph was the earliest widespread example of digital, or on-off, electrical communication.

to hear it. Telegraph signals did not have noise in the background, as did early phones.

Bits, Bytes, and Codes

When electronic computers became more common, digital technology developed to take advantage of digital signaling in communications. Digital means two states, on and off. Information that is represented digitally is shown as a series of on or off signals called **bits**. Bit stands for **B**inary dig**IT**, or one of two digits, and it is generally represented by either a zero or one (0 or 1). A group of bits that represent a character is called a **byte**. One byte is usually eight bits long.

Information that is represented digitally must first be encoded. A **code** is the use of one set of characters to represent another set of characters. For example, a code could be created in which each letter of the alphabet was replaced by a number: A = 1, B = 2, C = 3, etc. The number sequence 2–1–12–12 would then be the code word for the English word, "ball", Figure 20–5.

telegraph signaling described in Chapter 17. In telegraph signals, the time period that a voltage is present is either long or short, followed by a total absence of voltage. The voltage can be used to sound a buzzer or tone. The series of long and short voltage pulses are interpreted by a telegraph operator as a series of letters or numbers, Figure 20–4.

In this simple telegraph system, the sending operator is the encoder, the key and battery make up the transmitter, the wire between the two stations is the channel, the buzzer is the receiver, and the receiving operator is the decoder. Because a sending signal was either large enough to activate the buzzer or it was not, the operator was either able to hear a signal or was not able

LETTER	CODE CHARACTER	LETTER	CODE CHARACTER
A	1	N	14
B	2	O	15
C	3	P	16
D	4	Q	17
E	5	R	18
F	6	S	19
G	7	T	20
H	8	U	21
I	9	V	22
J	10	W	23
K	11	X	24
L	12	Y	25
M	13	Z	26

FIGURE 20–5 This simple code uses numbers in place of letters.

NUMBER	BINARY CODE
0	00
1	01
2	10
3	11

2–BIT CODE

NUMBER	BINARY CODE
0	000
1	001
2	010
3	011
4	100
5	101
6	110
7	111

3–BIT CODE

NUMBER OF BITS IN CODE	NUMBER OF DIGITS THAT CAN BE REPRESENTED
1	2
2	4
3	8
4	16
5	32
6	64
7	128
8	256
10	1,024
12	4,096
16	65,536
32	4,294,967,296

FIGURE 20–6 The number of bits in a binary code character determines how many different letters, numbers, or symbols the code can represent.

Similarly, one set of numbers can be coded to represent another set of numbers. Binary numbers made up only of ones and zeros can be used to stand for a full set of letters and numbers. Simple codes made up of two bits and three bits are shown in Figure 20–6. The number of bits used in each code character determines how many different characters the code can represent. The number of characters that can be represented by different length codes are also shown in Figure 20–6.

The **ASCII** code is the code most commonly used to represent letters, numbers, and punctuation (alphanumeric characters), Figure 20–7. ASCII stands for the American Standard Code for Information Interchange. The ASCII code can be either seven or eight bits long. Seven-bit ASCII can represent all the letters of the alphabet, numbers from zero to nine, and punctuation. Eight-bit ASCII, sometimes called Extended ASCII, is used to include added symbols and functions not found in ASCII.

LAST BITS \ FIRST BITS	000	001	010	011	100	101	110	111
0000			SPACE	0	@	P	`	p
0001			!	1	A	Q	a	q
0010			"	2	B	R	b	r
0011			#	3	C	S	c	s
0100			$	4	D	T	d	t
0101			%	5	E	U	e	u
0110			&	6	F	V	f	v
0111			´	7	G	W	g	w
1000			(	8	H	X	h	x
1001			)	9	I	Y	i	y
1010			*	:	J	Z	j	z
1011			+	;	K	[	k	{
1100			,	<	L	/	l		
1101			–	=	M	]	m	}
1110			.	>	N	∧	n	~
1111			/	?	O	__	o	DELETE

SPECIAL CONTROL CHARACTERS

EXAMPLE: G = 100 0111
FIRST LAST
BITS BITS

FIGURE 20–7 The American Standard Code for Information Interchange (ASCII) is one of the most commonly used binary codes.

Converting between Analog and Digital Information

In addition to letters and numbers, digital codes can be used to represent analog voltages. Thus, information that is naturally analog, such as telephone signals, can be represented digitally and sent digitally over long distances. The advantage in doing this is that digital transmission can be made virtually noise-free. Communication paths can be designed so that there is no error or question about whether a received signal is a zero or a one, and no error about the received code word. The only real noise in such a communication link is the noise introduced when the signal is changed from analog to digital at the transmitting end, and from digital back to analog at the receiving end.

At the transmitting end, a device called an analog-to-digital converter, or **A/D converter**, changes the analog signal into a binary, or ditigal, code. The more bits that the A/D converter uses, the more accurately the analog signal will be represented. Each code word stands for an actual voltage at a specific instant of time. The A/D converter generates a continuous stream of code words that describe the voltage as it changes with passing time. The rate at which the code words are generated is called the **sampling rate**. The sampling rate must be matched to the fastest speed at which the analog signal may change in order to give a true representation of the signal.

At the receiving end, a device called a digital-to-analog converter, or **D/A converter**, performs the reverse process. The D/A converter changes the digital signal back into an analog signal. The number of bits handled by the D/A converter should match the number of bits used in the code to ensure that all of the information contained in the code is properly used. As with an A/D converter, the more bits available and used by a D/A converter the more accurate the recreation of the analog signal. The more bits a converter can handle, the more accurate its **resolution**, Figure 20–8. If only one bit is used, then the resolution is only "on" or "off" (one or zero), so the resolution error is 50%. If two bits are used, the resolution error is twice as good, or 25%. Resolution indicates how closely the digital signal can

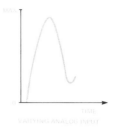

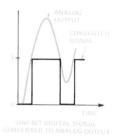

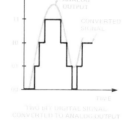

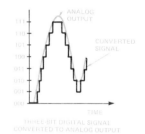

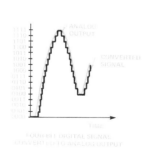

A/D, D/A CONVERSION

NUMBER OF BITS	RESOLUTION (PERCENT)
1	50
2	25
3	12.5
4	6.3
5	3.1
6	1.6
7	0.8
8	0.4
10	0.2
12	.05
16	.01

FIGURE 20–8 The more bits used in an A/D or D/A converter, the more closely it can represent a signal, or the better its resolution.

FIGURE 20–9 D/A converters are often made on a single integrated circuit. *(Courtesy of Burr-Brown Corporation)*

resemble the analog signal. The more bits, the smaller the resolution error, and the better the digital and analog signals match.

The communication from an A/D converter to a D/A converter (Figure 20–9) is at a **data rate** made up by the number of bits used in the conversion and the number of conversions performed each second (the sampling rate). Multiplying these two together gives the data rate in bits per second. Thus, if an A/D converter creates an eight-bit word at a sampling rate of 1000 samples per second, it has an output data rate of 8000 bits per second, or 8 kbps (''k'' stands for *kilo*, the Greek word for thousand).

One of the most common uses for A/D and D/A conversion is in the telephone system. When you talk on the telephone, it is likely that your speech is converted to digital form by a special A/D converter called a codec. In most modern systems, the sampling rate is 8,000 samples per second, with 8 bits per sample. This results in 64 kbps for speech. At the receiving end, the digital signal is converted back to analog. The use of digital transmission keeps the signal free of noise and makes it easier to switch.

Parallel and Serial Communication

For sending data over short distances, the most direct method is to use a group of wires to connect the transmitter with the receiver. Using the example of the two converters in the previous section, eight wires could be used to connect the A/D converter to the D/A converter, and eight bits would pass between the two 1000 times each second. The data rate is 8000 bits per second. This type of transmission is called **parallel communication** because the individual bits are sent on parallel pieces of wire, Figure 20–10. Because of the number of wires used, parallel communication is usually limited to short distances, such as from a computer to its printer located nearby. The advantage of using parallel communication is that very high data rates can be achieved simply.

When data is sent over longer distances, it is usually changed from parallel groups of bits to one continuous, higher data rate stream of data. This is called **serial communication**, Figure 20–11. The conversion from parallel to serial data is often performed in one special purpose inte-

FIGURE 20–10 Printers are often connected to computers by a parallel communication cable. This allows communication to be cost effective at a high data rate. *(Photo by Robert Barden)*

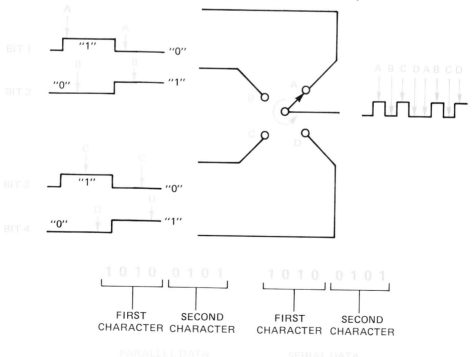

FIGURE 20–11 Serial data is formed when parallel lines are sampled sequentially, one after the other, before the data can change. After the last bit line is sampled, the data changes and the process is started again.

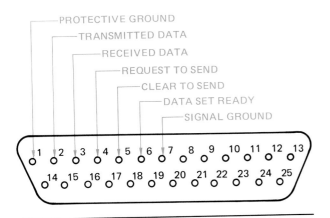

FIGURE 20–12 These are some of the more commonly used pins in the 25-pin EIA RS-232C connector.

grated circuit. In the conversion, the individual bit lines are sampled, one after the other, at a rate fast enough so that they can all be sampled before the information changes. The rate at which the bit lines are sampled is called the clock rate. The serial data stream is thus at a higher clock rate than the parallel rate, even though the data rate (the number of bits per second) has not changed.

One of the most common standards used in serial communication is the Electronic Industries Association Recommended Standard 232C (**EIA RS–232C**). This standard gives voltage levels, connector pins, and recommended control signals for serial communication between two devices, Figure 20–12. It is widely used throughout the world for connecting computers to serial printers, modems, terminals, and other devices, and for communication between any two automated devices.

Clocking: Synchronous and Asynchronous Communication

In order for data communication to be successful between a transmitter and a receiver, the receiver must ''know'' when to look at the data, so that the data is stable (not changing). This is done by sending some type of **clocking**, or **timing**, information along with the data. If an actual clock signal is sent along with the data, the transmission is called **synchronous** transmission because the data is locked, or synchronized, to

the clock. Higher data rate communication is most often synchronous, as this gives a very positive and efficient way of receiving continuous streams of data correctly.

Lower data rate communication is often done without a separate clock, character-by-character. In this kind of **asynchronous** transmission, the start of each character provides the receiver with the timing information needed to properly recover the data, Figure 20–13. Asynchronous communication is simpler, but less efficient, so it is usually limited to lower data rates.

Protocols

To understand some of the complexities of data communication, it is helpful to think of each of the steps involved in placing a telephone call (see Chapter 17). The first step is to lift the handset from the telephone. This sends a signal to the

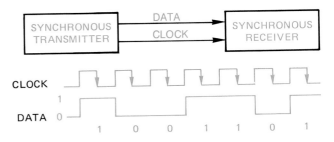

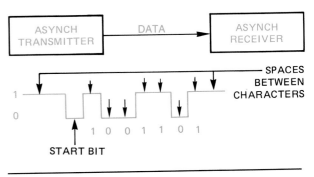

FIGURE 20–13 In synchronous communication, the receiver samples the incoming data at specific instants identified by a timing clock sent with the data. In asynchronous communication, the receiver starts its own timing at the start of each character.

telephone central office indicating that you wish to place a call. The central office responds by sending back a dial tone when it is ready to accept your call. You then dial or use push buttons to select the number of the party you wish to call.

Then you wait while the phone at the other end rings, until it is answered by the person being called. Information is exchanged only when you start speaking, and only if you both speak the same language. When the call has been completed, the handsets are replaced, setting up the telephones for another call.

In data communication, many of the same things have to happen automatically before a connection can be established. In addition, each transmitter and receiver has to observe a rigid set of rules to keep one end from transmitting when the other end is not prepared to receive. The set of rules that is used to establish calls, end connections, and keep the data flowing smoothly during the connection is called a **protocol**. The transmitting and receiving devices at both ends must observe the same protocol for successful data communication. In addition, the transmitter and receiver must be capable of operating at the same data rate and communication type (synchronous or asynchronous).

The communication protocol rules are often stored and implemented by a single chip computer specially programmed for this task. In desktop computers, a separate printed circuit board is often used that performs only communication-related tasks. Sometimes, the data communication protocol is so complex that the computer and support circuits needed on this communication board have more processing capability than the computer into which it is plugged, Figure 20–14.

Telephone Modems

When sending data over long distances, it is often most convenient and economical to use telephone lines rather than special data lines. As described in Chapter 17, most telephone lines appear to the user to be analog lines, even though some form of digital transmission may be used

FIGURE 20–14 Communication boards in computers contain special computers, support circuits, and programming for implementing the communication protocol. The processing power of some communication boards exceeds the capabilities of the computers into which they are plugged. *(Courtesy of MICOM INTERLAN, a division of MICOM Systems, Inc.)*

FIGURE 20–15 Modems are used to send data over telephone or other analog communication circuits. *(Courtesy of CXR Telecom)*

along the path. The device used to send digital data over an analog line is called a **modem**, which is short for **MO**dulator-**DEM**odulator (see Chapter 18 for a description of modulation). In a modem, the data signal is modulated onto an

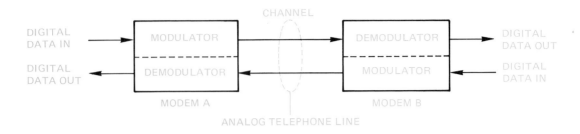

FIGURE 20–16 A modem modulates the data to be sent onto an audio carrier signal. At the receiving end, another modem demodulates, or removes the data from the audio carrier.

audio frequency carrier signal that can be sent over telephone lines, Figure 20–15. At the receiving end, the demodulating modem removes the data signal from the audio carrier signal, Figure 20–16.

Modems are used to connect personal computers to information stored in data bases hundreds of miles away, to allow large computers to exchange information with other large computers, to allow executives to retrieve electronic mail from their office automation systems, and to connect automatic teller machines (ATMs) to the central computer at a bank. Modems are used whenever data must be transmitted over telephone lines.

The relatively small bandwidth of a telephone line (see Chapter 19) limits the amount of information that can be sent over it in digital form. The amount of digital information sent over a line per unit of time is called the data rate, measured in bits per second, or bps. One thousand bits per second is one kilobit per second, or kbps. One million bits per second is one megabit per second, or Mbps. Telephone line modems used on dial-up circuits commonly carry 300 bps, 1.2, 2.4, 4.8, or 9.6 kbps. At 9.6 kbps, it would take only two seconds to send all the words on this page, but it would take almost two minutes to send a detailed multicolor graphic image.

High Data Rate Communication

In order to send large amounts of data more quickly, higher data rate transmission lines are needed. With special and relatively expensive modems, data rates of slightly more than 9.6 kbps can be achieved over dial-up lines. To get real improvements in data rates, however, special telephone data lines are used to give data rates of 56 kbps, 1.5 Mbps, and even 45 Mbps. At 1.5 Mbps, the detailed multicolor graphic image that takes two minutes to send over a dial-up line can

FIGURE 20–17 Some nationally distributed newspapers use high data rate digital transmission to send copy to regional printing plants in other parts of the country. *(Courtesy of Scott Malay/Gannett Co. Inc.)*

FIGURE 20–18 Often, high data rate communication equipment is made up of special purpose digital computers programmed to perform communications functions. *(Photo by Robert Barden)*

be sent in less than one second. People who have very large amounts of data to send, and who care about how quickly it is sent, have to install these special high data rate lines, Figures 20–17 and 20–18.

FIGURE 20–19 Some companies use satellite links as part of their private digital networks. This one is used for data transmission and for digitized video and voice transmission. *(Photo by Robert Barden)*

Some companies that have large amounts of data to transmit from one site to another install their own data communication circuits or have specialized communication companies provide such circuits, Figure 20–19. These are called private data networks, Figure 20–20, and can be city-wide or world-wide in size. Often, they use fiber-optic transmission links, satellite links, microwave links, and even undersea links.

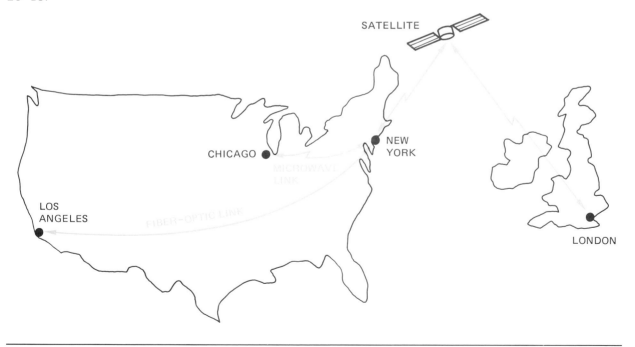

FIGURE 20–20 Private networks often use a variety of transmission media.

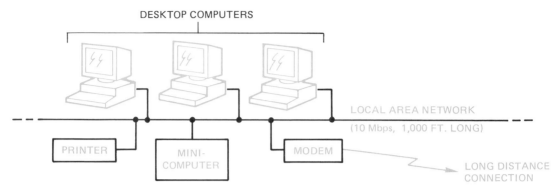

DESKTOP COMPUTERS

LOCAL AREA NETWORK
(10 Mbps, 1,000 FT. LONG)

PRINTER

MINI-COMPUTER

MODEM

LONG DISTANCE CONNECTION

FIGURE 20–21 Local area networks are high data rate networks that connect desktop computers, printers, and other office automation equipment in the modern office environment. *(Photo courtesy of American Electric)*

Local Area Networks

As office automation has become more widespread, the need for high rate data communication around an office has grown. Small travel agencies, insurance agencies, and other small companies have joined large offices in having a desktop computer on every desk, FAX machines, and printers all connected by a data communication network within the office. Such small networks are called **local area networks**, or **LANs**, Figure 20–21.

Connecting all these machines together with a local area network allows a person operating one desktop computer to retrieve a document stored electronically in another desktop computer to inspect or change the document. One or two printers can provide all the printing needs of a dozen or more desktop computers, rather than needing one printer for each computer. If the office needs information from a large data base stored on a large computer in another city (such as a travel agent checking airline schedules), then only one modem need be connected to the network to serve all the people in one office.

Because they are often used to send large documents from one machine to another in the office, local area networks usually operate at a very high data rate. The two most common LANs use data rates of 4 Mbps and 10 Mbps. At these data rates, large documents can be sent from one machine to another in a second or less.

It is important to keep system delays small. Studies have shown that system delays lead to less productive workers. Delays of 0.2 second or less have been shown to reduce the productivity of office workers, even though such delays can hardly be noticed. Because we now live in the information age, improving the productivity of information workers has become a very important task, Figure 20–22. This need, coupled with fast-developing new technology, has led to many new jobs in the field of data communications.

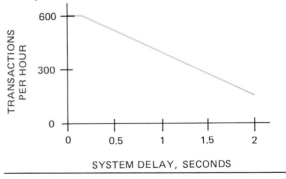

FIGURE 20–22 As communication and other system delays increase, information worker productivity goes down. One way to reduce delays is to increase communication data rates.

Data Communication at Work

Data communication is in use all around us everyday. These are only a few of the uses for data communication.

Data communication is used to transfer data from this computer to the plotter. *(Courtesy of Hewlett-Packard Company)*

The telephone modem attached to this student's computer allows him to carry on data communication over his telephone line. *(Courtesy of Racal-Milgo, international designers and suppliers of data communications and data security)*

This computer-aided design workstation gathers information from several remote sources using data communication. *(Courtesy of Moniterm)*

Data communication connects this unattended automatic teller machine (ATM) to the bank's central computer, 24 hours per day. *(Photo by Robert Barden)*

Data communication allows this automatic fire alarm system to gather information from many sensors. *(Photo by Robert Barden)*

This technician is using data communication test equipment to remotely repair telephone circuits. *(Courtesy of Fiber Optic Services)*

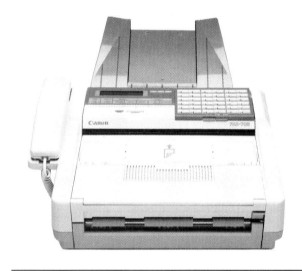

Copies of documents can be sent anywhere in the world in a few minutes, using data communication over the telephone lines. *(Courtesy of Canon U.S.A., Inc.)*

Summary

Information in audio, graphic, or machine readable form can be represented by a series of electrical signals. Electrical signals that are smoothly varying, following the natural signals they represent, are called analog signals.

All information, including analog informa-tion, can also be represented by digital, on-off signals. One principal advantage to using digital transmission, as opposed to analog transmission, is that digital transmission is not as susceptible to noise as analog transmission.

The smallest unit of information in digital transmission is called a bit. A bit is represented mathematically by a one or a zero, and by an on

Data communication allows these financial traders to have the latest prices from all over the world displayed instantly. *(Courtesy of Positron Industries, Inc.)*

or off voltage. A group of bits is used together to stand for characters. Each character can be used to represent information of another type, using a code that allows one to be translated into the other. The most common code is the ASCII code, which uses seven bits to represent 128 letters, numbers, punctuation, and other characters.

Analog signals can be changed into digital data by devices called analog-to-digital converters. Digital data can be changed back to analog information by digital-to-analog converters.

Digital information can be sent from a transmitter to a receiver by parallel or serial transmission. In parallel transmission, each bit is sent over a separate wire. In serial transmission, the bits are sent sequentially, one after the other, over one wire, before the information can change.

A data communication receiver is either instructed to examine the data it is receiving by a clock signal that is sent with the data (synchronous transmission), or is retimed by the start of each character as it is received (asynchronous transmission). Usually, asynchronous transmission is used at lower data rates, while synchronous transmission is used at higher data rates.

A protocol is a detailed set of rules that governs how a data communication call is established, how the data is sent from each transmitter to its corresponding receiver, and how the call is broken. Both the sending and receiving ends must observe the same protocol, data rate, and transmission type for successful data transfer.

A modem is a device used to send digital data over an analog telephone line. The transmitting modem modulates an audio signal with the data signal. The receiving modem removes the data signal from the audio carrier signal. Modems commonly used on telephone lines provide data communication at rates up to 9.6 kbps.

Higher data rate communication uses special transmission lines available from the telephone company or from other specialized communications companies. Often, a large company will install a private network, using dedicated fiber-optic, satellite, or microwave transmission links at data rates of 56 kbps, 1.5 Mbps, or 45 Mbps.

Local area networks are used in offices to provide very high data rate communication among office automation equipment, including desktop computers, printers, modems, and other devices. LANs allow workers to work more efficiently, with lower equipment cost than if each person used stand-alone equipment.

REVIEW

1. What is the principal advantage of digital data transmission over analog transmission?
2. Which of the following are examples of analog information and which are examples of digital information? Morse Code telegraph; speaking into a telephone hand set; storing information on a computer floppy disk; sending information in ASCII code; a television signal; turning a light switch off.
3. Using the ASCII table shown in Figure 20–7, write out the binary code for "Data."
4. How many bits are needed to represent a group of twenty different characters?
5. Draw a simple diagram of a system that uses an analog device to measure pressure in a pipeline, and then uses a digital transmission line to send pressure information to a computer located one hundred miles away using a modem.
6. An analog-to-digital converter is used to create 10 bits of information with each conversion. It is operated at a rate of 8000 conversions (samples) per second. What is the total data rate at which it sends data?
7. A printer is connected to a computer by a parallel communication cable. The cable carries 8 bits of data that change 1200 times per second. If the operator wants to change the cable to a serial communication cable (and change the printer and computer interfaces as needed), what data rate should be used for serial communication?
8. Why is synchronous transmission used for sending high data rate communication?
9. What is the function of a communication protocol? Can devices that are programmed for one protocol communicate with devices that are programmed for a different protocol?
10. What is the purpose of a telephone line modem? Draw a simple data communication system that uses modems.
11. Why are modems used for only relatively low data rates (up to 9.6 kbps)?
12. List three different kinds of transmission links that give high data rate transmission.
13. Give two examples of when high data rate transmission is needed.
14. What is a local area network? Why is it used in modern offices?

Recording Systems

KEY TERMS

amplifier	high fidelity	playback head
camcorder	magnetic tape	record head
cartridge	mechanism	reel-to-reel tape
cassette	medium	stamper
compact disk (CD)	monaural	stereo
digital audio tape (DAT)	mother	stylus
Dolby™ noise reduction	motor	video cassette recorder (VCR)
erase head	phonograph	video laser disk
heads	pickup	

Introduction

An electronic recording system stores electronic information for reading back at a later time. The information may be in analog or digital form (see Chapter 20). The part of the system that stores the information is called the **medium**. Common recording media include magnetic recording

FIGURE 21–1 Examples of the recording media described in this chapter.

tape, phonograph disks, optical disks, and floppy disks, Figure 21–1.

Typically, a recording system can be divided into five major parts. These include the recording medium to store the data, the transducer to encode (record) or decode (playback) the data to and from the medium, the amplification system to amplify all signals to a usable level, the mechanics to move the media past the transducer, and an output device to allow the user to see or hear the recorded data.

Magnetic Tape Recording

The principle of recording information on a material with magnetic fields has been with us for a long time. The first successful recording with a magnetic field was done in the late 1800s by a Danish engineer named Valdemar Poulsen. Poulsen's experiments were not done with the same type of magnetic tape we use today. He recorded on a steel wire. Wire recorders, as they were called, were very popular for recording voice data but they were not very good for music. These recorders were used by secretaries in offices of all kinds as late as the 1960s.

Something better than steel wire was needed for music and other **high fidelity** (music quality) uses. The flexible magnetic tape and a method of transferring information to it, similar to what we have today, was developed by a German scientist around 1930 and has been continuously improved.

Until the mid-1950s, all tapes were recorded **monaural** using one microphone and one speaker to play back. The music seemed to come from one spot. In the late 1950s and early 1960s,

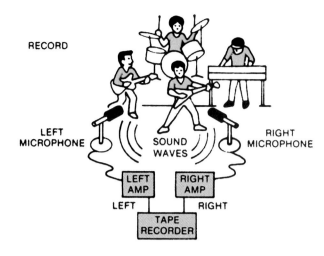

RECORD

LEFT MICROPHONE

RIGHT MICROPHONE

SOUND WAVES

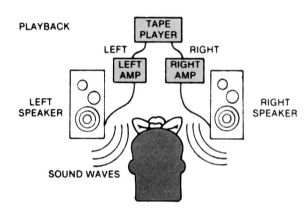

PLAYBACK

TAPE PLAYER

LEFT

RIGHT

LEFT AMP

RIGHT AMP

LEFT SPEAKER

RIGHT SPEAKER

SOUND WAVES

FIGURE 21–2 Stereo tape systems record two independent channels and play them back separately to give the listener a more realistic sound.

stereo recording was introduced. In stereo, two microphones are placed in different locations to record sounds, and two different recordings are made at the same time on the same tape. When the tape is played back, two different speakers are used, giving the listener the feeling of being at the performance, Figure 21–2.

Another improvement that occurred in the late 1960s was the **Dolby™ noise reduction** system. Dolby™ noise reduction improves the quality of the sound by reducing the background hiss that was noticeable on previous recording systems.

Also in the late 1960s and 1970s, video tape recording became economically feasible for widespread use.

Electromagnetism

Almost everyone knows what a magnet is and how it is attracted to ferrous materials like iron and steel. Most of us have been exposed to magnets even if we just played with them on a table top to see how they attract and repel each other.

You also learned in Chapter 16 that there is a relationship between magnetism and electric current flow. Two early experimenters, Andre Ampere and Michael Faraday, discovered this relationship in the mid-1800s. Their tests showed two very important facts that later helped others develop magnetic recordings. In one series of experiments it was discovered that when an electric current was passed through a conductor there was a magnetic field developed around that conductor, Figure 21–3. If the conductor was formed into a coil the magnetic strength was increased and magnetic poles were formed. If the intensity of the current was changed it caused the strength of the magnetic field to change. In another series of experiments it was discovered when motion

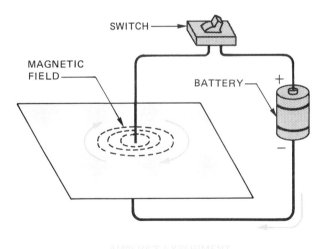

SWITCH

MAGNETIC FIELD

BATTERY

AMPERE'S EXPERIMENT

FIGURE 21–3 Ampere's experiment demonstrated that when current flows through a conductor a magnetic field is created around that conductor. This is sometimes called the Left Hand Rule for Straight Conductors.

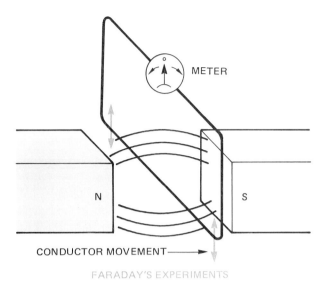

FARADAY'S EXPERIMENTS

FIGURE 21-4 Faraday's experiment showed that moving a conductor through a magnetic field created a current flow in that conductor and the direction of movement of the conductor controlled the direction of the current flow.

FIGURE 21-5 Typical eight-track cartridge mechanism. Note there is only one endless-loop reel of tape.

exists between a magnetic field and a conductor there will be a current flow created or induced in the conductor, Figure 21-4. If the conductor was wrapped into the form of a coil the current induced would be much greater. And finally if the strength of the magnetic field was changed, the amount of current induced would also change.

These principles enabled the development of transformers, electric motors, and generators (see Chapter 16). They are also the basis for the development of magnetic recording.

The Recording Media

One type of magnetic recording medium is the flexible **magnetic tape**. All tapes have similar construction: a plastic tape base covered with a magnetic coating. Low cost tapes use a thinner tape base and a magnetic coating that does not have many magnetic particles per square inch. By increasing the magnetic particle density, changing the type of magnetic media material, and making the tape base thicker, the quality of the tape will be improved (and the cost in-

creased). Improving the tape will improve the quality of the stored information.

Magnetic tape is packaged on reels or in special containers for ease of handling and storage. In the early days all tapes came as **reel-to-reel** tapes. These tapes had to be threaded by hand through the recorder/player. Reel-to-reel tape equipment is still popular for high-quality work in the audio field but it is mainly limited to studio work for video uses.

The next package to have widespread use was the eight-track **cartridge**, Figure 21-5. This is a continuous-loop tape wound in a plastic cartridge. Tape is pulled out of the middle of the winding, and wound back on the outside. These were very easy to load into a player/recorder and could be played in just about any position. Audio tapes could be taken almost anywhere. Eight-track cartridges were not very reliable for long periods of operation, however. This was due to the way they were made and loaded into the player/recorder. The cartridge and other parts wore down from constant inserting and extracting. This caused tape slippage and sound distortion.

The tape package in most widespread use today is the **cassette** cartridge. The cassette cartridge is a self-threading, reel-to-reel tape enclosed in a plastic case. This type of cartridge

FIGURE 21-6 Cassette tape mechanism showing the pressure pad and tape rollers.

Tape Recorder Components

In a tape recording system, the **heads** are the transducers that record the signal onto the tape, and that recover information from the tape during playback.

An **erase head** (Figure 21-7) is used to do just what the name implies. It erases the information stored on a magnetic tape. It does this by actually rearranging the magnetic particles on the tape to a position that would produce no audible sound. It is generally controlled by the erase oscillator. In some less expensive recorder/players the erase mechanism is simply a small magnet.

Record and **playback heads** change electric signals into varying magnetic fields and vice versa, so that information can be stored on the tape and retrieved from it, Figure 21-8. Most of

eliminated most of the wear/slippage problems of the audio eight track. The video cassette tape cartridge also made threading easier and tape damage occurred less often, Figure 21-6.

FIGURE 21-7 The white square in the center of the picture is a typical cassette deck erase head.

FIGURE 21–8 The silver square in the center of the picture is a typical cassette deck record/playback head. The red and white wires connect the head to the electronic circuits.

today's magnetic recorders have a combination record/playback head instead of separate heads.

The **amplifier** amplifies, or makes larger, electrical signals from the source of information (i.e., microphone) that are to be recorded on the tape or amplify the signals picked up from the tape to be sent to an output (i.e., speaker).

The **motors and mechanism** move the tape past the heads in order to record or play back the information, Figure 21–9. To be able to interchange between different players and to have excellent quality sound reproduction (high fidelity), the tape must be moved past the heads at a precise speed. This speed must be constant throughout the length of the tape. In modern tape equipment this is done by having the motor turn a metal spindle called a capstan which, in connection with the pinch roller, pulls the tape from the source reel to the take-up reel, Figures 21–10 and 21–11. The speed is regulated by high-quality compact motors and electronic motor speed controls.

FIGURE 21–9 Erase head (left) and record/playback head (right) on a reel-to-reel tape recorder. Notice the mechanical guides that position the tape on the heads and the spring tension screw used for the head adjustment.

The Tape Recording Process

The music, voice, or other information to be recorded is changed into an electric signal by a microphone or other device. The electric signal

FIGURE 21–10 The pinch roller and capstan that are used to move the tape past the heads in a cassette deck. Notice the lack of mechanical guides as in Figure 21–9 of the reel-to-reel tape recorder.

FIGURE 21–11 Pinch roller and capstan of a reel-to-reel tape recorder.

is amplified and passed through the record head, causing a varying magnetic field in the head. The head, being in contact with the tape, causes the magnetic particles on the tape to be repositioned, Figure 21–12. The position of these particles is related to the strength and polarity of the magnetic field in the record head. The new position of the particles now represents the recorded information, Figure 21–13. In an audio system, this recorded data occupies only a small portion of the tape.

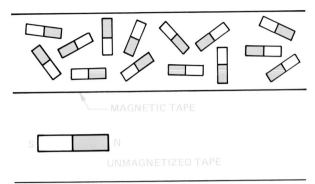

FIGURE 21–12 The magnetic particles are shown as tiny bar magnets. Notice there is no predictable pattern to the particles.

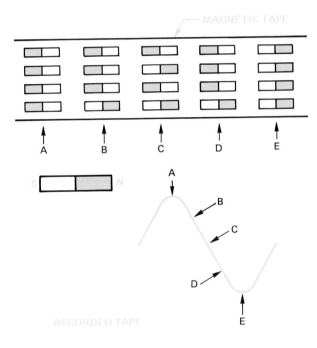

FIGURE 21–13 The magnetic particles are aligned to represent the variations of the recorded information. The letters below the magnetic tape correspond to the positions indicated on the sine wave.

The Playback Process

The playback process is the reverse of the recording process. As the tape, with its aligned particles, passes by the playback head, a varying current flow is created in the head coil. The current produced is proportional to the polarity and density of the magnetic field on the tape.

FIGURE 21–14 A high-quality videocassette recorder is used with a television receiver to view and listen to video programming. *(Photo courtesy of Sony Corporation of America)*

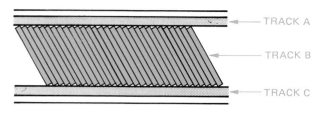

TYPICAL VIDEO TAPE INFORMATION

FIGURE 21–16 The three major portions of a recorded video tape. Track A contains the audio information. Track B contains the diagonal video information stripes. Track C contains the control data.

This current is then amplified and reproduced by the output device (i.e., speaker).

The previous description was based mainly on the principles of audio recording and playback. Although there are definite similarities, it would be inaccurate to say the same thing happens in a video system.

A **video cassette recorder** (**VCR**), Figure 21–14, uses almost the same principles and many of the same circuits as the audio recorder but most of these circuits are more sophisticated. Major differences in the audio and video recorder systems are the amount of information stored on the tape and the mechanical differences in the head/tape contact.

A video picture, stored on a tape, contains much more information than the sound on an audio tape, Figure 21–15. (You may recall from Chapter 19 that a video signal takes up 1,000 times more bandwidth than a telephone signal.) So much information is contained that the data cannot be placed in a straight line along the length of the tape. The data must be broken down into ''stripes'' which are then recorded diagonally across the tape. These stripes contain the video data as well as the color (chroma) and synchronizing signals needed for video reproduction. Audio tracks are also recorded to complete the television display, Figure 21–16.

The other main difference is in the mechanics of the recorder. In a video recorder the tape moves past the heads in a diagonal direction and the heads are not stationary. Video heads spin at a high speed. You can see there is a dual motion here—the tape moves and the heads move. The reason for this is to get more information placed on the tape in less tape length. In modern VCRs, in order to get even more information on a tape, the tape speed can be slowed down. The SP-LP-SLP switch controls this function. As the tape is slowed down, the diagonal stripes are pushed together and therefore more data is placed in a shorter space, Figure 21–17. As in audio systems, if we want to be able to replay a tape or exchange a tape with someone else, the tape speed, head speed, and the tape position must be carefully controlled on all video recorders/players.

In a video recording system, the information recovered from the tape is played back on a television set or video monitor. The sophistication of video playback systems continues to improve at a fast rate. Remote control devices keep many functions at the fingertips of the viewer.

FIGURE 21–15 The internal mechanics of a VHS cassette tape.

FIGURE 21–17 The record/playback head of a video recorder/player. One of the heads is located just to the left of the pencil point. Notice the angled position of the head drum. This allows the recording of the diagonal stripes and at the same time keeps the tape in a straight line.

Examples are forward-backward and slow motion. Black-and-white and color still prints can be made with video printing equipment. This makes it possible to select images that are shown on a television receiver and make prints similar to photographs. Currently, the quality of these prints is poorer than photographic prints, but improvements are being made rapidly.

The source of the video can be an off-the-air TV program, or a TV camera or **camcorder** (**cam**era-re**corder**), Figure 12–18. Camcorders are easy to use and give the operator several options.

Besides recording sound, camcorders allow the operator to take pictures with the lens in any position from wide angle to telephoto. A small motor attached to the lens moves the lens back and forth in a smooth motion. This gives the camcorder operator the chance to show close-ups, regular, or wide-angle views of selected scenes. One or more lights can be attached to the camera thus allowing dark or night scenes to be recorded on tape.

FIGURE 21–18 A camcorder is easy to use. *(Photo courtesy of Sony Corporation of America)*

Specialized Video Cameras Used for Research and Tracking Needs

The challenge of finding out why something works or does not work cannot always be determined by just looking at it with unaided eyes. The lenses of one or more video cameras are often used to record events in manufacturing, science, and nature. This makes it possible to study the photographed events at a later date.

High-speed video cameras are used to capture the events being photographed. Once the happenings are recorded on videotape, motion analysis equipment is used to view the taped image at any selected speed. Stop action control allows the researchers to view each scene as long as they wish.

Underwater video camera equipment makes it possible to record sea life at very great depths. Cases that house video cameras must be strong enough to withstand underwater pressures for breakage, be waterproof, and be convenient for operating the controls of the camera including the lamps.

Eye movement recorders allow the wearer to look in the direction of interest and photograph what is being viewed. This type of special equipment is useful for military surveillance and research in which it would be impossible or inconvenient for the photographer to use a hand-held camera.

Video photography is a useful tool for all forms of research where images will need to be studied. The immediate playback feature is of significance because time is often limited when decisions must be made.

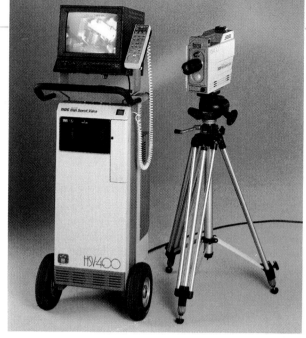

A high-speed video camera and motion analysis equipment which allows the viewer to see the taped image at any speed. *(Courtesy of Instrumentation Marketing Corporation)*

Underwater videocamera equipment. *(Photo courtesy of Sony Corporation of America)*

An eye movement recorder *(Courtesy of Instrumentation Marketing Corporation)*

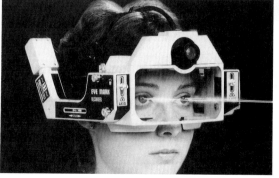

DAT

A new technology, **digital audio tape (DAT)**, may be as revolutionary to magnetic tape recording as the compact disc was to the phonograph record. DAT combines magnetic recording with digital technology.

In digital audio tape technology, music that is to be recorded is first converted to digital data by a digital encoder, or analog-to-digital converter (see Chapter 20). The digital data is then stored on the tape and recovered later. The digital data recovered from the tape is then converted back to analog music by a digital-to-analog converter, Figure 21–19. Because the information (music) is stored in digital form (1s and 0s) there is no hiss or tape noise on the recovered signal.

DAT Components

Because digital information requires a great amount of stored data to represent the recorded

sound, special high-density tape had to be developed. This tape has many more magnetic particles per inch than traditional audio tape. The size and shape of the particles is slightly different when compared to conventional audio tape. The tape lengths are also longer to accommodate faster tape speeds.

In order to store the huge amounts of data just described, it was necessary to develop special record and playback heads. These heads can record the information on a thinner track. This allows more tracks of information on a single tape. More tracks simply mean more stored data. Also, it is necessary to record and play back certain controlling data. This is data needed by the electronics circuits to play back the recorded information while checking for errors that would show up as noise in the speakers, among other things. DAT recorders also have separate record and playback heads.

The electronic circuits in a DAT recorder include an analog-to-digital converter for recor-

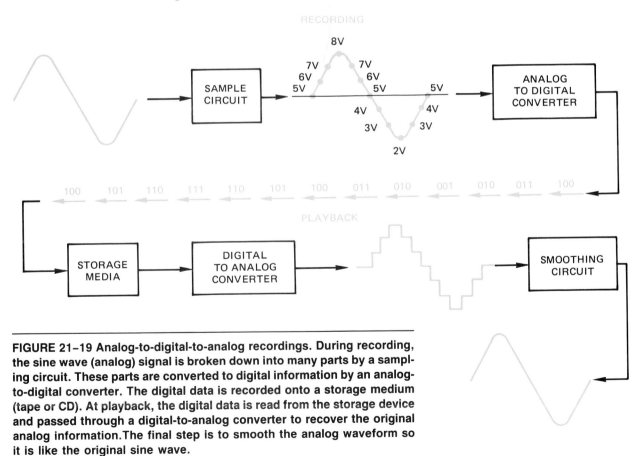

FIGURE 21–19 Analog-to-digital-to-analog recordings. During recording, the sine wave (analog) signal is broken down into many parts by a sampling circuit. These parts are converted to digital information by an analog-to-digital converter. The digital data is recorded onto a storage medium (tape or CD). At playback, the digital data is read from the storage device and passed through a digital-to-analog converter to recover the original analog information. The final step is to smooth the analog waveform so it is like the original sine wave.

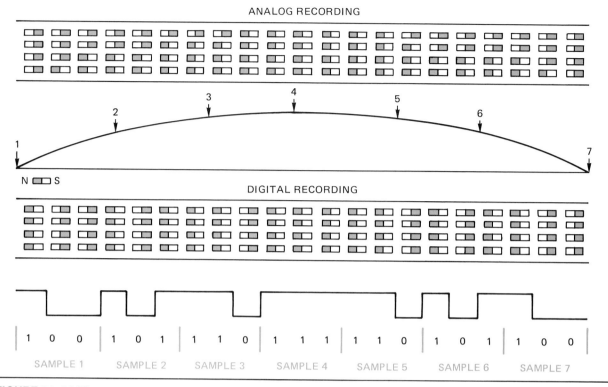

FIGURE 21–20 The top tape shows the typical magnetic particle arrangement of a recording of one half of a sine wave. The bottom tape shows how a three-bit digital recording of the same sine wave might appear. Of course, there are many magnetic poles on a tape track, while only four are shown here.

ding, and a digital-to-analog converter for playback in addition to the amplifiers normally found in a tape recorder. Other electronics are needed for the special information that will be used to check for errors in the recording. These errors may be problems such as bad spots on the tape. There is also data that will be used to be sure the recording is in sync or in time with the electronics circuits. That data must be processed by added circuits. Because of the number and complexity of all these circuits a microprocessor is needed to control their functions.

In DAT equipment, the tape must move faster and the position of the tape from record to playback is critical. These exact requirements mean the construction of the unit is more precise than the typical cassette player/recorder.

Digital Recording

One difference in a digital recording is the way the magnetic particles are arranged on the tape, Figure 21-20. In a conventional audio magnetic recording, the magnetic particles are arranged in a pattern that represents the varying audio signal. If we could look at the particles we would see they all point one way, then they are mixed, then they all point the other way.

In a digital recording, all the magnetic particles point the same way if a one (1) is represented. They all point the opposite way if a zero (0) is represented. This is sometimes referred to as saturation recording.

Another difference between analog and digital recording is in the amount of data that must be stored (bandwidth required on the tape) in order to reproduce the audio signal. Digital recordings are stored in the form of bits (0s and 1s). In a typical DAT recording, the analog-to-digital converter samples the music more than 40,000 times each second, converting it to a 16-bit code in each sample. The resulting data rate is more than 640 kbps (see Chapter 20).

The Advantages of DAT

If DAT recorders need all these special circuits and mechanics, why would they be so desirable? A DAT recorder has one big advantage: it reproduces sound with great clarity and no noise. It can do this because of the digital recording and error-checking circuits. Just as the digital transmission circuits used for telephone signals do not introduce noise (see Chapter 20), the digital tape does not introduce noise as music is stored on it or read from it. The end result is a sound quality that cannot be obtained with conventional audio magnetic recordings.

Other Digital Recording

There are other methods of magnetic digital recording in use today. These include computer disks and floppy disks (see Chapter 2) and direct-to-disk cameras (see Chapter 22). While the principles used are the same, the shape of the media and the drive mechanisms are different.

Phonograph Records

Thomas Edison developed the first practical **phonograph** about 1877. Of course Edison's phonograph was primitive and needed improvements to be practical. An amazing thing is that Edison's first phonograph and the modern record players of today are alike in many ways. The Edison phonograph did not have an electronic amplifier, of course, but the function of the major parts has not changed too much.

A phonograph record is a flat, vinyl disk with a groove cut into its surface that spirals from the outer edge to the center of the record. The groove is not smooth; it has irregular sides. These irregularities are formed to correspond to the audio information to be reproduced. To simplify this we could say the irregularities that are far apart will produce low-frequency (bass) sounds, and the irregularities that are close together will produce high-frequency (treble) sounds. Further, those irregularities that are shallow will produce faint sounds and deeper irregularities will produce louder sounds.

When modern records are made, the music is first recorded on high-quality audio tape (often

FIGURE 21–21 A highly magnified portion of a phonograph record. A close inspection of the center of the picture reveals the irregularities of the tracks. A scratch is clearly visible in the lower portion of the picture.

digitally recorded). The producer, artist, and engineer work with it until they are satisfied by the sound. A smooth record platter with no grooves in it is then placed on a machine that cuts grooves into it using a sharp stylus. The stylus vibrates as it is driven by the music from the master tape, creating irregularities in the grooves. The resulting master disk, or **mother**, is used to create a ''negative'' disk called a **stamper**. The stamper has hills in it instead of grooves. The metal stamper is then used to press grooves into smooth vinyl plastic disks that become finished records, Figure 21–21.

The Stylus and Pickup

The **stylus** (also called a needle) and the **pickup** (also called a cartridge) are used together to decode (translate) the record grooves into an electrical current and voltage. The stylus rides in the record groove. As the stylus passes through the irregularities, it is vibrated at different rates and intensities according to the size and shape of the groove irregularities. These vibrations are transferred to the pickup. The pickup changes these vibrations into a varying electrical current that corresponds to the vibrations.

There are two general types of pickups, crystal and magnetic. The crystal type works on the piezoelectric effect. The piezoelectric effect happens when a special crystal material is vibrated. That crystal then produces an electrical voltage in direct relationship to the vibration's frequency and intensity. Crystal pickups are inexpensive, but they do not have the best frequency response, and their operating characteristics are subject to heat and humidity.

A magnetic pickup has a coil, a stylus, and a magnet, Figure 21–22. In this pickup, the stylus is fastened to a small coil of wire. This coil is then placed in a magnetic field. As the stylus is vibrated from the record groove, it causes the coil to move within the magnetic field. The motion of the coil in the field causes a very small varying electrical signal in the coil. This varying electrical signal is directly related to the record groove variations. Magnetic pickups are more desirable for the listener because they have excellent frequency response (that is, they reproduce high and low frequencies equally

FIGURE 21–23 A rubber idler wheel and motor shaft.

well). They do, however, need extra amplifier circuits to boost up the small electrical signals to usable levels.

Regardless of the type of pickup used, electrical signals produced by the pickup must be amplified and presented to the listener through speakers or headphones.

The Drive Mechanism

All phonographs need a motor and drive mechanism to turn the turntable/record combination. The methods used to do this fall into three categories. Early phonographs used an idler wheel to drive the turntable and control the speed of the turntable. In this method the motor shaft turned against the rubber idler wheel which

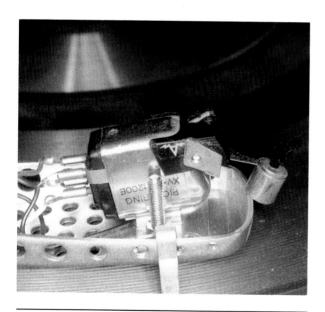

FIGURE 21–22 A magnetic cartridge and stylus.

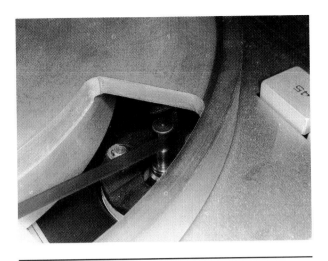

FIGURE 21-24 A belt drive with motor shaft.

in turn rubbed against the inside of the turntable, Figure 21-23. This method worked but was subject to wear and slippage which caused speed changes, thus causing distortion in the reproduced sound.

Manufacturers then developed a belt drive system to spin the turntable, Figure 21-24. As the name implies, a belt is used from the motor to the turntable. This method is less subject to speed changes due to wear.

The third drive method is the direct drive turntable. In this type of drive, the motor is built into the turntable, Figure 21-25. There is no slippage or wear with this type.

Compact Disks

In the early 1980s a new system for storing and retrieving data became available to consumers. The **compact disk**, or **CD**, was recognized right away for its quality of reproduced data. Sound that could not be reproduced on other forms of recording devices was now clearly heard. The CD's big brother, the **video laser disk**, was having equally amazing results in the video world. But what was not as obvious to the average user was the vast amount of data that could be condensed onto these disks. Another plus was the reliability of the data retrieval over long periods. These disks did not wear out. They were not subject to the wear through use as were all earlier methods of recording. The disks were also immune to external magnetic fields.

CD Player Components

A compact disk is a plastic disk onto which very tiny pits are formed. The surface of the disk has a very shiny coating of aluminum which is used to reflect the laser light, Figure 21-26. This aluminum layer is covered with a coat of lacquer

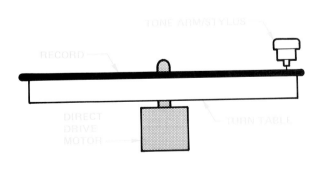

FIGURE 21-25 This drawing shows how a direct drive turntable is connected to the motor.

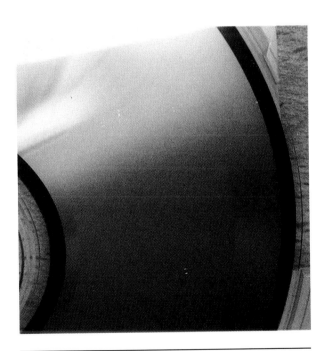

FIGURE 21-26 The surface of a compact disk. Notice the tracks are visible in the reflected light.

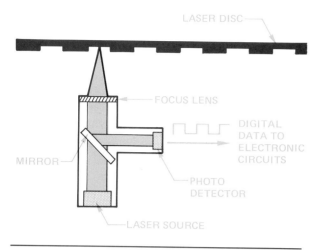

FIGURE 21–27 A typical CD laser pickup assembly. The internal mirror reflects the returning laser light into the photodetector where the changing light intensities are changed to a digital signal. This signal is processed by the electronic circuits.

to protect the fragile surface. The tiny pits are digitally encoded data. The length of the pits and the distance between them are read by the laser light and are decoded (translated) into digital 0s and 1s.

As in DAT, analog information such as music or voice is converted into digital format by an analog-to-digital converter before recording. The standard sampling rate for audio compact disks is 44.1KHz (44,100 samples per second), and each sample is converted into 16 bits, resulting in a data rate of more than 700,000 bits per second. On playback, the digital data is converted back to analog in a digital-to-analog converter.

A CD used for a home stereo has more information stored on the tracks than just audio. A functioning CD unit must also have error checking information, information for the correct channels, and front panel displays.

The CD uses a laser beam as the pickup device for reading data from the disk, Figure 21–27. The laser light is beamed towards the disk surface and is reflected back to a photodetector. Because the surface of the disk is pitted, the reflected light will vary in intensity. This is because the light does not reflect as well from the bottom of the pits as from the top of the

surfaces between the pits. The varying light intensities are amplified and turned into digital signals.

Basic Operation

Using the previous explanations of the disk and laser pickup we may be tempted to discuss the ways a CD player is like a record player. The CD spins and the pickup reads the data. The data is amplified and provided as output to the user. It is important to describe the differences between a CD player and a record player.

One big difference is the layout of the tracks of information on the CD. The first track of CD is on the inside, near the center of the disk. Its last track is at the outer edge of the disk. Another big difference is that a CD does not spin at a constant speed. On the first track (near the center) the disk spins fast and at the last track (near the edge) the disk spins slower. Finally, the information read from the disk is in digital form (on-off pulses) and must be changed into analog form (i.e., a sine wave) by the use of a digital-to-analog (D/A) converter if it is to be heard by the human ear.

All of the control and error checking data must be read and processed to complete the operation. This operation is like the descriptions given in the previous section on DATs.

The previous description was centered on audio uses. Recently a new use for the CD has been found. As you remember, the data stored on a CD is in a digital format. A computer directly uses digital information. Data from a CD can be directly fed into a computer without any digital-to-analog conversions. Now huge amounts of data can be available for retrieval from a single CD.

Summary

An electronic recording system stores electronic information for reading back at a later time. The information may be in analog or digital form. The part of the system that stores the information is called the medium. The medium may store information in magnetic form (magnetic recording tape, computer floppy disks), optical form (compact disks, laser video disks), mechanical form (phonograph records), or other forms.

When motion exists between a magnetic field and a conductor, there will be a current flow created or induced in the conductor. If the conductor is wrapped into the form of a coil, the induced current is greater. If the strength of the magnetic field is changed, the amount of current induced is also changed. These principles form the basis for magnetic recording.

Magnetic recording tape is made of plastic base tape covered with a magnetic coating. It is packaged on reels or in special containers (cartridges or cassettes) for ease of handling and storage. Cassettes are the most commonly used tape packages today.

In a tape recording system, the heads are the transducers that record the signal onto the tape and recover information from the tape during playback. An amplifier makes the signal from the playback head large enough to drive a speaker (audio tape) or TV monitor (video tape). The motors and mechanism move the tape past the heads.

In audio tape recording, information is recorded lengthwise along the tape in thin tracks. In video tape recording, information is recorded in diagonal stripes across the tape. The stripes are created when the tape is pulled across the head at an angle. In an audio recorder, the tape is pulled across a stationary head, while in a video recorder, the tape is pulled across a rotating head.

Digital audio tape (DAT) recorders convert analog music and voice to digital format and record digital data on tape. DAT players recover the digital data from the tape and convert it back to analog form. The use of digital recording on the tape yields very clear recordings with no tape noise.

Phonograph records store information mechanically in the irregularities pressed into spiral grooves in the record's flat surface. As the record spins, a stylus sitting in the groove vibrates, generating a varying electric voltage in the pickup. The small voltage is amplified to drive a speaker that recreates the recorded music.

Compact disks store digital audio information in optical form. Laser video disks store digital video information in optical form. Digital bits are stored on the disk in the form of pits of varying length. A laser beam is aimed at the surface of the disk as it spins, and the reflections from the surface and the pits are detected to give the digital information stored on the disk. The digital data is then converted back to analog form for playback. As in DAT, the digital storage gives very clear recordings with no noise added.

REVIEW

1. Name three types of recording systems.
2. Outline a brief history of magnetic recording.
3. Describe two principles that Ampere and Faraday discovered about magnetism and current flow.
4. Name three main parts of an audiotape recorder/player.
5. Name three types of audio tape packages.
6. Describe the main differences between an analog magnetic recording and a digital magnetic recording (DAT).
7. Draw a picture showing the magnetic particles on an analog audio magnetic tape and the particles on a digital audio magnetic tape.
8. Name three parts of a record player and briefly describe their operation.
9. Describe briefly how an audio signal is obtained from a phonograph record.
10. What are the major advantages of a CD over a phonograph record?
11. Describe the construction of a CD disk.
12. Explain how the laser is used as a pickup in a CD player.

SECTION ACTIVITIES

MULTIMETER APPLICATIONS

OVERVIEW

In this activity you are going to use one of the critical testing instruments for electronics. That instrument is the multimeter. With the multimeter you will check DC voltage, AC voltage, diodes, and resistance. You will measure values of various components and explore the common practice of electricians in trouble shooting problems. You will be able to correctly set up for making each of these measurements with a multimeter.

By observing the situation you will have the chance to predict the results, and then use the meter to see if you predicted correctly. You will be able to set the meter to the appropriate settings, and read and record the results.

MATERIALS AND SUPPLIES

To complete this activity, you will need the following equipment:

- digital multimeter with test leads
- DC batteries
- resistors
- diodes
- transformer with output between 5 and 15 VAC, resistors, and diodes

PROCEDURE

1. Step one for DC voltage measurement. Place the red or positive lead into the red V connector, and the black and negative lead into the COM connector.

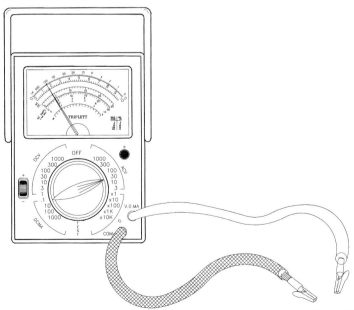

2. Set the function switch to the V = DC position.

3. Now touch the red lead to the positive and the black lead to the negative end of each battery and you are set. Wait for a few seconds for a steady readout and then record the value of each battery. Turn the multimeter off.

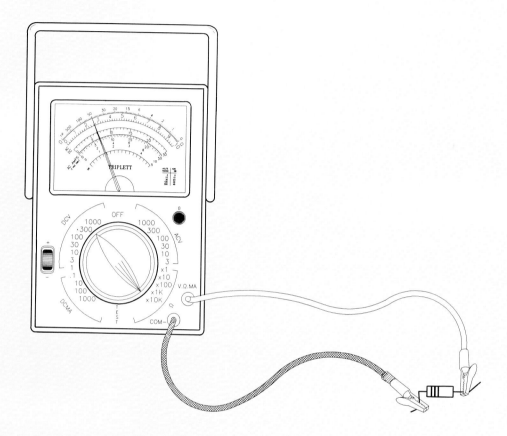

4. Measure AC voltage. With the leads in the same place, turn the meter selector switch to the V = AC position.

5. Touch the two test leads to the output side of the transformer. Read and record the voltage. Be sure to turn the multimeter off when you finish.

▶**CAUTION:** There is a dangerous high voltage present at the input side of the transformer and at the wall socket. Do not touch either of these points!

---

6. Measure DC current. With the test leads disconnected, turn the selector switch to the highest current scale.

7. Break the circuit and hook the test leads into the circuit as shown. If the reading is very small, change the selector switch to successively lower scales until you can read the meter accurately. Record the reading. Disconnect the test leads and turn the meter off.

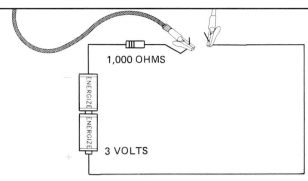

1,000 OHMS

3 VOLTS

8. Check a diode. Remember diodes only allow current to flow in one direction. Set up the leads so that the red lead is in the read V connector of the meter and the black lead is in the COM connector.

9. Turn the meter selector switch to the R × 1 ohms scale.

10. Connect the leads to the diode you wish to test. The black lead goes to the cathode end (color band end). The red lead hooks to the anode (other) end. The reading will now show on the meter. Turn the meter off and record your results.

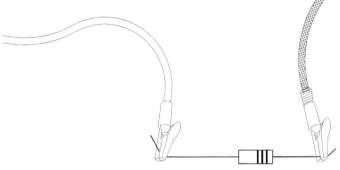

11. Record and measure the resistance of various resistors. Remember resistance is measured in ohms and each of the resistors is color coded with color bands to tell you the resistance value.

12. Use the color code chart to figure out and write down the resistance of each of the resistors.

FIRST SIGNIFICANT DIGIT

SECOND SIGNIFICANT DIGIT

MULTIPLIER

TOLERANCE

COLOR	FIRST AND SECOND STRIPES DIGIT	THIRD STRIPE MULTIPLIER	FOURTH STRIPE TOLERANCE
BLACK	0	1	
BROWN	1	10	
RED	2	100	
ORANGE	3	1,000	
YELLOW	4	10,000	
GREEN	5	100,000	
BLUE	6	1,000,000	
VIOLET	7		
GRAY	8		
WHITE	9		
SILVER			±10%
GOLD			±5%

13. Using the meter, turn the meter selection switch to the ohms position.
14. Touch the test leads together to ensure the meter reads zero ohms.
15. Place leads of the multimeter on either end of the resistor and wait for the stable reading. Compare your reading with the calculated resistance of the resistor. Remember to turn the power switch off when you are finished.

FINDINGS AND APPLICATIONS

Through these activities you have found the voltage, current, and resistance values of several of the major components used in the electronics industry. You have had the chance to perform trouble shooting functions, and measure and observe results.

ASSIGNMENT

1. Record the results of each of these activities and hand them to your instructor.
2. In your own words write the definitions of the following terms: ohms, volts, DC, AC, diode.
3. Ohm's Law describes the relationship among voltage, current, and resistance in a circuit. It states that voltage = current × resistance (E = IR). Using the measurements obtained in Steps 6 and 7, show whether Ohm's Law confirms what you have observed.
4. On a half sheet of paper list other courses you may wish to take that would be useful if you were to choose a career in electronics.
5. Using the multimeter, test other circuits and resistors, and record the results.

 ## BURGLAR ALARM

OVERVIEW

In this activity you will use the materials provided to design a burglar alarm using A) a laser as a light source and B) an electronic circuit. The laser used for this activity is a helium-neon laser (called HENE).

While this activity is not exactly the way most burglar alarms are made, it does give you a chance to make your own burglar alarm and use laser light in a different way. The second part of this activity is the actual electronic fabrication of a device to receive the laser light.

You will want to try to predict the effects of the different parts. What would happen if you used larger resistors, for example, or a greater distance for the laser beam to travel?

MATERIALS AND SUPPLIES

To complete this activity, you will need the following equipment:

- circuit board
- resistor
- light-sensing diode
- battery
- buzzer
- simple switch
- helium-neon laser
- concave lens
- heat sink

PROCEDURE

►**CAUTION:** In this activity you will be using the laser beam. Care must be taken to control the beam so that it will not be projected into the eyes of other students or yourself. Always wear safety glasses. Never look into a laser beam, even if it is reflected! Warning signs should be posted that state that a laser is in use; caution should be taken not to stare into the beam.

1. Collect all of the apparatus. Begin by assembling the receiver as noted on the schematic.

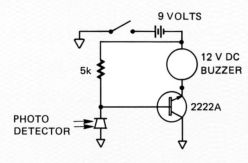

BUZZER SCHEMATIC

2. Depending on whether you want the receiver to be permanent or not, you may wish to solder the connections onto a simple circuit board. Should you choose not to, then the apparatus can be disassembled for other purposes.
3. Arrange mirrors, lenses, etc., so that the classroom door can be monitored by your burglar alarm. While it has not been too popular with students, the apparatus works well to find out who is late for class!

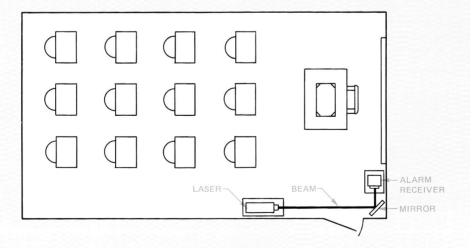

4. You can refine your design by arranging lenses to reflect the light beam and installing a switch in the circuit. (The apparatus can be made without a switch; however, it becomes a bit annoying in the setup stages if the buzzer continues to buzz until the light can be focused into the diode receiver.)

FINDINGS AND APPLICATIONS

In this activity you found that it is possible for light impulses to be transformed into sound. You also found that you could make a burglar alarm that may have a number of uses. Consider running a metric 500 using the burglar alarm as the line behind which all students need to stay while the cars are being run.

ASSIGNMENT

1. Complete the activity as designed and explained. Have your instructor grade you when you are finished.

2. List as many uses as you can for the apparatus you have made.

3. What could you do to improve on the design of this activity? Could a computer be attached to this device? What uses could you put such a combination to? Take about one-half page to explain your solution.

 ## SIGNAL MODULATION

OVERVIEW

Modulation is the process of putting information onto a carrier signal by changing its frequency (FM), amplitude (AM), or phase. In this activity you will investigate the possibility of overmodulaton (modulation in excess of ± 100%) of local AM (amplitude modulation) radio broadcasting stations. For this activity you will connect an AM receiver with independent IF (intermediate frequency) and/or RF (radio frequency) and AF (audio frequency) volume controls, to an oscilloscope. This connection will allow you to observe and measure the percentage of positive and negative peak modulation of the AM radio signal.

In this activity try to predict the results of increased overmodulation on the display screen of the oscilloscope. What happens when a signal is overmodulated? Why are the signals controlled by federal restrictions?

MATERIALS AND SUPPLIES

To complete this activity, you will need the following equipment:

■ AM receiver with independent IF/RF and AF volume controls.
■ oscilloscope

PROCEDURE

As you work through this activity, remember that all equipment must be properly grounded. When you have your equipment set up, check with your instructor *before* you turn on the power. Use caution with all voltages.

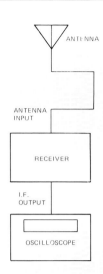

1. Arrange the equipment as shown using the following sequence:
   a. antenna connected to the receiver input
   b. receiver intermediate frequency (IF) output into the oscilloscope

2. With the receiver's automatic gain control (AGC) turned off, tune in a fairly strong AM radio station. A local station will be best to avoid signal fading.

3. Turn the automatic gain control on and adjust the oscilloscope so the peaks almost fill the display. Now turn the automatic gain control off.

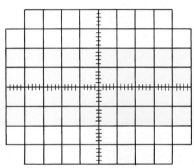

0% MODULATION

4. With the receiver radio frequency (RF) or intermediate frequency (IF) gain controls, set the signal level of the unmodulated signal (dead time when no voice or music is present) to fill two graticules centered about the horizontal centerline of the oscilloscope.

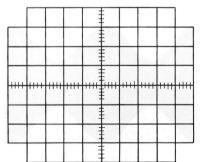

50% MODULATION

5. Set the sweep rate so that it will match the expected range of modulating frequencies. Adjust the amplitude so that it can accommodate an instantaneous increase in modulation of slightly more than 100%.

6. You are now looking at a graphic representation of the audio signal being transmitted by the radio station you selected.

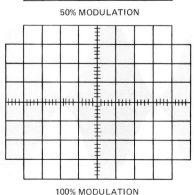

100% MODULATION

FINDINGS AND APPLICATIONS

With the above settings, the amplitude of the displayed waveform changes with the instantaneous changes in the degree of modulation. Positive modulation was noted when the waveform rises above the unmodulated level, and negative modulation is observed when the waveform falls below the unmodulated signal.

The activity provided for 100% positive and 100% negative modulation, the amplitude raises to more than twice the unmodulated carrier level. For more than 100% negative modulation, the display has a gap along the Y-axis, as indicated by a single bright horizontal line. When this negative modulation exceeds 100%, distortion in the audio signal can be observed.

In practice, all broadcast stations are required by federal law to maintain signal quality within certain guidelines. The percentage of time that a transmitted signal may display negative modulation measurement is severely limited. By using a form of modulation measurement such as that you have just performed, the broadcast station staff can monitor their own signal. This ensures that their transmitted signal falls within specified federal regulations. By doing this, they ensure higher quality programming for your listening pleasure, and restrict the possibilities of their "splattering" onto adjacent channels.

ASSIGNMENT

1. Perform the set-up described in this activity and have your instructor check the results.
2. Describe the reason for federal control of broadcasting signals.
3. Visit a radio broadcasting studio. Find out how and why they observe these federal restrictions.
4. Prepare a report describing how these signals are monitored.
5. Would you think that overmodulation may also be a problem with citizens band (CB) transmission? Please explain your answer.

 LASER LIGHT

OVERVIEW

Laser light is different from most regular light sources. This activity will give you a chance to look at laser light, and determine the differences and similarities between it and other types of light. You will use the helium-neon laser for this activity. You will also have the chance to find out the effects of some of the optical equipment used with light manipulation.

As you complete the exercise in this activity, you will develop a feel and an appreciation for this new technology. You will be able to describe some of the main properties of laser light as well as determine focal length and depth of field for simple lenses.

As you work through these activities, try to predict what the answer will be before you do the actual experiment. You will be surprised at some of the outcomes.

MATERIALS AND SUPPLIES

To complete this activity, you will need the following equipment:

- helium-neon laser (class 2)
- selected lenses to include a -10 centimeter negative lens and a +30 centimeter positive lens
- lens mounts

- one set of color/Polaroid filters
- mirrors
- laser receiver

PROCEDURE

1. To begin these activities with laser light, your first task is to become familiar with the safety precautions needed for directing the laser beam. Ask your instructor for instruction materials and complete the laser safety test.

►**CAUTION:** In this activity you will be using the laser beam. Care must be taken to control the beam so that it will not be projected into the eyes of other students or yourself. Always wear safety glasses. Never look into a laser beam, even if it is reflected! Warning signs should be posted that state that the laser is in use and caution should be taken not to stare into the beam.

---

2. Set the laser flat on the table pointing away from any of the students in the class. With the laser turned on, sprinkle or pat chalk eraser dust through the beam along the length of its course. This works best in a darkened room. As the chalk dust flows through the beam, what happens? Why do you think this happens?

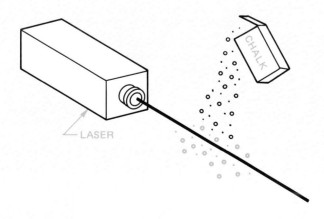

3. Set the laser so that it sends the beam through a prism and onto a plain sheet of paper. What colors are projected onto the paper? When sunlight is projected through the prism onto a sheet of paper, what colors are projected on the paper? Is there a difference? Why?

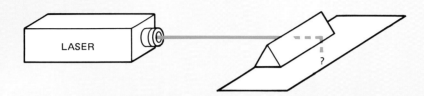

4. Position the laser 10 feet from the wall or beam stop. Lay a large sheet of plain paper 4 inches in front of the wall. Mount the +30 centimeter positive lens in its holder and set it down 4 inches from the wall. Turn on the laser. Move the lens slowly away from the wall. When the beam appears at its smallest on the wall, stop and mark the position of the +30 centimeter lens on the paper. Now continue to move the lens further from the wall until the beam is at its largest. Mark this position on your paper. Using a ruler, put a third mark on the paper exactly midway between the two marks. Measure the distance between the wall and the third mark (which was the last one we marked) in inches. The resulting distance is the focal length. The distance between the nearest and farthest marks will also give you the depth of field.

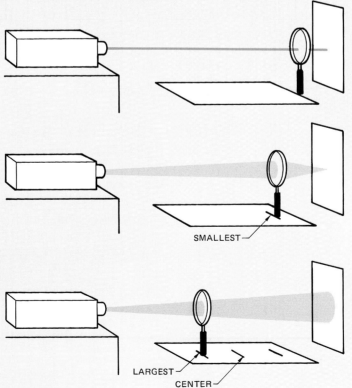

5. Arrange the mirrors around your room, being careful not to reflect the beam toward any other students. The beam can be directed from the laser to the first mirror, to the second mirror, to the third mirror, to a prescribed point or target located on the wall. The major difficulty you will find is in fine tuning each mirror to allow for the reflection you desire.

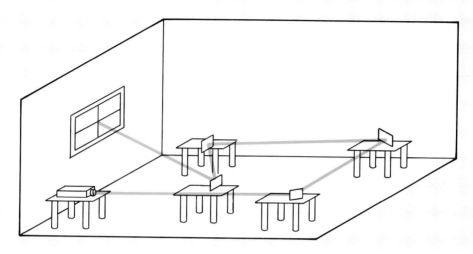

6. Split the laser beam by shining the beam through a clear piece of glass. Then, direct both beams to a prescribed target located on the wall using reflective mirrors.

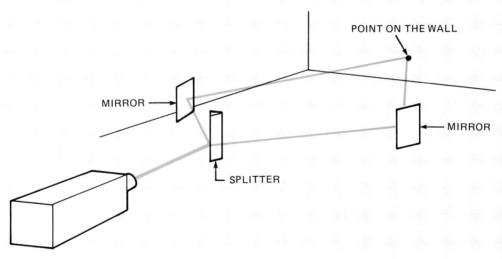

FINDINGS AND APPLICATIONS

During these processes you have had the chance to direct a beam to a variety of positions and observe what the laser beam does. In class discussions, you may be able to discover many uses for the laser in today's technology.

ASSIGNMENT

1. Pay attention to the news, newspaper, or magazine articles that mention or describe uses for lasers. Bring copies of the news articles or short summaries from radio or TV and give them to your instructor. (Makes great bulletin board material!)

2. Turn in to your instructor the results of the focal length and depth of field activity.

3. On one-half page, describe how the helium-neon laser works to produce a laser beam.

4. Describe what happens when a beam is divided by shining it through a clear piece of glass. Include your estimate and the actual number of beams resulting from this activity.

5. If you could do this activity again, what would you do differently?

LASER TRANSMISSION OF SOUND AND FIBER OPTICS

OVERVIEW

In this activity you will use a simple helium-neon laser and receiver to transmit sound. You will also use fiber-optic cable and perform the same activity. Students will be able to observe the effects of distance and interference in an open beam transmission as well as the quality and uses of fiber-optic sound transmission.

During this experiment you will estimate effects of various types of interference and distance on the quality of the signal. Questions will be answered such as: Would you expect an increase in distance to affect the quality of the signal? If so, how? Is there a way to at least partly resolve the effects of distance? Could you suggest a way this may be resolved? Regarding fiber-optic cable: How does fiber-optic cable work? What are the two major problems with using fiber-optic cable?

MATERIALS AND SUPPLIES

To complete this activity, you will need the following equipment:

- modulated helium-neon laser (class 2, 0.5 milliamp works well)
- receiving unit
- tape recorder or radio
- fiber-optic cable
- convex lens

PROCEDURE

▶**CAUTION:** In this activity you will be using the laser beam. Care must be taken to control the beam so that it will not be projected into the eyes of other students or yourself. Always wear safety glasses. Never look into a laser beam, even if it is reflected! Warning signs should be posted that state that the laser is in use and caution should be taken not to stare into the beam.

The first activity will be to transmit basic sound using a modulated laser beam.

1. Review related safety information. Be sure you are able to pass your instructor's laser safety test. Remember, exposure to the laser beam can cause serious eye damage.

2. Assemble the apparatus so that the tape recorder or radio output jack is connected to the input jack on the modulated laser.

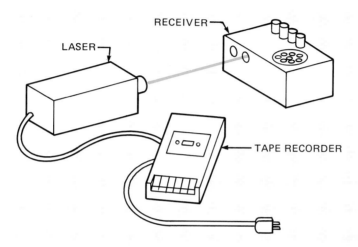

3. Turn on the tape recorder. Turn on the laser beam.

4. Set up the receiving unit so that the beam is directed right into the input hole on the side of the receiving box.

5. Set the receiving unit about 10 feet away from the laser. At this point if you are receiving sound but it pops, you are overdriving your unit. You can resolve the problem by turning down the volume on your recorder or radio. If everything lines up and you still get no sound, check the batteries on your receiving unit and be sure it is turned on.

6. After you have gotten the sound to work, try sprinkling chalk dust through the beam. Notice what happens to the transmission or the sound we hear coming from the receiving unit.

7. Try moving the receiving unit a greater distance away from the laser, perhaps 20 ft. Observe the sound at the greater distance. Also observe the size of the beam around the input hole in the receiving unit.

8. Now refocus the beam using the convex lens (magnifying glass). Focus the beam into the hole. Does this increase the volume?

9. Now try it using fiber-optic cable. To center the fiber-optic cable in the laser and also in the input hole in the receiving unit, a centering device is needed. By drilling a hole through a small cork just large enough for the cable to fit through, and then inserting the cork into the laser, you are able to adjust the beam to its brightest by observing the other end of the fiber-optic cable. Then, drill a hole into a still smaller cork, and insert the other end of the cable into the small cork. Insert this into the receiving hole on the receiving unit.

10. Now turn on the recorder or radio, the laser, and the receiving unit. Adjust the focus of the beam by turning the cork slightly in the input hole of the receiving unit. Sound should be clear and not affected by any external factors such as dust.

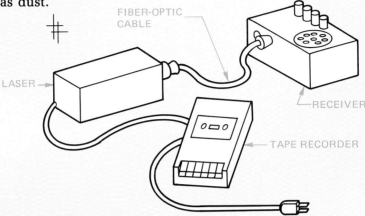

FIBER-OPTIC CABLE

LASER

RECEIVER

TAPE RECORDER

FINDINGS AND APPLICATIONS

During this activity you discovered that sound can be transmitted over light waves and can be affected by dust and other impurities. It is also affected by distance as the impurities in the air itself refract the beam. You also had a chance to refocus the receiving beam. You were able to experiment with a rather crude form of fiber-optic transmission and see what the fiber-optic cable can do.

ASSIGNMENT

1. Look into the uses of lasers and fiber-optic cable in the medical field. Submit a one-page summary of the information you find to your instructor.
2. List any places where fiber-optic cable or lasers affect you in your life.
3. Do you know people who work with lasers in their job?
4. What classes in school do you suppose would help if you were interested in a job dealing with lasers, communications, or fiber optics?
5. Can you run the beam though water and still get the sound? Try it and explain what happens.
6. If you could do this activity again, what would you do differently?

 AUDIOVISUAL TECHNOLOGY

OVERVIEW

In this activity you will have the chance to write, direct, and produce a commercial from a suggested source of information. If you prefer, you may develop a commercial for an original product or service, or you may develop a safety video referring to a certain situation or machine in your lab.

As you complete your video activity, take time to review it and decide if it needs to be redone. Try to predict and plan for a quality video.

MATERIALS AND SUPPLIES

To complete this activity, you will need the following equipment:

- several story board forms
- video camera
- videotape
- necessary props
- lighting
- pencil
- microphone

PROCEDURE

This activity works well with small groups or pairs of students. The first task is to choose a topic or article you wish to develop for your video.

1. Obtain from your instructor the time limit for the tape of your activity.
2. Remember that both audio and video are available. As a result, develop your script to include sound, and also a sketch to indicate a video portion of your commercial. This is done on story board forms. The story board allows a small space for a quick sketch and the related text that will be put together in the production.

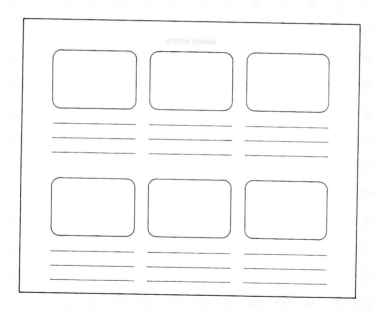

3. Once you decide on a topic and outline your production on the story board, you are ready to select and organize the stage for your production. Check that you have enough lighting and control of the environment or the area around where you are shooting your video.
4. Decide who is to be in the video and practice your production.

5. Shoot your production.

►**CAUTION:** Video equipment is fragile. As a result, it can easily be damaged. It is suggested that the camera be mounted on a tripod to help prevent accidents.

---

6. If you have access to the proper equipment, it's possible to dub music to add another dimension to your video.
7. Once your video is made, review it and decide if it needs to be redone. If you are comfortable with it as it is, you are ready to submit it with a copy of your story board and the name of all students in your group to your instructor for review and a grade.

FINDINGS AND APPLICATIONS

During this activity, you have seen that commercials are constantly being used to sell certain products to us, the consumers. You have also had the chance to review exactly what is involved in making a commercial. You also are more aware of commercials and their importance in industry today.

You have developed a video commercial and reviewed some of the things involved in making such a commercial.

ASSIGNMENT

1. Turn in to your instructor materials you have developed, including the story board planning sheets and a copy of your video.
2. Record the number of commercials in your favorite one-hour movie. Also record the length of these commercials and the product being sold. Write a one-line summary of the process that was shown to sell you the product.
3. Develop a video of the activities and opportunities in your communications class to be used to show other students about technology education.
4. If you could do this activity again, what would you do differently?

 SATELLITE TECHNOLOGY

OVERVIEW

This activity is divided into two parts. First, it is necessary to make a simulated receiver to determine when the appropriate signal is being received. The second part of the activity allows you to simulate the uplink and downlink portions of a satellite transmission. You will have a chance to use a schematic drawing to assemble the various parts of the receiver. You will also learn the basic principles of satellite technology.

Consider the similarities between your simulated communication system and those you notice in your community. If you were to put an object between the mirror representing the satellite and your receiver dish, you would expect the beam to be blocked and the apparatus not to work. Is this similar to setting up your microwave dish behind the house with the house between your dish and the satellite? What would happen if someone built a high-rise building adjacent to your home that blocked the direction of your receiver dish?

MATERIALS AND SUPPLIES

To complete this activity you will need the following equipment:

- parts list for the circuit (handout)
- schematic diagram of the circuit
- flashlight
- two reflective mirrors

The apparatus will be put together to represent the receiver dish and will use the electronic circuitry you have put together.

PROCEDURE

1. Assemble all of the pieces for the receiver station. This includes electronic parts as well as the simulated dish, base, and pieces of PVC pipe.

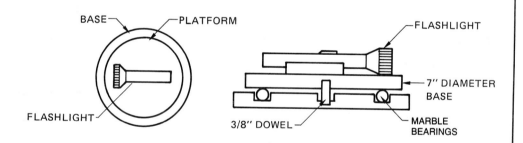

2. Assemble the circuit as in the schematic drawing. Test your circuit with a flashlight to be sure it works. (A 9-volt power supply will save replacing batteries.)

3. Arrange the mirrors to simulate satellite stations. You are to beam the light from the flashlight at one location, to a mirror which will reflect the beam directly into the receiver that you have made.

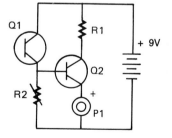

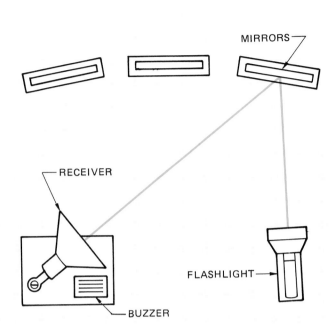

4. Without moving the first mirror, arrange the second mirror to simulate the second satellite. (Communication satellites are all located in a special place directly over the equator called the Clarke belt.) You have arranged both mirrors so that the flashlight beam can be rotated from one to the next and each will reflect the beam to the receiver. Where is the uplink? Where is the downlink?

5. Try moving the mirrors to locations further away from the receiver dish and flashlight. What happens as you do this?

6. After you have arranged the mirrors, the flashlight for the uplink, and your receivers for the downlink, and have it so it works and receives the beam and the buzzer sounds, ask your instructor to evaluate your work. Your instructor will be looking for quality and neatness in the assembly, and to see if you have arranged your satellites so that they work.

FINDINGS AND APPLICATIONS

When you have completed this activity, you have successfully made your own Clarke belt (the mirrors), your own transmitting station (the flashlight), and your own receiver (the microwave dish). Some things are easier in this activity than others. List those parts of the activity that you felt were the most fun. Also note those that seemed to give you the most trouble. Add to this your estimate of the number of receiver dishes you expected to see, and the actual number that exist between your school and your home.

As you worked through this activity, you began to notice that this form of communication is much more common than you expected. You also noticed that there are many different sizes and shapes of receiver dishes in your community. What else did you find interesting about this activity?

ASSIGNMENT

1. On your way home from school today, count the number of receiver dishes you see between school and home. Before you begin counting, note your estimate of the microwave dishes you expect to see.
2. You have had the chance to develop a simulated exercise. Now list other uses that are made of the satellite communication system that we enjoy in our homes and communities today.
3. Turn in the required written assignment to your instructor.
4. If you could do this activity again, what would you do differently?
5. See if a laser beam can replace the flashlight for a signal beam.

 ## MODEM COMMUNICATION

OVERVIEW

Modem communication has become a common part of communication. The ability for one computer to communicate with another using inexpensive telephone networks is an important communication development. In this activity you will review the process and actually transmit messages through the computer to students in another school. You will be able to communicate questions and information about yourself and your school program.

In this activity you will want to develop the information to be transmitted to make you look as intelligent and knowledgeable as possible. This is easy to do as you are able to edit and develop your message before actually sending it.

MATERIALS AND SUPPLIES

To complete this activity, you will need the following equipment:

■ a computer with a telephone modem hookup and telephone line with appropriate software to use the modem and develop basic word processing information.

(Reprinted from LIVING WITH TECHNOLOGY by Michael Hacker and Robert Barden, © 1988 by Delmar Publishers Inc.)

PROCEDURE

For this activity to be most effective your instructor will need to find out where other programs exist that have modem capabilities. Now you can begin:

1. Identify the person with whom you wish to communicate.
2. Develop the message you plan to send.
3. Turn on the modem and send the message through the computer.
4. Receive and respond to any messages that are returned.
5. Properly shut down the system. Make sure you don't leave the phone connected and on line.

FINDINGS AND APPLICATIONS

The International Technology Education Association is part of a network system accessed by modem which links participating schools across the country. There are other similar networks, both commercial and private. Uses for this communications system are limted only by your imagination. You might consider cooperating with another school to develop a manufacturing activity using JIT (Just in Time) material management. You may also wish to share information about club activities or upcoming statewide events dealing with student organizations.

ASSIGNMENT

1. Develop a message, transmit it, and receive information from a student in another school.
2. Develop a communication network of students in your school and at least one other for the purpose of sharing information, ideas, and activities.
3. Prepare a graphic plan or working drawing and transmit it to another school using the modem.
4. Organize a manufacturing project with another school using the computer and modem to communicate and cooperate in the manufacturing process.

SECTION SIX

THE FUTURE OF INTEGRATED COMMUNICATION SYSTEMS

As in many other technological fields, more and more different technologies are being combined to provide communication of all types. Computers and electronics are increasingly being used in technical drawing, printing, and photography. In electronic communication, radios are being combined with telephones, and computers are found in nearly all forms of communication.

Communication and computer processing are being increasingly combined, and in some systems, it is difficult to tell whether the process is communication or data processing. The combining of computing and communication is producing a new field called information technology. Many corporations have recognized this trend, and established the Manager of Information Systems as a corporate officer.

This section examines some of the trends that will shape communication in the coming years.

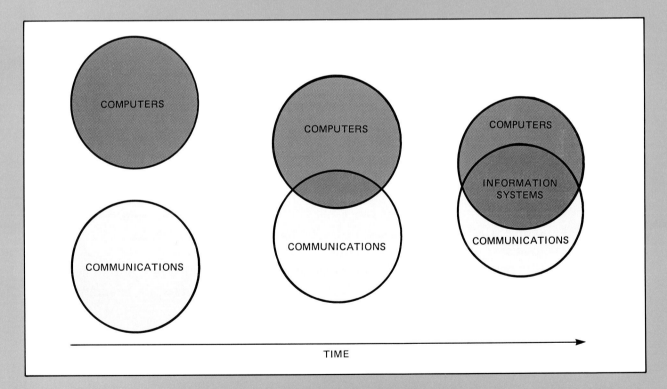

THE CONFLUENCE OF COMPUTING AND COMMUNICATION

Chapter 22 ■ The Future of Communications

The Future of Communications

OBJECTIVES

After completing this chapter you will know that:

- In the future, more and more jobs will require the ability to manipulate and process information.
- Electronic still cameras (ESCs) will use small floppy disks to store images digitally, rather than film.
- Ion printers may be used in the future.
- Advances in electronics will continue to make more complex computers available at lower cost.
- In the future, a variety of communication services will be provided over the Integrated Services Digital Network (ISDN).
- More and more people will be able to live in their favorite environment and telecommute to work, as communications get better.

KEY TERMS

artificial intelligence (AI)
biometrics
cell
digital audio tape (DAT)
direct broadcast satellite (DBS)
electronic still camera (ESC)
expert system
follow focus

gateway
high definition TV (HDTV)
inference
integrated networks
ion printer
parallel processing
protocols
quartz shutter

slow scan video
telecommunities
telecommute
teleworkers
trace
voice recognition system
waferscale integration

Introduction

Imagine being able to read a newspaper that only has articles that interest you. Each time you read an article, the newspaper learns more about your interests. The next time you read it, the newspaper (a computerized tabloid called Newspeek) displays articles even more customized to your reading preferences.

FIGURE 22–1 A videodisk tour of Aspen, Colorado *(Courtesy of MIT Media Laboratory)*

FIGURE 22–2 Soon, musicians will play along with computerized accompaniments. *(Courtesy of MIT Media Laboratory)*

How about taking a walking tour of a city without leaving your own home? While viewing video images of city streets, you can make up your own tour by touching the computer screen. You can indicate the direction in which you wish to walk. You can even visit the insides of buildings.

How would you like to be able to sing a song and have a computerized piano adjust to your tempo and play along with you?

In places like the Media Lab at the Massachusetts Institute of Technology, Figure 22–1, these and other futuristic technologies are being developed. By linking computers to communication media, a whole new world of communication technology is opening as we move into the 21st century, Figure 22–2.

We now live in the information age. The information age is an era when communication technology satisfies many business and societal needs. In the future, more jobs will require the ability to manipulate and process information. The economies of industrialized nations will depend upon the use of advanced means of communication for the rapid exchange of information.

The Future of Printing and Typography

Production printing is going high tech! Publishers are searching for ways to produce their printed materials faster and more economically. Publishers will be linking their authors and editors more directly to the phototypesetters. Authors will submit their work on floppy disks which will be edited and coded for phototypesetting by an editor. The editor will telecommunicate the text via modem to the phototypesetting house. This procedure can cut the time it takes to produce galleys from original manuscript from three weeks to one week.

Another development involves using workers in other countries to do word processing. These workers may work for lower wages than their counterparts in the United States. These foreign workers telecommunicate the text back to the home office computers by facsimile machines and satellite. Facsimile machines will be able to transmit a page of text in five or six seconds. An entire book can be transmitted in less than an hour.

New varieties of computer printers are being developed. One such device is called an **ion printer**. This device uses technology similar to that used by laser printers, but is more durable and less expensive. At the heart of the ion printer is a flat cartridge, about 2″ wide and 12″ long. The cartridge has three layers. The top two layers form an electrical grid. The layers are made of electrical wires, and are at right angles to each other. The bottom layer is a metal screen containing over 2,000 tiny holes. Each hole is positioned under one of the intersections of the wire grid.

When electricity flows through the wires, the air around the wires ionizes. The negatively charged ions speed through the holes in the screen onto a printing drum. Toner is attracted to the ions and adheres to the drum. As the drum rotates, the toner is transferred to paper, Figure 22–3.

Photography: A Picture of the Future

There is a bright future for photography. Amateur and commercial interest in photography is rising, and technology applied to camera equipment will result in new features that will enhance photographic techniques.

Electronic Still Photography

Future cameras will be filmless! **Electronic still cameras (ESCs)** are electronic marvels that are similar to camcorders in their operation, Figure 22–4. Like camcorders and videocameras, electronic still cameras use charge-coupled devices (CCDs) as light sensors. The output of the CCD (and an electronic image) is stored on a 2″ × 2″ standard-size video floppy disk which spins at a speed of 1800 RPM. These disks can store between 25 and 50 images. The disk can either be played back directly from the camera, or can be inserted in a player, similar in concept to a videocassette player. The image can then be seen on a TV screen, or outputted to a color printer to give an instant print.

ESCs will have capabilities beyond those of ordinary cameras. For example, the shooting

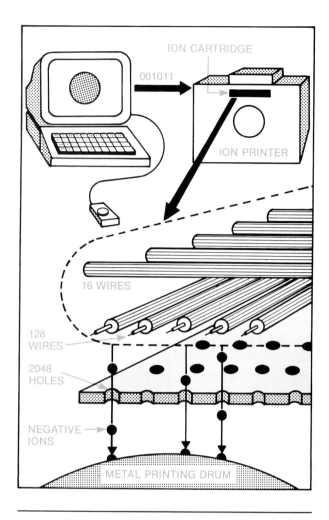

FIGURE 22–3 As electricity flows through the wires, the air space is charged (negatively ionized). These ions are shot through holes in the screen and form an electronic image on the rotating drum.

speed can be fast enough (20 images-per-second) so that the camera can be used for time-and-motion studies.

Additionally, since the photographs will be in digital form, they will be able to be transmitted over telephone lines almost instantaneously to remote locations. Devices called "transceivers," which are similar to computer modems, will be used.

One of the most novel possibilities offered by ESCs will be the ability to store an image, alter it, and print the new version. Since the image has been converted to digital bits of information, the information can be input to a computer and

science and electronic technology. **Quartz shutters**, which have no moving parts, would use a thin slice of quartz as a window in front of the film. When electricity is applied to the quartz plate, it would change from transparent to opaque. Enormously fast shutter speeds could be attained, up to a millionth of a second. Artificially intelligent cameras could provide a **follow focus** feature which would allow the photographing of an object that is rapidly moving. Only when the object is in sharp focus will the picture be taken.

One spinoff from the space program is the development of optically pure glass which is produced in zero gravity. This glass will be used to produce lenses of near-perfect quality. Another new feature will be image size selection. The photographer will be able to automatically select a full-height, waist-up, or head-and-shoulders image.

FIGURE 22–4 The Nikon QV–1000C camera is an example of an electronic still camera (ESC). *(Courtesy of Nikon Inc.)*

manipulated. For example, architects will be able to add a photograph of a model of a new building to an existing photograph of a downtown city scene. They would be able to show clients what the impact of new construction might be on a city skyline. Or, artists would be able to easily change the color of certain parts of the scenery in a photograph.

One must consider the ethical implications which are raised by this new technology. Because there are no film records, it would be easy to manipulate an image and claim it is the original. People might be shown to be in locations where they have never been. In criminal cases, this could be entered as evidence. Some newspapers have already ruled that photographs that have had changes made to their content may not be used. Digital security code technology has been proposed as a way of providing ''proof of originality'' of digital photographs produced by ESCs.

Other Future Camera Developments

Additional features influencing future camera design are the result of advances in material

FIGURE 22–5 Kodak's Ektar film has very low image graininess providing excellent image quality. *(Reprinted courtesy of Eastman Kodak Company)*

Film Improvements

Amazing new color films are in our future. Colors will be even more vibrant, and the sharpness of prints and slides will be fantastic. Companies like Kodak, 3M, and Agfa are developing films with ''T-Grain'' technology. These films position particles of photosensitive material parallel to the film surface. In this way, light absorption during exposure is maximized. An example of this new film is the Ektar film produced by Kodak, Figure 22–5. The Ektar film has very low image graininess providing excellent image quality. The film will be made up of nine layers of plastic, five of which will be image-recording. The new films will be able to be push-processed to provide four times the original film speed.

Future Electronic Communication

New developments in electronics continue to occur at an increasing rate. These new developments are also driving down the costs of building electronic parts, making things that are expensive today much less expensive in the future. The impacts of these new developments reach to all phases of communication, but none as widely as electronic communication.

Trends in Electronics and Computers

The first integrated circuit, which was developed in 1959, held only one single transistor. Chips have already been designed to include millions of components, Figure 22–6. Most integrated circuits of today are made of silicon. Other kinds of materials, such as gallium arsenide, are being more widely used to increase processing speed and reduce cost.

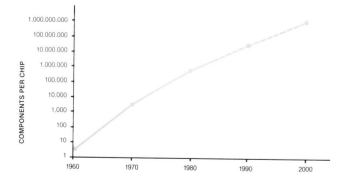

FIGURE 22–6 By the year 2000, it is projected that a chip will contain one billion components.

A new chip construction technique, called **waferscale integration**, holds the promise of greatly increasing chip capabilities. Currently, many integrated circuits are made on thin, semiconductor wafers that are several inches in diameter. Each integrated circuit is cut out of the wafer, and mounted in a protective package that greatly increases its size, and slows down its operation. In waferscale integration, the individual circuits would be left on the wafer and interconnected. Several wafers could then be stacked up before being mounted on a printed circuit board. One application of this technology is to shrink billions of bytes (Gbytes) of computer memory that currently occupies ten cubic feet (and is very expensive) into a few cubic inches of inexpensive semiconductor material.

Interesting things will happen in the computer world as we move into the decades ahead. Because so many technological processes involve computers, a great deal of research is devoted to improving their capabilities. Both computer hardware and computer software will become much more powerful.

New kinds of computer programming will greatly change computer capability. **Parallel processing** allows computers to work faster on problems that require the same calculations over and over again. Instead of performing sixteen, thirty-two, or more calculations one after the other, parallel processors perform them at the same time, giving the result in a fraction of the time. Future computers will be structured with more parallel parts, often connected by special networks. These machines will use parallel processing hardware and software.

In addition, **artificial intelligence (AI)** programming gives computers **inference** rules to arrive at conclusions. These conclusions, just as with human intelligence, are not always right, but **expert systems** have the ability to learn from their mistakes, once the mistakes have been pointed out. Artificial intelligence systems will become more widespread in some kinds of occupations.

Computers are now able to recognize vocabularies of several hundred words routinely. At the Thomas J. Watson Research Center of the IBM Corporation, engineers have developed an ex-

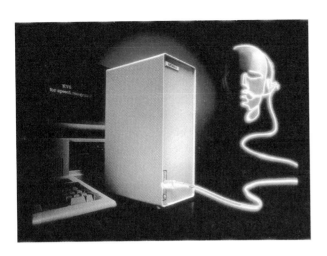

FIGURE 22–7 An experimental voice recognition system. *(Courtesy of Kurzweil Applied Intelligence, Inc.)*

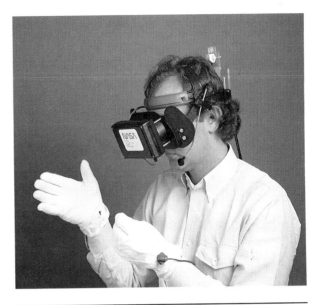

FIGURE 22–8 This helmet is equipped with a position and orientation sensor. By moving the head, the user can give instructions to the computer. *(Courtesy of NASA)*

perimental **voice recognition system** that has a vocabulary of 20,000 words. That is about 38 percent of the average person's spoken vocabulary. Voice recognition technology will change the way people work and live. Spoken commands will control machines in the factory and appliances at home. Typewriters and word processors will be spoken to, rather than keyed, Figure 22–7.

Using special interfaces, computers will be able to recognize human gestures. With sensors that monitor head position, the image on a computer screen can be made to shift as the observer changes the viewing angle, Figure 22–8. NASA has developed a new helmet that astronauts could wear. It would project a visual picture of a scene outside the spacecraft. As the astronaut's eyes move, a robotic camera outside the spacecraft moves in the same direction, Figure 22–9.

Hand movements can be interfaced to computers as well. Wearing a glove made with special tactile-feedback sensors, a user can do a set of mechanical operations that can be duplicated by a robotic device at a remote location. Fiber-optic cables, carrying light, run through the

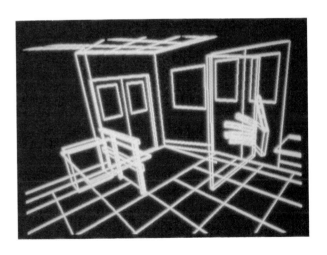

FIGURE 22–9 This is a picture that is projected on the inside of the helmet shown in Figure 22–8. By turning the head, the user can actually view the computer-generated picture from a different angle and cause it to change. *(Courtesy of NASA)*

glove. When the fingers bend, light escapes from the cables. The more the fingers move, the more light escapes. Light sensors detect the amount of light and thus monitor the hand position, Figure 22–10.

For financial transactions or security applications, sensors have been made that can recognize a person's fingerprints or the pattern of blood vessels in a person's eye (retina scanning). These are more positive forms of identification than code words or numbers that can be forgotten or stolen. The field of computer measurement of personal characteristics is called **biometrics**.

Telephones of the Future

Telephones and telephone lines will play a much more important role in the future of communications. Telephone lines will be used to support on-line communication of all types. New telephone services will be commonplace, as well. Among these will be an increase in the use of cellular telephone technology.

Cellular telephones are now found mainly in cars. These telephones are really radio transmitters that broadcast a signal to a receiver in a geographic area called a **cell**. The signal, once received, is connected directly into the telephone network (see Chapter 17). Thus, cellular telephones can be used to communicate with any other telephone in the world, Figure 22–11.

The new trend in cellular phones is to use small portable telephones that can be hand-held. Portable phones give power outputs of about half a watt, compared to mobile phones which produce about 3 watts of power output. Portable phones are still rather expensive. The least expensive phone sells for around $1,500.

In England, a new system is under development that will be made up of tens of thousands of small transmitter/receiver units located in public buildings and along roadways. These will be used as base stations for even smaller, lower power, and less expensive telephones that will easily fit in pockets or pocketbooks.

Because the telephone system will be used for so many types of communication, the sales of portable telephones will most likely rise

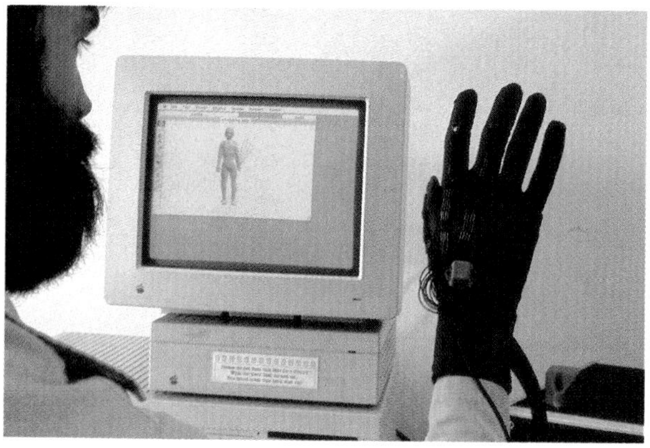

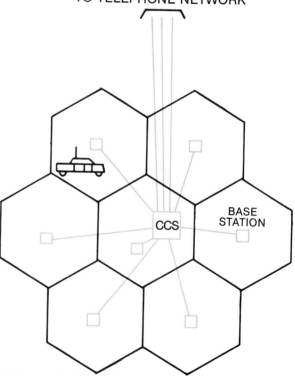

FIGURE 22–10 The DATAGLOVE, developed by VPL Research, uses fiber-optic cables, with an LED at one end, and a phototransistor at the other end. As the fingers flex, the light falling on the phototransistor changes. (Courtesy of Ann Lasko Harvill)

FIGURE 22–11 Diagram of a cellular system, showing separate cells with their own antennas. A central control system (CCS) determines when one cell should hand-off the signal to another cell.

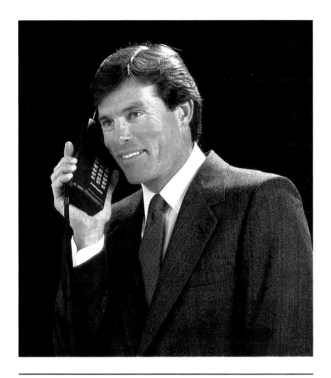

FIGURE 22–12 As prices drop, hand-held portable phones will become more and more popular in the years ahead. *(Courtesy of Mitsubishi Electric)*

steadily, Figure 22–12. With a portable computer and a portable telephone, users will be able to access on-line data bases from almost anywhere. This means that whenever you have need for information, it is available on demand.

The new era of telephones will be able to provide video as well as audio signals to users. A video camera built into the telephone will send video pictures as people talk to each other on the phone. The picture phone concept has been around since the 1964 World's Fair. In the future, this two-way video telephone system will become common.

Some proposed systems transmit black-and-white **slow scan video**. This means that the video camera scans the person to be televised at a slow rate. The picture that is displayed is a still picture of the person at the other end which takes about 5 seconds to create, Figure 22–13. Other systems can transmit full-motion video.

Telephone companies are experimenting with new technologies that will expand services

FIGURE 22–13 These video phones display still video images. *("Visitel" is a trademark of Mitsubishi Visual Telecom Division) (Courtesy of Stiller Public Relations Inc.)*

to customers. One service allows many different callers to talk to each other on a conference call. This conference service has been available to business users for some time. Using these new services, strangers are able to simply call a local number and join a telephone conference where the discussions may center on any topic of interest to the participants.

Another new service involves being able to **trace** calls made to your telephone number. Using a device already available to telephone customers in some parts of the country, the number of the calling party is displayed before you pick up the telephone, so you know who is calling before answering the phone. This service has raised objections from groups who think that personal privacy may be sacrificed by the system.

Soon, telephones that translate your voice into other languages will be available. When you speak into your telephone, a computerized voice synthesizer will digitize your voice, store it, and be able to translate it into any number of foreign languages on command, Figures 22–14 and 22–15.

The kinds of numbers that you can dial in the future will include a whole new variety of phone calls. You will be able to dial new numbers that start with 700 to find out information, play games, and purchase tickets. Other numbers will

FIGURE 22–15 New telephone technology will allow people in foreign countries to hear an immediate translation of your voice in their own language.

allow you to get children's stories, travel information, jokes, and sports scores.

Future TV Viewing

Because of changing technology, the television industry is undergoing many changes. In its formative period, it was dominated by major networks of stations that carried a relatively small selection of programs at any one time. The spread of cable TV systems and satellite broadcasting in the 1980s have resulted in a much wider choice of programs for viewers. In some cities, more than 100 channels are available. This trend will continue in the 1990s.

The introduction of **Direct Broadcast Satellite (DBS)** systems was delayed because satellites couldn't be launched for several years after the tragic explosion of the Challenger. Launching of satellites for this service is imminent now, however. DBS satellites broadcast programs

FIGURE 22–14 Software is being developed to translate technical terms from a number of languages into **Japanese.** *(Courtesy of Mitsubishi Corporation)*

Television on the Telephone Line

CODEC stands for coder-decoder. It is a device that converts analog information into digital information, and vice-versa. CODECs that operate on video signals enable video to be sent over special telephone lines. This has been difficult to do in the past since the bandwidth needed by full-motion video signals is greater than the telephone lines could handle.

CODECs work by compressing the video signal from a high bandwidth to one low enough to be sent over special telephone lines.

The technique is based on the fact that the whole video image does not have to be trans- mitted continuously; only motion that causes change in the image must be updated. The background of a scene, or a person's shirt collar would not change, for example, but moving lips would. The number of pixels on a television screen image that have to be updated is therefore less than the whole picture. Compression technology checks for changes in motion. The whole screen does not have to be refreshed over and over again.

CODEC technology compresses a video signal so that it can be sent over telephone lines. *(Courtesy of Concept Communications, Inc., Dallas, TX)*

(Courtesy of Concept Communications, Inc., Dallas, TX)

directly to home receivers that are connected to small satellite antennas mounted on roofs or chimneys. These special satellite antennas are less than three feet in diameter.

Stereo TV is now available in some cities across the United States, and it will become universally available during the 1990s. In stereo TV transmission, two different audio channels are sent with the picture. The receiver decodes the two separate channels and produces the audio in two speakers. The sound is more lifelike than conventional TV with monaural, or single-speaker sound.

High-definition TV (HDTV) has been demonstrated by several manufacturers, and transmission standards are being developed so that HDTV can be universally available. You may recall from Chapter 18 that conventional TV uses 525 scan lines to produce a picture. As the picture is enlarged to fill a large screen, the individual lines become visible, giving a grainy look to the picture. One HDTV demonstrated by SONY Corporation uses 1125 scan lines, producing a much smoother, more detailed picture. In addition, the picture is wider, making it look more realistic to the eye.

Integrated Networks

Today, most types of communication occur within separate networks. For example, the nationwide telephone network and the various cable TV networks exist side by side. Each is independent of the other.

In the future, a variety of communication services will be provided over an **integrated network**. In several different countries, a project is underway which attempts to send voice, text, data, and video over a single network. This network, called the Integrated Services Digital Network, or ISDN, combines these services and transmits them over a single fiber-optic system. Countries including France, Great Britain, Italy, Germany, and the United States are developing ISDN technology.

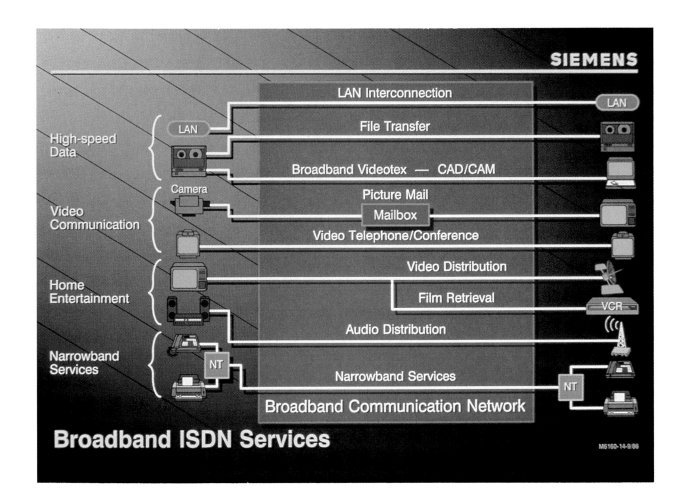

FIGURE 22–16 Broadband ISDN services. *(Courtesy of Siemens Public Switching Systems, Inc., Boca Raton, FL)*

With ISDN technology, the telephone becomes a computer which sends digital information over the phone lines. Since digital information is sent, modems will no longer be necessary. (Modems change digital data to analog data, and vice-versa.) Basic ISDN service will be made up of two channels that send data at the rate of 64 kilobits per second. These are referred to as B channels. A third channel (D channel) will carry data at a rate of 16 kilobits per second. The two B channels can be used for voice, text, data, and graphics transmission. The D channel will be used for signaling and control. As an example, a user could carry on a conversation on one B channel, send data on the second B channel, and receive an electronic mail message on the D channel.

One of the current problems the communications industry faces is the lack of standardization of the way computers communicate. For example, when Apple Computer, Incorporated and IBM first introduced their personal computers, they could not easily communicate with each other. More recently, a great deal of work has been to standardize **protocols**, the rules used by computers to communicate with each other (see Chapter 20). In addition, computers using one protocol can communicate with computers using different protocols through devices called **gateways**.

Protocol standardization and the futher development of gateways, coupled with ISDN, will result in an international network including telephone, facsimile, data, and television transmission. This network could allow you to use a single terminal in your home or office to send and receive audio, video, and data to locations all over the world, Figure 22–16.

On-line data networks are those that give users a means of accessing huge amounts of information by dialing into a central source. The central source is usually a mainframe computer where data can be constantly updated to give current information on thousands of topics to its users. Some of these on-line communication services are Bibliographic Retrieval Service (BRS), CompuServe®, The Source, and Genie. These data bases will continue to grow in number

as the number of households with personal computers increases.

As people rely more and more on on-line data bases for information, systems will be developed to coordinate a person's use of the different systems. The system would direct the user to the on-line data base that meets his or her needs. For example, a system called UNISON new acts as an intelligent network. If you wanted to check your electronic mail, the system would display the mail you have received in all the networks to which you subscribe, Figure 22–17.

In the information age, people will continue to need more instant access to vast amounts of information for personal and business reasons. For example, the whole contents of encyclopedias can be retrieved. By using on-line services, people can turn their home terminals into a total communication and information system. Newspapers from all over the world can be read. Financial information from international stock exchanges is available. Airline schedules are accessible at once.

On-line networking also allows people to directly communicate with each other. In France, the Minitel network connects over three million users who make more than 46 million telephone calls each month.

Students use the system to get help with their homework from professors by connecting to a conference called SOS-PROFS. Expert instructors then give answers to their questions. Lonely single people can connect to the MINITEL ROSE conference to look for romance. Should a user need legal advice, he or she need only connect to the SOS JURDIQUE (legal) service. Users can also access banks, travel agencies, railroads, medical hotlines, and shopping services.

Minitel was first developed as a means of eliminating the need for telephone books. Using the Minitel terminal, a subscriber can access telephone numbers from all over France. The service provides the same amount of telephone information as hundreds of telephone books.

In the United States, U.S. Videotel has patterned its service after Minitel. Videotel claims that it will be "the first low-cost desktop communications network." The system also offers home shopping, entertainment, news and sports,

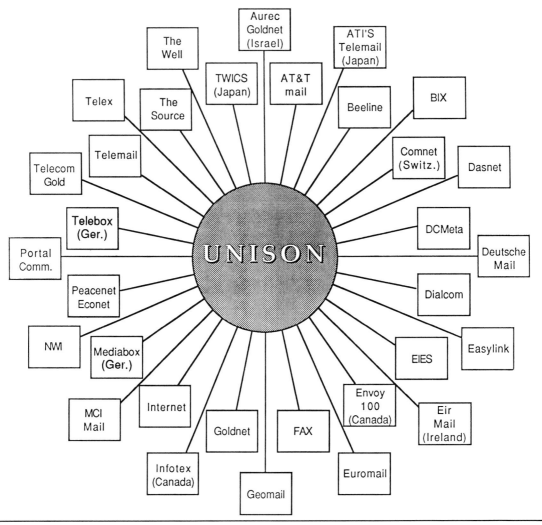

FIGURE 22–17 UNISON's Universal Mail System™ connects these networks together. *(Courtesy of Unison Telecommunications Services)*

and information on subjects as diverse as home and garden, health, community activities, business, and travel, Figure 22–18.

One caution! On-line communication can become addictive. Once people start using the system to talk to other people, telephone bills can skyrocket. Some users have monthly phone bills of more than $1,500.

Telecommuting

As the world keeps shrinking, people will not be limited to working in conventional offices. In the past, people needed to live close to their work places so that they could commute easily. In the

future, people will be able to live in their favorite environment and **telecommute** to work. Through telecommuting (exchanging information and conducting business via computer, modem, and telephone), living and work space can be anywhere in the world. A writer who likes to ski, for example, could live in the snow-covered Rocky Mountains, Figure 22–19, and transmit novels to publishers by computer and modem, over the telephone line.

Besides being able to choose a preferred living environment, people may wish to telecommute for other reasons. A mother or father might prefer to stay at home with a newborn baby. Physically disabled people might find it more convenient to work from home. Some people

FIGURE 22–18 People can access the U.S. Videotel network through compact terminals that plug into ordinary telephone lines, or through their personal computers. *(Courtesy of U.S. Videotel)*

may simply prefer the home environment to that of the typical office. A survey done in Great Britain showed that by the year 2010, 20 percent of the labor force would work from home. Companies such as Blue Cross—Blue Shield, Xerox, and British Telecom are already encouraging some workers to telecommute from home. It is quite clear that millions of people will become **teleworkers** in the future.

A variety of **telecommunities** have sprung up which give complete telecommunication capabilities for residents. One such community is Mesa-Z, near Telluride, Colorado, Figure 22–20. This 700-unit, 2,200 person community will be finished by 1990. It would give people who live there access to fiber-optic and local-area net-

works which connect to large information data bases. Residents could live in beautiful surroundings and not worry about driving to work.

Another idea for an electronic village is Eaglecrest®. Eaglecrest® is a carefully planned community nestled in a beautiful forest, but it is designed for the information age. Each home will be a center of an electronic network. This network will control appliances, lighting, electrical, and alarm systems. A household management and communication center will provide for teleshopping, financial planning, bill paying, and electronic mail, Figure 22–21. The electronic network will also provide leisure activities and entertainment, educational activities and a computer station for work-related activity.

FIGURE 22-19 Telecommuting will allow people to work from their homes in remote locations. *(Courtesy of Siemens Capital Corporation)*

Future Recording Techniques

Recording in the future will become more and more digital. **Digital audio tape (DAT)** for the home will be introduced commercially and become the preferred method of tape recording in the early 1990s (see Chapter 21). In addition, compact disks that can be recorded upon in home audio or video recorders will be introduced in the same time frame. These also will be used for large memory systems in personal computers.

As integrated circuit memory technology progresses, it will replace tape and disks in some current computer applications. This also will be true in some audio and video recording equipment. At some time, small integrated circuit memory packs may replace tapes and disks for storing music.

Summary

The information age is an era when communication technology satisfies many business and societal needs. In the future, more jobs will require the ability to manipulate and process information. Increasingly, the economics of many nations will depend upon the use of advanced means of communication.

FIGURE 22-20 Plan for Mesa-Z. *(Courtesy of John Lifton)*

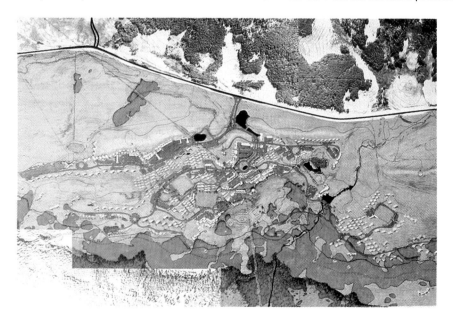

FIGURE 22–21 Some companies already market a complete line of equipment for teleworkers. *(Courtesy of Sharp Electronics)*

Electronic still cameras (ESCs) will replace film cameras in many applications, just as video camcorders have replaced motion picture cameras for everyday use. ESCs store images digitally on floppy disks. A computer can be used to change the digital images to modify colors, shapes, or positions of people or objects shown in the pictures.

Printing will be enhanced by electronics, which will provide faster typesetting and new ways of printing, such as ion printers.

Improvements in electronic construction techniques will continue to make more complex and more powerful computers available at a lower cost. Computers will be built to mimic the ways that people solve problems. Computers will

become easier for people to use through the use of voice recognition, and gesture recognition circuits and programs.

The increasing use of digital technology will result in telephones with more advanced features. The use of portable and mobile phones will become more widespread.

Home TV viewing will offer a wider choice of programming with better quality pictures and sound. Direct broadcast satellites will bring more programs to homes. Stereo TV and High Definition TV (HDTV) will be adopted to improve viewing quality.

The development and installation of the worldwide Integrated Services Digital Network will provide superior telephone, data, and video communication. This will enable many more workers to live in the environment of their choice, telecommuting to work (working from home with data sent over communication lines to their offices).

Recording technology will continue to shift toward digital recording. Digital Audio Tape (DAT) and compact disks will both become available for home recording of audio, video, and computer information.

REVIEW

1. What jobs may be created or outmoded as voice recognition technology develops?
2. Explain the difference in operation between electronic still cameras and traditional film-based cameras.
3. Why are new varieties of computer printers, such as the ion printer, being developed, when laser printers produce such high-quality output?
4. What are some of the ethical implications inherent in the use of electronic still camera technology?
5. What improvements in film and camera technology do you expect to see in the future?
6. How will the telephone system play a more important role in the future of communications?
7. What is the major advantage of using ISDN technology?
8. What is appealing or unappealing to you about the possibility of telecommuting to work?

SECTION ACTIVITIES

 TOMORROW'S INVENTOR

OVERVIEW

As you did the activities in this book, you became aware of numerous technological developments. Many developments are in response to a specific need. Examples include satellite communication, remote communication systems such as cordless telephones and microphones, fiber optics, computer software, and lasers. In this activity you are to review the communication technology developments, and suggest an improvement to an existing product. You are to illustrate your idea using a computer or hand sketching.

As with any new idea, you must consider what the negative and positive impacts of this new idea will be.

MATERIALS AND SUPPLIES

Apparatus for this activity will vary with the concept you choose to develop. You may plan to construct a prototype, or a simple poster or sketch to illustrate your idea.

PROCEDURE

As you develop your invention, you will want to convince your instructor and others in your class that the idea is worthy of your time and effort. You may consider producing a way to stabilize the hologram construction so that vibration from the building does not ruin your hologram, or consider generating a new use for holography in today's world. Other considerations may be for new uses of photography or new filming practices for video instruction. Any of these ideas may benefit others in our society.

Design a way to make any communication method more effective.

1. Select a topic, and sketch or outline its development and use.
2. Refine the design and construct a prototype to share with others.

FINDINGS AND APPLICATIONS

As was suggested earlier, it is important to find the applications and implications of your idea or invention. Many times it is possible to develop new things that help others in our society. What applications could be derived from your invention?

ASSIGNMENT

1. Complete the process outlined in producing your own invention utilizing all of the knowledge and resources studied in this unit.
2. Prepare a presentation of your idea using those technologies you have studied throughout this course. You may prepare a slide tape show, video, poster, computer demonstration, or any other method to present your idea.
3. Look around your home and list as many examples of communications methods as possible, including television, voice, nonvoice, garage door openers, security systems, etc. Complete this list and turn it in to your instructor as assigned.

GLOSSARY

A/D converter　Analog-to digital converter. An electronic circuit that converts an analog (continuous) voltage into a series of binary (digital) numbers.

adhesive binding　A book-binding method using hot-melt, flexible, plastic glue to join signatures or individual pages together.

alternating current (AC)　An electric current that periodically changes polarity (as opposed to direct current, DC).

amplifier　A device that increases the amount (amplitude) of a voltage, current, or other quantity.

analog　Continuously (smoothly) varying, as an analog voltage. Signals, such as sound and light, that occur in nature are generally analog in nature.

animation　A technique that gives a viewer the illusion that an object is moving, due to many small changes in position of the object shown to the viewer quickly in sequence.

antenna　A wire that is specially designed to launch a radio frequency signal into the air, or recover a radio frequency wave from the air and convert it to an electric current.

aperture　In a camera, the amount of opening of the shutter that lets the light in. The larger the opening, the more light is allowed in to expose the film.

applications program　A set of instructions that enables a computer to perform a well-defined task.

artificial intelligence　A programming technique that allows computers to use a set of rules to arrive at conclusions even if complete information is not known about a problem.

ASA　American Standards Association. In photography, the sensitivity rating of a film.

assembly drawing　Drawings that show how to put component parts together to make a product.

bit (Binary Digit)　The basic binary element. A bit is generally represented by an on or off signal, or a ''0'' or ''1.''

boldface　In printing, a heavy (dark) typeface.

broadcast　A radio or TV transmission that is intended to be received by a large, unrestricted audience of receivers.

bubble diagram　A rough sketch showing the relationship of a structure's parts as a group of circles.

byte　A group of bits organized to represent a character or other unit of information. Bytes are eight bits long.

CAD　Computer-aided drafting. The use of a computer and special software to generate, store, edit, and plot drawings.

CADD　Computer-aided design and drafting. The use of a computer and special software to generate, store, edit, and plot drawings, and to assist in the analysis and actual design of the object.

CAM　Computer-assisted manufacturing. The use of a computer to control an automatic machine that makes parts.

carrier　A radio frequency signal that is used to carry information. The information is carried through a process called modulation.

cartography　The branch of technical drawing that creates maps.

casebound　A method of binding a book together using a hard cover (case) to enclose the signatures and pages, which are sewn together.

cellular radio　A mobile radio telephone technology that uses many base stations, each covering a small area (cell) and connected to the others through telephone switches. As a mobile telephone unit moves from cell to cell, the connection is maintained because the switches hand the call from one cell base station to the next.

central processing unit (CPU)　The part of the computer that controls the flow, storage, retrieval, and operations on data.

channel　The route or path a message takes in a communication system.

CIM　Computer-integrated manufacturing. The use of a computer to control many of the management functions of a production line, including inventory monitoring and replenishment, work scheduling, parts delivery, etc.

circuit　A collection of electronic components that are interconnected to perform a specific function.

clip art　A set of standard illustrations that may be used in printed material, usually for a price paid to the clip art provider.

code　A series of characters that are substituted for letters and numbers. Codes can be used to make it easier to send a message over a channel, or to disguise the message so that unintended receivers will not find out its contents.

collating　The process of combining individual pages into the proper order.

color separation　The process of separating a photograph or other multicolor image into the three process colors (magenta, cyan, and yellow) and black.

communication　The exchange of data or information between at least two parties. The two parties can be people, animals, or machines.

424

component   A discrete electronic part that performs a specific function.

comprehensive layout (comp)   An artistic representation of exactly what a final printed product will look like.

conductor   A material that offers little resistance to the flow of electric current.

continuous tone photography   Photography that uses the full range of darkness tones from white to gray to black, as opposed to halftone photography.

D/A converter   Digital-to-analog converter. An electronic circuit that changes a series of digital (binary) signals to a smoothly varying analog voltage.

darkroom   A light-tight space in which exposed film can be processed into photographs.

data communication   Communication between two devices using digital signaling. The most common forms are between computers and terminals, printers, and each other.

decoding   Converting a signal recovered from a communication channel into a form that is easily understood by the recipient.

depth of field   In photography, the range between the closest point and the farthest point in the picture that are in acceptable focus.

desktop publishing   The combination of a desktop computer, special software, and a high-quality graphics printer to form a system that can be used to create, edit, and produce a master original for printing operations.

developer   In photography, an alkaline solution that turns exposed image areas on a film dark.

digital   Composed of a series of binary (discrete) bits (as opposed to analog).

digital audio tape (DAT)   A tape recording system that records music and voice in digital form, eliminating noise and distortion on playback.

direct broadcast satellite (DBS)   A satellite system that provides direct satellite broadcasts of programming to millions of homes equipped with small satellite antennas.

direct current   An electric current whose polarity does not change with time. The current may be steady or it may vary, so long as it does not change polarity.

document oriented   Software that allows the text to flow from page to page is document oriented (as opposed to page oriented).

downlink   The portion of a satellite communication channel from the satellite down to the earth station.

earth station   The transmitter/receiver/antenna portion of a satellite communication system that is located on the ground.

EIA RS–232C   Electronic Industries Association Recommended Standard 232C. A widely adopted standard for the interchange of data communication among terminals, computers, and other digital devices.

electromagnet   A device that uses electric current flow to generate a magnetic field.

electronic mail   The sending, receiving, and storing of memos or mail-like documents among a group of people using computers and terminals rather than paper.

electronic publishing   The combination of a computer, special software, and a high-quality graphics printer to form a system for the creation, editing, and production of camera-ready master originals for printing processes.

electronic still camera (ESC)   A camera that captures images in electronic form, and stores them on a small floppy disk or other digital memory.

electrostatic printing   A system of nonimpact printing using a toner that adheres to the paper only in areas that have been given a static electric charge. The toner is then heated to form a permanent image on the paper.

elevation   An architectural drawing that shows a room or a structure from the side to illustrate heights of various features above the floor.

embossing   The printing process of compressing a sheet of paper between a male and female die, leaving a raised impression of the image to be printed.

encoding   Converting a message into a form that is easily sent over a communication channel; preparing a message for transmission.

enlargements   Photographic prints that are several times larger than the negatives from which they are made.

exposure mode   The method selected to adjust the camera for shutter speed, aperture, and light control. Some methods available on modern cameras include programmable, automatic, manual, and flash.

FCC   Federal Communications Commission. The U.S. governmental agency that regulates telephone, television, radio, and other communications.

feedback   In a closed-loop system, the process of sampling the actual output of the system, comparing the actual output with the expected output, and modifying the system based on the result.

fiber-optic cable   A flexible glass fiber or plastic cable that guides light along a path with little loss. Fiber-optic cable is often used for communication using light as a carrier.

fixer   In photography, a solution that makes the image permanent by dissolving the unexposed silver halide crystals on the film, and hardening those that were exosed and developed.

flexography A form of relief printing using a flexible rubber or polymer image carrier that conforms to an irregular or fragile surface to be printed.

floor plan An architectural drawing that illustrates the placement of furniture, walls, counters, and other interior features in a view from above the room.

frequency The number of complete cycles of a periodically changing signal (electricity, light, or other) that occur in one second.

f-stop In photography, a measure of the aperture opening of a camera. The higher the f-stop, the smaller the aperture opening.

galley A long strip of phototypeset paper with type in the proper size, style, and measure to be printed.

geosynchronous Moving in synchronism with, or at the same speed as, the earth. A satellite is in a geosynchronous orbit if it is above the equator and moves around the earth once in twenty-four hours, the same time that it takes the earth to turn once on its axis. An object in a geosynchronous orbit will appear to be fixed in the sky when observed from the earth.

graphics frame A special area created in the middle of text displayed on a computer screen that can be used for the creation and display of graphics.

gravure printing A printing process that uses an image carrier with etched or sunken areas that retain ink, and then transfer it to the paper to be printed.

halftone photography Photography that converts whites, grays, and blacks into a pattern of dots, more dense for darker shades, and less dense for lighter shades.

high definition TV (HDTV) A new TV format that provides greater resolution (more detail) in pictures presented on the screen.

hologram A three-dimensional photographic image made using a laser light source.

icon An image used on a computer screen to indicate an action or command instead of typing the command itself. An example is having a computer operator point to a picture of a file cabinet on the screen rather than typing SAVE DOCUMENT on a keyboard.

image assembly Paste-up. The process of combining different elements of the document to be printed.

image generation The production of type and illustrations in a printed document.

image preparation In printing, the combined processes of photo conversion, film assembly, proofing, and image carrier production.

ink jet printing A form of electronic printing in which a fine jet (stream) of ink droplets is sprayed on the paper to be printed.

insulator A material that offers very high resistance to the flow of electricity or heat. Plastic, rubber, or ceramic are usually good insulators.

integrated circuit (IC) A complete electronic circuit that includes multiple components built at one time on a single semiconductor chip.

ISDN Integrated Services Digital Network. A worldwide communication network standard that provides voice, data, and visual image transmission using digital technology.

ISO International Standards Organization. In photography, a rating of the sensitivity of film.

italic A slanted typeface used in printed material.

justified In printing, aligning type so that the left- and right-hand sides of the typeset columns form a straight vertical line.

laser Light amplification by stimulated emission of radiation. A device that converts electrical energy into a coherent, highly pure and narrow beam of light.

LED Light-emitting diode. An electronic semiconductor that provides light when an electric current flows through it.

lens In photography, the part of a camera that focuses the image on the film.

letterpress Relief printing. A printing process that uses an image carrier with a raised surface to transfer ink to the paper to be printed.

lithography A printing process that uses water to repel ink from non-image areas of the image carrier, allowing ink to exist only in areas transferring images to the paper to be printed.

local area network (LAN) A high data-rate digital communication network that is usually small (contained in one building or on one floor of a building), used to connect many desktop computers to shared memory devices, printers, modems, and larger computers.

mainframe computer A large computer used by a company, government agency, or university for administrative work, such as making out payroll checks, keeping personnel records, or doing routine scientific or research work.

mass communication Communication with very large numbers of people at one time. Examples are radio, TV, magazines, and newspapers.

memory The part of a computer that stores data for later use. Various types of memory include read only memory (ROM), random access memory (RAM), and mass storage memory.

microcomputer A computer built around a microprocessor (computer on a single integrated circuit) and associated support circuits.

microprocessor A single integrated circuit that contains all the functions of a complete computer.

microwave The region in the radio frequency spectrum between 1 GHz and 100 GHz. The microwave spectrum is used for microwave radios, radar, satellite communication, microwave ovens, and other uses.

modem modulator-demodulator. A device used to send digital signals over analog transmission lines. The most common example is a modem used to send data signals over telephone lines.

modulation The act of changing a signal, usually to superimpose information onto a carrier signal.

Morse code The most commonly used code for sending telegraph signals, invented by Samuel F. B. Morse.

multiplex To send more than one information signal over a single transmission line by either time-sharing or frequency-sharing the line.

network A group of computers, modems, telephones, or other communication devices connected together so that they can communicate with each other.

offset printing A printing process that uses an intermediate image carrier (often a rubber blanket) to transfer the image from the metallic carrier to the paper.

operating system Computer software (a set of instructions) that allows the user to control a computer's memories, printers, and other attached devices (peripherals).

optical disk A round, flat disk that stores digital information in a series of pits on its surface that are read by reflections of laser light shined into the pits.

orphan In typesetting, when the last line of a column of type ends with the first line of a paragraph that is continued on the next column or page.

page imposition The assembly of several pages of print master onto a single large sheet (signature) that, when folded properly, will give a sequence of pages in the final printed product.

page oriented Software that allows text to flow from column to column only on one page is page oriented, as opposed to document oriented.

panning Moving a camera to follow an object in motion while taking its picture.

PBX Private Branch Exchange. A telephone switch installed on a customer's premises. Modern PBXs are digital and offer many advanced features to the customer.

perforating The process of making a series of small holes or slits in a paper to make it tear easily and accurately along the perforation line.

persistence of vision The ability of the eye to hold an observed image for a split second before another image comes into view.

phototypesetting The production of type and illustrations to be used in creating a printed document, consisting of input, editing, and output stages. See also **image generation**.

pica In typesetting, a measure of line length. One pica is equal to twelve points, and there are six picas to the inch.

platemaking The making of an image carrier by exposing a photosensitive plate to high-intensity light through a film master of the image to be printed.

point A measure of type size. There are 72 points in one inch.

print In photography, a photograph developed on a piece of paper.

process colors The three colors (magenta, cyan, yellow) used in combination with black to make all other colors of the spectrum in printing.

program A set of instructions used to operate a computer. See also **operating system** and **applications program**.

projector A device that shines a powerful light through a transparent film containing images, projecting the images on a screen or viewing surface some distance away.

propagation In radio communication, the spreading of a radio signal through the air or through space.

protocol A set of well-defined, mutually agreed-upon rules that enable two data communication devices to successfully communicate with each other.

radio wave An electromagnetic wave that has been lauched by an antenna, usually for the purpose of communication (radio, TV, etc.) or sensing (radar).

random access memory (RAM) Computer memory that can be written into or read from any memory location at any time. It is sometimes called a computer's main memory, and it is erased when the computer is turned off.

read only memory (ROM) Computer memory that is permanent (cannot be written over) and which remains after the power has been turned off. ROM usually contains the instructions used to set the computer up for use when it is first turned on (boot-up instructions).

receiver A communication device that recovers an information signal from a channel.

refraction The bending of light or other electromagnetic waves.

registration In printing, the alignment of different images that must be printed on the same paper at different times (for example, for different color inks).

relief printing Letterpress. A printing process that prints an image from a raised surface on the image carrier that deposits ink on the paper to be printed.

resistance are called conductors, while those with high resistance are called insulators.

saddle stitching   A method of binding signatures or pages together using wire staples.

screenplay   The story on which a motion picture is based.

screen printing   A printing process in which ink is pushed through a stencil onto the substrate (material to be printed upon). The ink is deposited in the pattern set by the stencil.

semiconductor   A material that offers moderate resistance to the flow of electric current. By specially processing semiconductor materials, electronic components with special operating capabilities, such as diodes, transistors, and integrated circuits, can be made.

shrink-wrap   A packaging method using a clear plastic wrapping that shrinks to the shape of the item being packed when exposed to heat.

shutter   In photography, the part of a camera that opens for a very precise amount of time to let light in to expose the film.

signature   A group of pages that is printed on one large sheet at one time, and then folded and cut as necessary to form a sequential group of pages in a printed document.

single-lens reflex camera (SLR)   A camera in which the same lens is used to aim and focus the camera and to take the picture.

slow motion   A motion picture technique in which an activity is photographed at a high speed, and then replayed at standard speed to examine the movements more closely.

specification   A statement of the exact requirements of a job, in as much detail as is necessary to ensure that the job is done correctly.

spectrum   The range of frequencies existing in a channel.

stop bath   In photography, a slightly acidic solution that is used to stop the action of the developer.

substrate   Any object that is printed on. For example, a substrate can be paper, a cardboard carton, or a tee shirt.

supercomputer   An extremely fast computer used to conduct very sophisticated research or to solve scientific, engineering, or other problems requiring many calculations.

telecommute   To use telecommunication to work from home rather than to travel to an office every day.

telegraph   A method of communication using on-off pulses of electricity in a coded form. Morse code is the most common code used in telegraphy.

telephone   A device that sends a voice signal from one place to another by electrical signaling.

telephoto lens   A camera lens used to photograph a distant object, making it appear closer than it actually is.

teleport   Communication centers used by businesses throughout a city for reliable, high-capacity communication with other locations around the world.

thumbnail sketch   A small, very quickly drawn pencil sketch of various designs that may be appropriate for the layout for a particular printing job.

tolerance   The amount of variation allowed between what was specified in a design and what is actually built as a product.

transistor   Short for **trans**fer-re**sistor**. A semiconductor device that allows a small amount of current into one terminal to control a larger flow of current to another terminal.

transmitter   A device that sends a message, or places it onto a channel.

twin-lens reflex camera (TLR)   A camera in which one lens is used to aim and focus the camera, and another lens is used to take the picture.

TVRO   TV Receive Only. A kind of satellite earth station that only receives signals from satellites, and cannot transmit signals up to satellites.

typesetting   The process of converting manuscript copy into typeset copy.

uplink   The portion of a satellite communication link that sends signals from a ground station up to the satellite.

video   An electronic signal that represents a visual image.

view camera   A camera in which the photographer can directly view the image that will be captured on film before taking the picture.

viewfinder camera   A basic camera that uses a separate viewfinder for aiming the camera.

voice recognition system   A system that analyzes a speaker's voice to identify words or the speaker.

voltage   A measure of electrical potential (electromotive force).

wavelength   A measure of the distance that one complete cycle of a periodic wave form occupies in a transmission channel.

wide angle lens   A camera lens used to photograph an object that is close to the camera, gathering image light from a wide field of vision.

widow   In typesetting, when the top of a column of type starts with the last line of a paragraph from the previous column or page.

WYSIWYG   ''What you see is what you get''; a term that refers to the display of a page of text and illustrations on a computer terminal in exactly the same form as the final printed copy. (Pronounced whizzy-wig.)

zoom lens   A camera lens that can be adjusted to change the viewing angle.

INDEX